Asset Management Excellence

This is the third edition of *Asset Management Excellence: Optimizing Equipment Life-Cycle Decisions.*

This edition acknowledges and introduces the many changes to the Asset Management business while continuing to explain the supporting fundamentals. Since the second edition, there have been many influences of change in asset management, society's expectations, and supporting technologies.

In this edition, the contributors have revisited the content and have updated and added insights and information based on the emerging influences in thinking and the continued evolution of applied technologies since the prior editions.

New in the Third Edition:

- Updates across each of the second edition chapters to align with today's insights
- Updates on technologies now available to support Asset Management, including related software packaging, the Internet of Things (IoT), Machine Learning, and Artificial Intelligence
- Insights on how Information Technology can step up to help an asset-intensive organization compete, drive to operational excellence and automation
- A chapter on sustainability and the influence Asset Management may have on this higher-focus priority
- A chapter on change enablement as the process and technology changes impact the various stakeholders of asset-intensive organizations.

The fundamentals of Asset Management are essential as Asset-intensive organizations look to technologies to help them compete. AI is becoming pervasive but must be confirmed and aligned with the fundamentals. This edition will provoke thought as each organization determines its next steps toward its new challenges in Asset Management.

Asset Management Excellence

Optimizing Equipment Life-Cycle Decisions

Third Edition

Editors

Don M. Barry
John D. Campbell
Andrew K. S. Jardine
Joel McGlynn

This Edition was edited by

Don M. Barry

CRC Press
Taylor & Francis Group
Boca Raton London New York

CRC Press is an imprint of the
Taylor & Francis Group, an **informa** business

Designed cover image: Getty Images

Third edition published 2024
by CRC Press
2385 NW Executive Center Drive, Suite 320, Boca Raton FL 33431

and by CRC Press
4 Park Square, Milton Park, Abingdon, Oxon, OX14 4RN

CRC Press is an imprint of Taylor & Francis Group, LLC

© 2024 selection and editorial matter, Don M. Barry; individual chapters, the contributors

First edition published by CRC Press 2001
Second edition published by CRC Press 2011

Library of Congress Cataloging-in-Publication Data
Names: Barry, Don M., editor.
Title: Asset management excellence : optimizing equipment life-cycle decisions / edited by Don M. Barry.
Other titles: Maintenance excellence
Description: Third edition. | Boca Raton, FL : CRC Press, 2024. | Revised edition of Maintenance excellence : optimizing equipment life-cycle decisions / edited by John D. Campbell, Andrew K.S. Jardine. New York : M. Dekker, [2001] | Includes bibliographical references and index. |
Summary: "This is the third edition of Asset Management Excellence, Optimizing Equipment Life-cycle Decisions. The fundamentals of Asset Management are essential as Asset-intensive organizations look to technologies to help them compete. AI is becoming pervasive but must be confirmed and aligned with the fundamentals. This edition will provoke thought as each organization determines its next steps toward its new challenges in Asset Management"-- Provided by publisher.
Identifiers: LCCN 2023041136 (print) | LCCN 2023041137 (ebook) | ISBN 9781032679396 (hbk) | ISBN 9781032679594 (pbk) | ISBN 9781032679600 (ebk)
Subjects: LCSH: Production engineering. | Product life cycle. | Production management.
Classification: LCC TS176 .M336 2024 (print) | LCC TS176 (ebook) | DDC 658.2/02--dc23/eng/20231107
LC record available at https://lccn.loc.gov/2023041136
LC ebook record available at https://lccn.loc.gov/2023041137

ISBN: 978-1-032-67939-6 (hbk)
ISBN: 978-1-032-67959-4 (pbk)
ISBN: 978-1-032-67960-0 (ebk)

DOI: 10.1201/9781032679600

Typeset in Times
by SPi Technologies India Pvt Ltd (Straive)

Contents

SECTION I   Maintenance Management Fundamentals

SECTION II   Managing Equipment Reliability

SECTION III   Optimizing Maintenance and Replacement Decisions

SECTION IV *Optimizing Maintenance and Replacement Decisions*

Foreword

In 1999, I had the privilege of being welcomed by John D. Campbell to embark on a sabbatical from the University of Toronto at his Centre of Excellence in Maintenance Management within PricewaterhouseCoopers consultancy. During this enriching period, amidst fulfilling my consulting responsibilities, John and I jointly formulated a concise course on Physical Asset Management for industry professionals, slated to be delivered at the University of Toronto. Simultaneously, we collaborated on the creation of the book "Maintenance Excellence: Optimizing Equipment Life Cycle Decisions," which found its publication under Marcel Dekker (now CRC Press) in 2001.

Regrettably, John succumbed to cancer in 2002.

In 2009, CRC Press invited me to prepare a 2nd Edition of our work, titled "Asset Management Excellence: Optimizing Equipment Life-Cycle Decisions." This Edition would feature an additional editor, Joel McGlynn, Vice President and the Global Leader for Enterprise Asset Management at IBM Business Consulting Services. Don M Barry, Associate Partner and leader of IBM Canada's Supply Chain and Enterprise Asset Management Consulting Practice, assisted Joel in this endeavor.

In this 3rd Edition, I am delighted to welcome back Don M Barry, who has undertaken a substantial overhaul of the content in most chapters. His efforts ensure that this Edition considers the significant developments in asset management that have transpired since the publication of the 2nd Edition in 2011. Don brings a wealth of experience to this Edition, supporting educational programs at the Centre for Maintenance Optimization and Reliability Engineering (C-MORE), Mechanical and Industrial Engineering, University of Toronto, for over 20 years.

The Physical Asset Management (PAM) course has consistently been in high demand at the University of Toronto and beyond. Over the years, the PAM program has leaned on Don's expertise and exceptional teaching abilities. Participants in our program consistently highlight the value of his practical insights and commend his teaching skills. Personally, I am grateful for the wealth of experience he has gained from his work at IBM, and I appreciate that Don has collaborated with me to disseminate that knowledge.

As we celebrate the 23rd anniversary of the original publication, I am thrilled to present this updated Edition, confident that it reflects the evolution of asset management practices and benefits from the expertise of dedicated contributors like Don M Barry.

This Edition emphasizes the increasing role of technology in transforming asset management practices. It highlights the availability of technology for automating decisions in asset management, contributing to operational excellence. The text advises caution in embracing technology without a solid understanding of the fundamentals being automated.

One of the new Chapters on Emerging Technologies emphasizes the convergence of Information Technology and Operations Technology, Industry 4.0, IoT, Big Data, Sensors, Mobile tools, and business risk management. The evolving landscape of Asset Management faces significant demands on IT, necessitating a proactive approach to stay abreast of these advancements.

Andrew K S Jardine, Professor Emeritus
University of Toronto

Preface

This is the third edition of *Asset Management Excellence: Optimizing Equipment Life-Cycle Decisions*. This edition acknowledges and introduces the many changes to the Asset Management business while continuing to explain the supporting fundamentals. As with the time between the first and second editions, there have been many influences of change in this space. The original authors of the past chapters and editions continue to be acknowledged for their sound information and principles related to working toward maintenance excellence at that time. In this edition, the contributors have revisited the content and have updated and added insights and information based on the emerging influences in thinking and the continued evolution of applied technologies since the prior editions.

Maintenance has emerged to be considered in the broader context of Asset Management. Asset Management exists to support the needs of Operational Excellence. For asset-intensive organizations, Operational Excellence is now expected for that organization to become and remain leading in their chosen markets.

Society 4.0 needs and global tolerances have created the notion of Industry 4.0 as technology has created the expectation that Information Technology (IT) needs to be an integral supporting partner in the success of an asset-intensive organization's pursuit of business growth, resilience, and sustainability goals. Today the "environmental sustainability" agenda is a component of any "corporate sustainability" agenda. As this edition is being released, the world has announced Society 5.0 goals which primarily suggest that success for asset-intensive organizations are expected to report their contributions to Green House Gas (GHG) emissions as responsibly as they would the financial results and forecasts. The expectation of accountability and responsibility for an organization's citizenry will only become a deeper requirement moving into the next decade. Waste is becoming a renewed focus frontier.

IT Technology has a role to play in pursuing Asset Management Excellence. The notion of connected devices through the Internet of Things (IoT) is now in everyone's awareness as most first-world people over 16 can access smartphones. Over the past decade, IT has enabled and converged with Operations Technology (OT). Access to large databases on agile Cloud infrastructures is the emerging norm. The IT Cloud can be multiple types or a hybrid of types. Data analytics has evolved from spreadsheet Pareto graphs to simple query capabilities leveraging Business Intelligence to leveraging Machine learning and artificial intelligence (AI) with what can be seemingly unlimited computing capacity with the now available quantum computing.

The notion of AI in almost everything today is exciting and concerning. The dependencies emerging on IT technologies bring concerns about business and personal data security. It also concerns whether the insights gained from every data analysis deserve immediate acceptance.

As the world's asset-intensive organizations work to compete, there is no doubt that technology will have an accelerating role in supporting its success. This success will likely depend on creating leading processes and practices and automating. Automation has existed for decades, and hyper-automation will become the norm for successful global organizations striving for Asset Management and Operational Excellence.

Foundationally the journey's IT enablement must be based on proven maintenance and asset management fundamentals. The need for understanding and executing the known fundamentals has never been more critical. Not ensuring leading practices are executed within the IT enablement and automation journey automates a poor path and is potentially disastrous for an organization, its reputation, and the environment.

This third edition will feature the same organizational format as previous editions, with some of the terms rebranded to better align with how it is taught to new organizations. Asset Management sponsor alignment helps to affirm the need for aligned strategy, organizations, and stakeholders to the mission of an asset-intensive organization. Acknowledging the asset management team's scope of direct influence confirms that maintenance execution, parts management, and workable maintenance tactics and process metrics are supported by engaged IT support systems. The next level of elements supporting Asset Management Excellence involved stakeholders working together beyond the management and direct asset-facing maintenance teams. Cross-functional stakeholder engagement is required across multiple business areas to fully execute Total Productive Maintenance (TPM) and risk and reliability positioning with Reliability Centered Maintenance (RCM) teams. The final level of Asset Management Excellence is now branded "Uber Change," given that incremental change will not be enough with all the influences suggested earlier in this Preface.

Change must happen for asset-intensive organizations to expect to be prosperous, resilient, and sustainable. Asset Overall Equipment Effectiveness is still essential but is now an expected outcome for asset managers working to contribute to Operations Excellence. Change must be more profound, broader, and IT-enabled. Asset Management supporting software vendors has recognized the business and societal needs. Change must happen. A chapter on Change enablement has been added to this third edition as the stakeholders involved will specifically need to be helped through the pending transitions.

Asset-intensive organizations from all asset classes and sizes will find this book insightful. The authors of these chapters recognize that there are different levels of Asset Management maturity, readiness for change, and IT's willingness/ability to support. What is needed in one organization may be well established in another. Understanding which fundamental leading practice principles and concepts are correct for your organization may or may not be correct for another at that time. This edition includes the leading concepts, trends, and new information on emerging areas such as technologies, sustainability needs, and change enablement.

It is recognized that asset management entities universally have varying needs in several key areas. Trends in using the many frameworks and models help organize and prioritize an organization's area of focus. This edition continues to provide such a framework. Identifying an organization's unique needs across diverse business departments and asset classes is supported throughout this book.

This third edition aims to provide a combination of practical and insightful information. The content, theories, and methods can be used by maintenance/asset management entities and their supporting stakeholders of varying sizes and maturity for their benefit or to generate thought for personal, group, or academic rigor.

   Once armed with the insights and information contained herein, an asset-intensive organization, team, or individual can take stock of the areas they need to address and organize for success. They can use models and frameworks to set a road map and prioritize improvement initiatives. They can apply what they have learned to evolve into something "new" or move to the next generation of asset management excellence.

Acknowledgments

Thank you to all who have contributed to this and the previous two editions of this important title and topic. I wish to acknowledge the past Edition editors, particularly:

John D. Campbell
Andrew K. S. Jardine
Joel McGlynn

The contributors and editors have recognized the many current and emerging standards, technologies, and commitments that drive Asset Management Excellence. Asset Management Excellence supports Operational Excellence and an asset-intensive organization's resilience and sustainability. This tenacious focus has been a constant and is impressive.

The fundamentals of asset management leading practices are essential as the world is supported and influenced by technologies, emerging operational and societal needs, and uber change, respectively.

Don M. Barry
3rd Edition Editor.

Editor

Donald M. Barry is a Principal Consultant with an Asset Management and Technology Services organization, supporting Risk and Reliability Strategies, ISO55000 and Asset Management Strategy consulting, Enterprise Asset Management, and Asset Performance Management solutions. Previously he was the Global Lead for IBM's Asset Management Center of Competency and an Associate Partner, leading IBM's Asset Management Practice (for 15 years).

Don M. Barry has over 40 years of experience in asset management-related service delivery support systems and application development, including three years in field service management and 15 years in business process development and supply chain management.

Mr. Barry's direct client list has included industries such as Upstream Oil and Gas, Pipelines, Power Generation and T&D Utilities, Mining, Forestry, Airlines, Electronics Manufacturers, Steel Manufacturers, and Federal, Provincial, and Municipal Governments.

He is a primary instructor of the Physical Asset Management program at the University of Toronto and has been teaching this class since 2014.

PUBLICATIONS

He was a prime contributor to the 2nd edition of *Asset Management Excellence: Optimizing Equipment Life-Cycle Decisions* published in 2011, CRC Press.

He is the author of *Maintenance Parts Management Excellence: A Holistic Anatomy*, published in 2023 by CRC Press.

FUNCTION AND SPECIALIZATION

- Asset Management strategy and change consultant, instructor, and Program Executive
- EAM (i.e., Maximo) and related systems implementation Program Executive
- Reliability Centered Maintenance Practitioner
- Maintenance Parts Management consultant and Program Executive
- Consulting Professional

Mr. Barry is a recipient of the Lifetime Achievement Award in Plant and Production Maintenance, awarded by Federated Press.

Contributors

Brett Barlow is the Asset Optimization Practice Leader for IBM Canada. He is a Reliability-centered Maintenance (RCM) trained professional providing clients' business and technical expert advice on transforming their maintenance organizations and implementing the supporting technologies. He is a trusted advisor to CIOs, CFOs, VPs, Directors, and Managers of Asset Management. Brett has presented at several of the IBM Asset Maintenance-related conferences (Think, Pulse, Edge) focused on Asset Management solution implementations. He has spent over 20 years delivering asset management solutions to clients in various industries (Transportation, Utilities, Oil & Gas, Aviation). He has helped clients transform their asset maintenance capabilities from rudimentary paper-based organizations to planned and predictive-based organizations. He is known for establishing trust with his clients and delivering with integrity.

Don M. Barry is a Principal Consultant with an Asset Management and Technology Services organization, supporting Risk and Reliability Strategies, ISO55000 and Asset Management Strategy consulting, Enterprise Asset Management, and Asset Performance Management system solutions. Previously he was the Global Lead for IBM's Asset Management Center of Competency and an Associate Partner, leading IBM's Asset Management Practice (for 15 years).

Don M. Barry has over 40 years of asset management-related service delivery support systems and application development, including three years in field service management and 15 years in business process development and supply chain management.

Mr. Barry's direct client list has included industries such as Upstream Oil and Gas, Pipelines, Power Generation and T&D Utilities, Mining, Forestry, Airlines, Electronics Manufacturers, Steel Manufacturers, and Federal, Provincial, and Municipal Governments.

He is a primary instructor of the Physical Asset Management program at the University of Toronto and has been teaching this class since 2014.

He was a prime contributor to the 2nd edition of *Asset Management Excellence: Optimizing Equipment Life-Cycle Decisions* published in 2011, CRC Press.

He is the author of *Maintenance Parts Management Excellence: A Holistic Anatomy*, published in 2023 by CRC Press.

Mr. Barry is a recipient of the Lifetime Achievement Award in Plant and Production Maintenance, awarded by Federated Press.

Andrew K. S. Jardine, Ph.D., C.Eng., P.Eng. FCAE, FIISE, FEIC, FISEAM (Hon.), was the Founding Director of the University of Toronto's Centre for Maintenance, Optimization and Reliability Engineering (CMORE).

During the period 1986–95. Dr. Jardine was Chair of the University's Department of Industrial Engineering.

He was the co-editor (with J. D. Campbell and J. McGlynn) of *Asset Management Excellence: Optimizing Equipment Life-Cycle Decisions*, published by CRC Press in 2010. His most recent book is the 3rd edition of his earlier work, *Maintenance, Replacement, and Reliability: Theory and Applications*, published by CRC Press in 2022 and co-authored with Dr. A.H.C. Tsang.

Professor Jardine has garnered an impressive array of awards, honors, and tributes, including having been the Eminent Speaker to the Maintenance Engineering Society of Australia (now the Asset Management Council), as well as the first recipient of the Sergio Guy Memorial Award from the Plant Engineering and Maintenance Association of Canada. In 2013, he received the Lifetime Achievement Award from the International Society of Engineering Asset Management (ISEAM).

In 2020 he was awarded Life Membership of PEMAC: Asset Management Association of Canada. Professor Jardine is listed in Who's Who in Canada.

He has been elected a Fellow of the Canadian Academy of Engineering, a Fellow of the Institute of Industrial and Systems Engineers, a Fellow of the Engineering Institute of Canada, and an Honorary Fellow of the International Society of Engineering Asset Management. Besides writing, researching, and teaching, Dr. Jardine has completed numerous consulting assignments with organizations worldwide, including mines, government agencies, power and transit companies, and others.

Brian Helstrom, Ph.D. is an Executive Consultant for Technical Architecture and Strategy, assisting clients in transforming their enterprise and operations by framing industry opportunities and challenges into specific strategic options, formulating actionable strategies that intersect business and technology, and accelerating implementation through tailored operations and change programs and bringing Business consulting and IT architecture perspectives to his clients, holding credentials from IBM as a Certified Consultant, Certified IT Specialist, Senior Certified IT Architect, and from Open Group as a Distinguished Certified IT Architect, Master Certified Specialist, and TOGAF9 Certified, with over 35 years of IT infrastructure architecture, applications architecture, solution architecture, Enterprise Architecture, and implementation management experience across multiple and various industries.

Helstrom's background is very diverse and includes Enterprise Architecture, SOA, application architecting, solution architecting and business integration, ITIL Service Management, methods for IT Strategic Planning, Asset Management, IT operational management, roll-out planning, and effectiveness assessments.

Atif Sheikh is IAM certified Asset Management practitioner and leader with a unique combination of plant maintenance leadership experience in petrochemical environments and EAM systems implementation. As an EAM principal, Atif has led complex asset maintenance and reliability-based solutions around the globe (Asia, the Middle East, Europe, Canada, and the Caribbean). As Plant Maintenance and MRO Supply Chain strategist, well versed in ISO55000 and GFMAM principles, Atif has helped Utilities, O&G, and Mining clients build EAM framework and CMMS (Maximo) Architecture, identifying ROI opportunities and formulating transformation road maps.

1 Asset Management Excellence*

Don M. Barry
*Previous contributions from Don M. Barry and
John D. Campbell*

1.1 INTRODUCTION

Asset Management excellence is many things done well. First, it's when a plant performs up to its design standards and the equipment functions as and when needed. Also, its maintenance costs tracking on budget, with reasonable capital investment. Next, it's high service levels and fast inventory turnover. It's motivated, competent trades. Finally, it is leveraging the Industry 4.0 era technology to assist asset management processes, improve data accuracy, and automate where practical to drive operational excellence from asset productivity. All of these things are required to achieve asset management excellence. Today, more and more of these elements are automated, and there is a distinct interest in asset management contributing to operational excellence, customer experience, and stakeholder value.

Achieving asset management excellence is complicated because much of what happens in an industrial environment is influenced by external forces. Our goal is operational excellence. However, the random nature of these dynamic variables does not make achieving excellence easier.

Maintenance management has evolved tremendously during the past century. About one hundred years ago, maintenance was primarily a run-to-failure response discipline. We have seen the progression of the steam age (now called Industry 1.0) in the late 1700s evolve to mass production assembly lines in the 1800s (in what is now called Industry 2.0). The late 1900s brought automation supported by numeric controls and electronics-enabled computers (Industry 3.0). Today, more than 20 years into the 21st century, we maintain all the previous production capabilities with cyber-physical systems, networks, and devices supported through the "Internet of Things" (Industry 4.0). Our expectations have changed with industry evolution, and now Industry 5.0 is emerging. Greenhouse gas (GHG) environmental, safety, and product quality issues are now as important as reliability at act as gates to be in business. Equipment changes have advanced maintenance practices, changing predominant tactics from run-to-failure to prevention, prediction, and eventually reliance. With an upgraded mission to integrate and drive "Return on Asset" and contribute to the operational excellence of an organization's perceived value and the evolution of technology, the Asset Management (and Maintenance Manager) role is more critical today than ever. The reliance tactic and the emerging dependence on technology focus on this edition.

DOI: 10.1201/9781032679600-1

Many organizations had developed a systematic approach to maintenance planning and control in the late "70s" and early "80s," abandoning it as we shifted to a global marketplace and significant economic harmonization. Today, we must start over. We must first rebuild fundamental maintenance management capabilities before proving the value of reliability management and maintenance optimization. We also must prepare for the technological advances that are rapidly becoming available in concert with our maintenance strategy maturity.

In this third edition, you will see introduced or reaffirmed some key concepts we believe to be fundamental in how a leading organization should look at asset management.

They are that:

- Five Asset Classes are emerging that, although they are distinct, they have much in common;
  - Real Estate and Facilities
  - Plant and Production
  - Mobile Assets
  - Infrastructure
  - Information Technology

More and more assets are being developed with electronic intelligence and IP addresses to allow the asset to communicate via the Internet. Some solutions have evolved to recognize that the maintenance of one asset class is primarily the same process as another.

The Maintenance Excellence Asset Management Pyramid that John D. Campbell introduced in the mid-1990s provides a fundamental and holistic approach to understanding where an organization is in its maintenance maturity and can act as the baseline for where they want to be. In this edition, the pyramid has been modified slightly to align with how it is explained and how the levels align with the groups that will be engaging.

Asset Management Excellence is structured with four levels in the stacked cylinder pyramid:

- Beginning with visible and effective **"Sponsor Alignment"**:
  - To establish the foundational guidance needed to align with the enterprise business goals, maintenance goals, and the related supporting asset management goals.

Real Estate and Facilities   Plant and Production   Mobile Assets   Infrastructure   Information Technology

FIGURE 1.1   Suggested five asset classes.[1]

- To organize the asset management goals aligned to support the enterprise goals so all stakeholders in the enterprise understand their role and contribution to the program's success.
- To ensure that leadership is "seen" leading and sponsoring the parts operation to support asset management excellence.
- Understand how "change enablement" initiatives may need to be injected into the foundational culture to empower staff and enable leaders to be lead and be successful.
- Then controlling the **"Managed Scope"** of the functions assigned explicitly to asset management:
  - To establish the basic working efficiencies of planning and scheduling of technician **Work Execution**.
  - To measure success with effective **Metrics** that work across all Business dimensions (levels) in support of asset management excellence.
  - To establish how **Parts Management** is executed to support the asset management and operational goals of the business.
  - To determine and support the various maintenance **Tactics** that may be applied to the functional goal of an asset and its components (preventive, predictive, corrective maintenance, as well as inspections and redesign tactics).
  - To support the activities, transactions, and analysis and provide a workable audit trail with Information Systems having a role at this level in **Data and Systems Support**. Preparing the systems to support the future also has a significant play in achieving asset management excellence.
- Some key activities supporting and advancing asset management excellence require multiple cross-functional support and guidance that cannot be managed by Maintenance staff alone. We have **"Stakeholder Engagement"** activities:
  - To establish the asset's **Risk and Reliability** tolerances and mitigations for each critical asset's operating context. Here an organization needs to establish each asset's support philosophy and required infrastructure, leveraging facilitated input from Design Engineering, Operations, and Maintenance staff.
  - To support a culture of "reliability by the operator," a joint operations and maintenance discipline called **Total Productive Maintenance** must be supported.
- Ultimately, the highest level is called **"Uber Change"** to influence the need for significant change to meet future needs and lead the charge in establishing the enterprise as a leader in its industry. Foundationally, Uber Change can include:
  - Re-thinking Process elements to allow for new methods, systems, or tools in achieving asset management excellence
  - The ability to predict and bridge the technology chasm with new algorithms and associated automation
  - Ensure that all asset management processes align with standard legal, financial, and business regulatory requirements
  - Re-thinking the current processes to prepare for the future.

FIGURE 1.2   Asset management excellence pyramid.[2]

The Asset Management Lifecycle Model, also initially introduced by John Campbell, should be considered to ensure each organization understands the full impact of an Asset purchase or disposition and the role maintenance can play in promoting the length and quality of the asset's life.

The challenge of maintenance excellence, and the goal of this book, is to provoke thought on strategic issues around maintenance and develop tactics that will minimize breakdowns and maximize the rewards of planned, preventive, and predictive work.

There are three goals on the route to maintenance excellence:

- **Strategic.** First, you must draw a map and set a course for your destination. You need a vision of what maintenance management will be like in three years. What is the plant condition, availability, maintenance cost structure, amount of planned work compared to unplanned reactive work, and the work environment? You must assess where you are today to get where you are going. This way, you will know the gap size to be closed. Finally, you must determine the human, financial, and physical resource requirements and a time frame to make your vision real.
- **Tactical.** You need a work and materials management system to control the maintenance process. The work transactions and metrics are managed in a computerized maintenance management system (CMMS), an enterprise asset management system (EAM), or a maintenance module in an enterprise resource planning system (ERP). Maintenance planning and scheduling are most important for job work orders, plant and equipment shutdowns, annual budgeting exercises, and creating a preventive and predictive program. Also, measure performance at all levels to effectively change people's behavior and implement lubrication, inspection, condition

FIGURE 1.3   Asset Lifecycle Model.[3]

monitoring, and failure prevention activities. Technology also has a role here. Now smartphone technologies are an expected maintenance process support tool. Data accuracy is a given focus, and risk and reliability disciplines are expected. Asset performance management (APM) tools (to complement EAM tools) collect and respond to dynamic data triggers from the risk and reliability thresholds.

- **Cross-Stakeholder Continuous improvement**. Cross-stakeholder engagement is needed to enable a continuous improvement culture. When you engage the collective wisdom and experience of your entire workforce and adopt best practices from within and outside your organization, you will complete the journey to systematic maintenance management. But continuous improvement requires diligence, consistency, and commitment across all entity organizations working toward a common goal (asset value.) You need a method, a champion, strong management, and hard work to make it work. Information technology (IT) can be significant in driving continuous improvement to help an organization prepare and compete. Sustainability influences drive each organization to "exist" in their future and be environmentally responsible; preparing to be agile and automated is becoming a functional requirement as they move toward Industry 4.0.

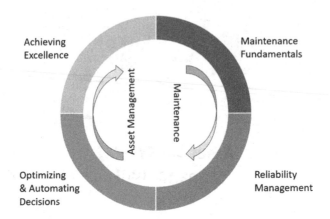

FIGURE 1.4   Phases toward achieving asset management excellence.[4]

1.2   ASSET MANAGEMENT (MAINTENANCE) EXCELLENCE FRAMEWORK

As you read on, you will learn how to manage your equipment reliability and optimize maintenance – the life cycle of your plant, fleet, facility, and equipment. The purpose of this book is to provide a framework. It is organized into four sections: (1) maintenance management fundamentals, (2) managing equipment reliability, (3) optimizing and automating asset management decisions, and (4) achieving asset management excellence. Throughout this book, the influences of technical support will be introduced.

1.2.1   Section I – Maintenance Management Fundamentals

Chapter 2 introduces a new fundamental understanding of how the asset classes have converged and how Asset Lifecycle dynamics have evolved.

Chapter 3 overviews the basic strategies, processes, and approaches for equipment reliability through work management and leading maintenance management methods and tools.

Chapter 4 explores the concept: if you can't measure it, you can't manage it. You'll learn how to monitor and control the maintenance process. We discuss the top-down approach, from setting strategic business plans to ensuring maintenance fully supports them. We also examine the shop floor's measures to manage productivity, equipment performance, cost management, and stockroom materials management. When you pull it all together, it becomes a balanced scorecard.

Chapter 5 discusses the latest computerized maintenance management and EAM systems. We describe the basics of determining requirements, then justifying, selecting, and implementing solutions that realize the benefits. Spare parts are the single most significant expense for most maintenance operations, so it's no wonder that poor maintainability is usually blamed on a shortcoming in parts, components, or supply.

Chapter 6 describes managing maintenance procurement and storing inventory to support efficient work management and equipment reliability. We show you how to wisely invest the limited parts inventory budget in yielding top service levels and high turnover.

1.2.2 Section II – Managing Equipment Reliability

Chapter 7 is about assessing risks and managing to international standards. We begin by looking at what equipment is critical based on several criteria, then develop methods for managing risk. Finally, we describe relevant international standards for managing risk to help you develop guidelines and easily assess an enterprise's risk state.

Chapter 8 summarizes the reliability-centered maintenance methodology as the most powerful tool to manage equipment reliability. This often misunderstood yet compelling approach improves reliability while helping to control maintenance costs sustainably.

Chapter 9 examines reliability by the operator and what many leading companies are naming their total productive maintenance programs. We describe how equipment management and performance is everyone's job, especially the operators.

1.2.3 Section III – Optimizing Maintenance Decisions

Chapter 10 provides the basics for using statistics and asset life costs in maintenance decision-making. We hear about a company's Six Sigma quality management program, but the link between "sigma," or standard deviation, and maintenance management is often subtle. This chapter shows you that collecting simple failure frequency can reveal the likelihood of when the next failure will happen. While exploring failure probabilities, we guide you through life cycle costs and discounting for replacement equipment investments.

With the fundamentals in place and ongoing effective reliability management, Chapter 10 begins the optimization journey. Here, we explore mathematical models and simulations to maximize performance and minimize costs over equipment life. You'll find an overview of the theory behind component replacement, capital equipment replacement, inspection procedures, and resource requirements. Various algorithmic and expert system tools are reviewed per their data requirements.

Chapter 11 focuses further on critical components and capital replacements. In addition, we include engineering and economic information in our discussion about preventive age-based and condition-based replacements.

Chapter 12 looks at how condition-based monitoring can be cost-effectively optimized. One of the maintenance manager's biggest challenges is determining what machine condition data to collect and how to use it. We describe a statistical technique to help make practical decisions for run-to-failure, repair, or replacement, including cost considerations.

Chapter 13 looks at a hybrid "case study" and how much of what is discussed in the preceding chapters can be analyzed and developed into a short-term and long-term business transformation road map.

1.2.4   SECTION IV – ACHIEVING MAINTENANCE EXCELLENCE

This section introduces some of the challenges of real estate and IT assets, summarizes concepts from this book, and prepares you for some of the evolving trends in the future of maintenance excellence.

Chapter 14 introduces the challenges driving Asset Management and Maintenance Excellence in Facilities and Real Estate assets.

This chapter focuses on "leading practice" approaches in the management of all aspects of corporate real estate and facilities management, notably including:

- Asset Strategy, Process, and Organization
- Lifecycle Delivery considerations
- Information and Communication Technology Systems
- Optimization.

Chapter 15 introduces the Information Technology Service Management Lifecycle.

IT people like to have their language and have adapted one that fits the nature of their delivery of services. However, when we take the time to analyze the IT languages and terms, it is remarkably similar to the maintenance strategies and services discussed throughout this book. This chapter will discuss some unique language used to describe IT Service Management while showing the parallelisms with other forms of maintenance and why it is part of this text.

Chapter 16 Information Technology Asset Management.

This chapter looks at what IT assets are and why we would want to manage them. We will spend some time here to understand why we include IT within the realm of maintenance excellence and why IT asset management has become a part of the enterprise asset used to deliver productivity to the organization.

Managing and optimizing IT assets is critical to delivering cost-effective support to the business. A multidisciplinary approach to managing these assets and processes allows IT organizations to minimize costs while maximizing return on assets and achieving targeted service levels of support to the business.

Chapter 17 shows how to apply the concepts and methods to the shop floor. This involves a three-step process. The first is determining your current state of affairs, the best practices available, and your vision. You need to know the size of the gap between where you are and where you want to be. The second step involves building a conceptual framework and planning the concepts and tools to execute it. Finally, we look at the implementation process and all that goes into managing change.

In Chapter 18, we look at "people" elements of achieving Asset Management Excellence. As in any process-intensive operation, a change enablement focus is required to expect a successful outcome. Leadership is the most significant influence on making successful change happen. Changing the culture from an existing state to one motivated to drive asset management to contribute to operational excellence is critical. Change enablement is even more relevant given the needed introduction of technology to automate many processes.

Chapter 19 will explore the evolution of Society 4.0, now being declared Society 5.0, and how that influences Industry 5.0 and the explicit requirement to consider greenhouse gas (GHG) environmental, safety, and product quality issues as important gates to be in business.

In Chapter 20, we will look at the evolution of the influence of Industry 4.0 toward asset management and the emerging technologies that may directly or indirectly influence the inevitable need to embrace. We will also look at what is needed to keep an enterprise competitive and the asset management staff essential to employ help from their IT colleagues.

1.3 THE SIZE OF THE PRIZE

We could debate an organization's social and business mandate with a significant investment in physical assets. But this is sure: it becomes irrelevant if these assets are unproductive. Productivity is what you get out of what you put in. Maintenance excellence is about getting exemplary performance at a reasonable cost. So what should we expect to invest in maintenance excellence? What is the size of the prize?

Let's look at capacity. One way to measure it is as follows:

$$\text{Capacity} = \text{availability} \times \text{Utilization} \times \text{Process Rate} \times \text{Quality Rate}$$

We have the required maintenance output if the equipment is available, running at the desired speed, and is precise enough to produce the desired quality and yield.

Now, look at the cost. This is a bit more difficult because the cost can vary depending on the working environment, the resources and energy required to accomplish capacity, equipment age and use, operating and maintenance standards, technology, etc. For example, a breakdown maintenance strategy is more costly than linking maintenance actions to the likely failure causes.

But by how much? Here's a helpful rule of thumb to roughly estimate cost-saving potential in an industrial environment:

$$\$1 \text{ Predictive, Preventive, Planned} = \$1.5 \text{ Unplanned, Unscheduled}$$
$$= \$3 \text{ Breakdown}$$

Accomplishing "one unit" of maintenance effectiveness will cost 1$ in a planned fashion, $1.50 in an unexpected way, and $3 if reacting to a breakdown.

In other words, you can pay now, or you can pay more later. Emergency and breakdown maintenance is more costly for several reasons:

- you must interrupt production or service
- the site isn't prepared
- whoever is available with adequate skills is pulled from their current work
- you must obtain contractors, equipment rental, and special tools quickly
- you have to hunt down or air-freight in materials
- the job is worked on until completed, often with overtime
- there usually isn't a clear plan or drawings

For example, if the total annual maintenance budget is $100 million, and the work distribution is 50% planned, 30% unplanned and unscheduled, and 20% breakdown:

50% × 1 + 30% × 1.5 + 20% × 3 = 50 + 45 + 60	155 "equivalent planned units"
Planned work costs 50/155 × $100 million	$32 million, or $.645 million per unit
Unplanned work costs 45/155 × $100 million	$29 million
Emergency work costs 60/155 × $100 million	$39 million

To compare the difference, imagine maintenance improvement yielding 60% planned, 25% unplanned, and 15% breakdown:

Planned work costs $.645 × 1 × 60%	$39 million
Unplanned work costs $.645 × 1.5 × 25%	$24 million
Emergency work costs $.645 × 3.0 × 15%	$29 million
Total	$92 million
Savings potential	$100 million − $92 million = $8 million

The focus on percent planned versus unplanned is still needed. So is the need for specific data elements to help an enterprise automate maintenance selection and frequency against critical assets. Industry 4.0, with the combined commitment from the asset management and IT staff, drives a focus from the percent planned to the percent predicted in an automated way. Imagine if you have this capability fully operating and the injection of machine learning (ML) and artificial intelligence (AI) how an enterprise can achieve operational excellence and enable critical assets to achieve the total "Return on asset" (ROA).

Imagine the "size of the prize" when an organization can utilize the gains they have achieved in executing planning and scheduling to a leading level and then, with the help of IT, leverage the available tools to move the focus from "planned maintenance" to "automated prescriptive maintenance." Data can be used to view the supply chain demand, asset, and operational constraints, and prescribe the operational risk of an organization's value chain.

Partnering with asset management and IT personnel is the "Art of the Possible" journey in asset management. Assets were acquired to drive value aligned with the enterprise's goals strategically. Asset Management optimization is achieved when successfully integrated with operations to optimize operational excellence. Change enablement (of stakeholders and culture), leadership, IT commitment to support, and the enterprise's asset management strategy will all be needed to achieve this success.

As you read through this edition, we hope you will continue to let it provoke thought on strategic asset management and maintenance issues and help you develop tactics that minimize breakdowns and maximize the rewards of planned, preventive, and predictive work. We also hope you will consider how the asset classes continue to evolve. IT is now a factor that will play a more vital role. We also need to ensure that the need for "green" environmental impact thinking across an asset's life cycle is included. In other words, it is not enough to effectively achieve asset capacity optimization. We must also consider the resource and energy impact the asset requires to reach capacity.

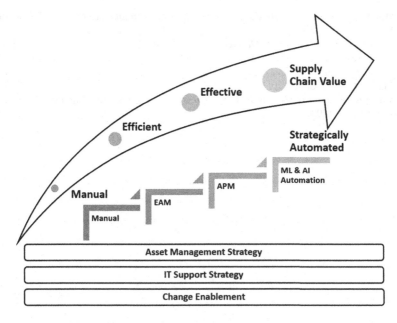

FIGURE 1.5 The asset management and IT continuum.[5]

PERMISSIONS

* Chapter adapted from pages 1–8, *Asset Management Excellence: Optimizing Equipment Life-Cycle Decisions*, 2nd Edition by Editor(s), John D. Campbell, Andrew K. S. Jardine, Joel McGlynn Copyright (2011) by Imprint. Reproduced by permission of Taylor & Francis Group.

1 Copyright pictures by D. Barry, Toronto, 2022.

2 Adapted from copyright Figure 13.1, *Asset Management Excellence: Optimizing Equipment Life-Cycle Decisions*, 2nd Edition by Editor(s), John D. Campbell, Andrew K. S. Jardine, Joel McGlynn Copyright (2011) by Imprint. Reproduced by permission of Taylor & Francis Group.

3 Adapted from copyright Figure 15.2, *Asset Management Excellence: Optimizing Equipment Life-Cycle Decisions*, 2nd Edition by Editor(s), John D. Campbell, Andrew K. S. Jardine, Joel McGlynn Copyright (2011) by Imprint. Reproduced by permission of Taylor & Francis Group.

4 Adapted from copyright Figure 1.4, *Asset Management Excellence: Optimizing Equipment Life-Cycle Decisions*, 2nd Edition by Editor(s), John D. Campbell, Andrew K. S. Jardine, Joel McGlynn Copyright (2011) by Imprint. Reproduced by permission of Taylor & Francis Group.

5 Adapted from copyright *Physical Asset Management Lecture Notes* (Part 1 and Part 2), UofT (2019), edited by (Barry). Reproduced by permission of Asset Acumen Consulting Inc.

BIBLIOGRAPHY

Don M. Barry. *IT and IoT in Asset Management Training*, Toronto, Asset Acumen Consulting Inc., 2021.

Don M. Barry. *Physical Asset Management Program Training*, Toronto, Asset Acumen Consulting Inc., 2022.

Don M. Barry. *Maintenance Parts Management Excellence: A Holistic Anatomy*, Boca Raton, CRC Press, Taylor & Francis Group, 2023.

John Dixon Campbell. *Uptime: Strategies for Excellence in Maintenance Management*, Portland, OR, Productivity Press, 1995.

J.M. Moubray. *Reliability-Centered Maintenance*, 2nd ed., Oxford, England, Butterworth Heinemann, 1997.

Section I

Maintenance Management
Fundamentals

2 Asset Classes and the World of Life Cycle Asset Management*

Don M. Barry
Previous contributions from Joel McGlynn and Frank "Chip" Knowlton

2.1 INTRODUCTION

In most corporate organizations today, tangible assets in real estate, manufacturing, transportation fleets, physical infrastructures, and information technology dominate the balance sheet. As a result, they are frequently one of the two highest overhead costs (after personnel and benefits). Our previous edition suggested that the financial investment in capital assets was growing aggressively.

Manufacturers in process industries, such as natural resources (mining, oil, and gas, etc.), utilities, transportation, and facilities management, are deeply asset-intensive and significantly depend on the value generated from their asset investments. As a result, maintaining and replacing capital assets can represent a significant portion of total operating costs and limit the ability to compete.

As an example, the typical plant maintenance cost ratio to Total Operating Cost in an enterprise by industry is estimated to be:

One of the organization's first challenges in determining the impact of assets on the bottom line is identifying and categorizing what constitutes an asset. For purposes of this book, we describe assets physically rather than from a financial portfolio perspective. As shown below, we have classified them into real estate and facilities, plant and production, mobile assets, infrastructure, and information technology equipment and networks.

Each asset class has unique characteristics to the assets typically found in that grouping.

Conversely, some distinct similarities in the overarching processes must be addressed regardless of the asset class. This fact attracts various Enterprise Asset Management (EAM) software providers to develop products that span all asset classes. Several examples include:

- **Asset management and configuration**
  - Track asset detail
  - Establish asset location and hierarchy
  - Monitor asset conditions

TABLE 2.1
Plant Maintenance Cost by Industry

- Steel mill = 29%
- Nuclear utility = 30%
- Bauxite mine = 52%
- Petrochem refinery = 28%

- Underground metal mine = 36%
- Non-ferrous smelter = 32%
- Pulp and paper mill = 26%

Hidden Costs can be:

- Safety
- Environment
- Legal compliance

- Service interruptions
- Shareholders

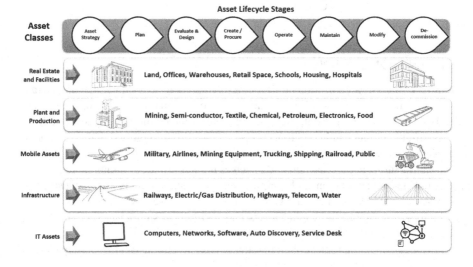

FIGURE 2.1   Asset classes and asset lifecycle stages.

Asset Class	What is Unique in some of these Asset Classes?
Real Estate and Facilities	- Asset hierarchies, value to stakeholders - Focus on location, construction, lease management
Plant and Production	- RCM, TPM, Production ROA focus - Configuration management
Infrastructure (Linear) Assets	- Asset hierarchies and data, by location - Depreciation and maintenance forecasting focus
Mobile Assets	- Asset configuration, regulatory compliance - Tracking mobile asset locations, timing of planned maintenance
IT Assets	- Asset Configuration, version management, change management

FIGURE 2.2   Unique requirements in some asset classes.

- **Work management**
  - Manage resources, plans, and schedules
- **Materials management**
  - Track inventory transactions
  - Integrate work management with materials management
- **Procurement**
  - Vendor management, vendor performance analysis, and KPIs
  - Event-driven purchasing
  - Enterprise-wide leverage in spend analysis
- **Contract management**
  - Manage vendor contracts
  - Manage alerts/notifications to optimize vendor SLAs
- **Service management**
  - Accept and manage new service requests
  - Manage Service Level Agreements

Companies increasingly face a competitive environment, requiring the development of more efficient and cost-effective operations than ever before. Many asset-intensive organizations are under intense globalization, shifting markets, outsourcing, and external regulation. These factors drive organizations to increase productivity, reduce costs, and improve product quality. A one percent improvement in performance can be worth millions of dollars annually for a manufacturer. In addition, service rates are often regulated, making business survival dependent on efficient management of capital assets using best practices and standards. As a result, organizations are now looking for ways to extend the capabilities of their existing systems. Some organizations have enjoyed, at times, double the Compound Annual Growth Rate (CAGR) in specific asset-intensive manufacturing sectors; optimizing the maintenance excellence of all the assets in each asset class will be a continued high priority for years to come. The fastest-growing asset type sectors are expected to be Equipment Assets and IT and Software Assets. EAM/APM/ Software and Services revenues from managing asset-intensive organizations are expected to grow between 5–12% CAGR, depending on the market dynamics and year. As the enterprise's asset management maturity grows, more and more investment is going to Asset Performance Management (APM) solutions.

In the past, asset management was described as maintenance management with an exclusive focus on the programs, procedures, and tasks necessary to optimize the uptime of an organization's equipment. Today, it requires active Lifecycle Management of the critical assets and components from design and inception to disposal to achieve a competitive edge against the competition. A more strategic view of asset management first requires a new consideration of which assets must be managed. In a traditional view, assets may only include items from a few categories, such as machines, factories, vehicles, or specific infrastructure. Or the responsibility for these items may have been lumped by their critical path to corporate values, job function, financing scheme, or procurement categories. This old approach has a few weaknesses. The company leaves its management to either chance or unstructured processes by

ignoring essential categories. The company may have difficulty prioritizing investment or cost savings decisions by not taking a complete portfolio view. In a traditional model, where different categories are managed separately, it can be near
impossible to weigh decisions against one another. For example, a cost-reduction
effort may be horrifically executed if the decision-maker can't balance equipment
cost and repair in the same analysis. It is easy to imagine internal turf wars and the
politics of asset category managers overriding decisions based on business strategy.
Imagine a forestry company mandated to cut costs. They improve margins by purchasing less expensive (and less reliable) equipment but not adjusting their repair
capabilities. At the same time, the logistics group under the same cost-reduction mandate reduces the number of vehicles available. In the short term, these decisions return
the mandated cost reduction, but soon an increased amount of equipment breakdowns
result in logging stoppages, and the lack of trucking capacity handicaps the company's ability to bring in backup equipment. The company then has to front money to
rent emergency replacement assets, pay repair teams huge premiums to work overtime, and perform "damage control" for angry customers missing shipments.
Production frequently stalls and the company misses revenue opportunities while its
field labor sits idle. The inability to analyze different asset attributes cripples the
company's ability to drive cost reductions when needed.

With an expanded view of asset classes, the asset manager can have a broader and
more complete influence over how the business spends and controls its prioritized
properties. This approach leaves fewer assets to be managed informally or by inconsistent procedures. By bringing more asset classes together (i.e., under a common
purview and portfolio), the asset manager can better support the business, including
investment decisions, performance decisions, or compromises across the entire
portfolio.

2.2 ASSET LIFECYCLE MANAGEMENT

An expanded view of classes brings new benefits to the completeness and rigor of
asset management. Similarly, an expanded view of the asset lifecycle provides a new
level of rigor and understanding. The practice of Asset Lifecycle Management (or
ALM) takes an expanded view of how assets are planned for, used, maintained, and
ultimately dispositioned. A traditional view often separates or ignores critical phases
within the asset lifecycle. For example, a procurement officer may buy new mobile
assets in an established company, such as planes, trains, buses, or ships. He is motivated (and probably measured) on specific criteria for success, most likely negoti-
ating low prices and meeting the needed number of vehicles. The maintenance of
these vehicles is managed by someone else whose job is to keep repair costs down.
Another manager may handle the financing and the disposition/liquidation by yet
another. While these job roles will always be needed, the company may have hurt
itself by not completely understanding the entire cycle. When managed separately,
we are inclined to ask: "Were repair costs factored in at the purchase?"; "Does the
company know the total cost of ownership?"; "Could smarter costing be possible if
finance and procurement worked with the entire portfolio?" Whether the company in

our example suffered from a lack of knowledge is unclear, but the fact that they may not find the answers demonstrates a primary shortcoming.

Shown below is the Asset Lifecycle Management framework that IBM formulated. It breaks down the lifecycle of assets into discreet phases of activity. In practice, companies should analyze their portfolio of assets (including the expanded view of asset classes) across the entire lifecycle to make decisions and define asset strategy. The framework comprises eight lifecycle phases of use and planning, each supporting financial management, and technology attributes to consider. See Figure 1.3 to view the Asset Lifecycle Framework.

2.2.1 Asset Strategy

Set an asset strategy that makes sense for the asset class and your company's business requirements. Activities may include assessing asset management practices, developing a comprehensive strategy, and developing a measurement program with key performance indicators (KPIs). Managers need to determine whether they own their primary assets or choose to access them "on-demand" or take a hybrid approach. For example, a chemical company might have a strategy where they own and maintain all equipment related to core manufacturing but decide that all customized product development be manufactured with leased infrastructure.

2.2.2 Plan

Clearly define asset targets, standards, policies, and procedures focusing on the asset management strategy delivery. Companies may wish to develop policies and standards and conduct Portfolio Asset Management Planning. This enables them to plan across the entire portfolio of assets. For example, a petroleum company able to plan land acquisition and equipment construction/repair simultaneously may be more agile in negotiations when purchasing equipment, better able to conduct discovery activities quickly, and better able to deal with emergencies.

2.2.3 Evaluate/Design

Evaluate the assets if purchased or design the assets that need to be created. This phase includes developing a capital program assessment model, which informs buying decisions. Computer-aided facilities planning can reduce the complexity of managing buildings, storage, and plants. For example, consider a pharmaceutical company ramping up to manufacture a new drug and building completely new manufacturing facilities and processes around the product lifecycle. The new product will require bioprocessing infrastructure (vats, bioreactors), manufacturing space, cold storage, shipping, new safety equipment, process monitoring technology, etc. By integrating the asset design plans with the product lifecycle, they will better understand their infrastructure spend regarding the overall product profitability and ensure that the asset management activities support the time frames of the product launch.

2.2.4 CREATE AND PROCURE

This phase involves creating, building, or procuring the planned assets. This phase may have one of the most visible impacts, as the first significant money is spent on asset management. New practices include Capital Project Management, e-MRO (i.e., automated/computerized materials resource optimization), and new procurement/ project delivery strategies. Imagine an asset procurement manager who can make purchasing decisions globally across all aspects of his production facilities with an integrated view. They can: make purchases with few redundancies and fewer shortages, negotiate with suppliers better, and manage the installation/delivery/deployment of assets in an integrated, coordinated fashion. They would also look for maintenance parts insights to support the asset maintenance strategy and the spare parts policies.

For example, when planning to install a new significant asset, a typical issue is understanding when the maintenance parts data will be identified and integrated into the enterprise's asset management toolset. Often, this is not well executed. The Design Engineer with Procurement needs to determine when the original manufacturer will receive the data and how that will integrate into the existing inventory placements, echelons, and policies. Questions that need to be carefully addressed include:

- What is the maintenance strategy for this new asset?
- What is the asset's planned lifecycle?
- Who will create and provide the parts data?
- How will this new asset be supported in the enterprise?
- Who signs off on the transfer or responsibility for design to operations?
- How will the needed parts support integration into the enterprise's inventory management policy?

The maintenance parts questions will be further discussed in the Maintenance Parts chapter (Chapter 6) of this edition.

2.2.5 OPERATE

Operate the assets per the strategy, using the standards, policies, and procedures with feedback into the ALM. The operation of assets is where performance is most affected (i.e., what value the assets deliver to the company). New practices in this area include formal IT Asset Management (ITAM), APM strategies, and Total Asset Visibility solutions. A mining company, for example, could track ore production to actual equipment ratios to understand which types of deployments are higher-producing. This operational data could be used when planning new asset acquisitions and deployments.

2.2.6 MAINTAIN

Maintain the assets supporting the strategy and targets using the standards, policies, and procedures with feedback into the ALM. Maintenance costs/resources can wildly alter the total cost of ownership, from repair costs to downtime. New practices

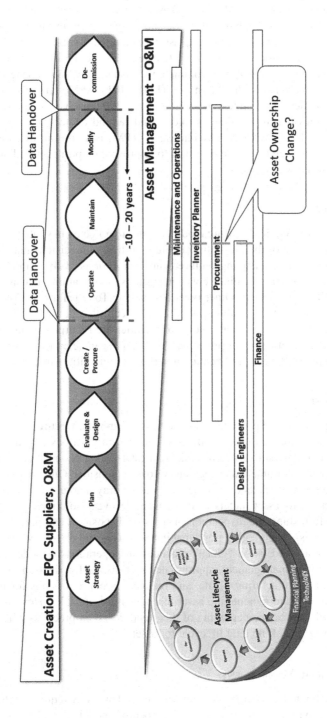

FIGURE 2.3   Asset management lifecycle and the data management dynamics.[1]

include conducting process improvement workshops with multi-disciplinary staff (i.e., users and technicians) and deploying EAM software systems. Predictive maintenance becomes a mainstay, based on understanding the past through failure databases, risk and reliability analysis, and asset tracking tools, ultimately lowering reactive maintenance allocations. EAM systems enable asset managers to track and manage assets across the enterprise, complete with centralized monitoring (even by a mobile device). IT sensors, the Internet of Things (IoT), APM solutions, and other "smart" technologies can be integrated with our assets themselves. Imagine factory robots or pipelines that report their problems and remind owners of their maintenance schedules. Total Productive Maintenance (TPM) is a cultural discipline that encourages operators and maintenance staff to improve the uptime and reliability of critical assets when well deployed.

2.2.7 MODIFY

Modify assets when required. Ensure modifications are reflected in strategy, policies, procedures, etc. Some of the most challenging modification decisions may come in IT-related assets, where changing requirements and options evolve rapidly. Many firms deploy strategies that facilitate constant systems modification, such as Service-Oriented Architectures (SOA). Other practices include Total Lifecycle Costing and Performance Improvement Analysis. Modification can also be crucial to the life extension of assets as machines are retooled, facilities are repurposed, and technology is adapted to facilitate newer processes. Medication can be either an asset modification or a support process modification in a risk and reliability environment.

2.2.8 DISPOSE

The strategy, policies, and procedures include asset disposition, retirement, or liquidation. The disposition can have significant financial implications beyond replacement. For example, market variations make real estate calculations in constant flux. Some assets have environmental or regulatory costs to consider. Other disposal strategies are finding new pockets of income from online gray markets. Other programs may include precious asset metals or component recovery solutions or initiatives focusing on refurbishing the disposition of parts or disposition equipment to minimize disposal costs. An emerging trend making headlines and driven by new regulations is the increased focus on "green" practices and operations. Practices such as sustainable facilities management, appropriate asset disposal, reduction of the carbon footprint at manufacturing plants, and reduced carbon emissions for the fleet are now becoming requirements the asset manager must consider. How assets are disposed of will only be the beginning of this trend, as green practices will need to move into every stage of the total lifecycle for assets.

2.2.9 FINANCIAL MANAGEMENT

Each phase has financial management implications and planning requirements. These are often most pronounced during the "create/procure" and "disposal" phases, but of great importance are also "operate" and "maintain," where financial performance is also affected. Maintenance, for example, can be a massive contributor to the Total Cost of Ownership (TCO), and "operate" performance can be a massive contributor to financial performance.

2.2.10 TECHNOLOGY

We refer to technology as an asset management tool (EAM, APM, AIP), not the asset itself (although the asset management system is indeed an asset). Technology can transform how each of these phases is planned for and executed. In an EAM system, models for planning and management are resident within a standard, centralized system. Active cataloging, monitoring, and measurement of assets are also tracked to aid repair actions, enable quick procurement/replacement decisions, and monitor performance. Technology also integrates the EAM/APM/AIP with other vital systems, such as accounting, procurement, and business performance management (BPM) dashboards. An EAM solution primarily tracks the asset and executes maintenance activity against it. A leading EAM solution would help the enterprise execute its maintenance tasks "Efficiently." An APM solution would typically need to be in place to hold the data and triggers for what maintenance to do. While the EAM helps execute maintenance "effectively," an APM helps manage the "efficient" portion of this relationship. The Asset Investment Planning (AIP) tool rounds out the defined scope of Asset Management Tools and is primarily used to help forecast capital planning dynamics for the enterprise.

2.2.11 ASSET-RELATED SUSTAINABILITY IMPACTS

With the evolution of Industry 4.0 moving to Industry 5.0, sustainability is becoming more and more critical and a gate that all large organizations must be seen to be addressing to operate. As a result, asset lifecycle planners must consider the environmental footprint of all significant assets and how they will contribute to the organization's environmental, social, and (corporate) governance (ESG) metrics. ESG standards are practices used to evaluate a corporate entity's social and environmental

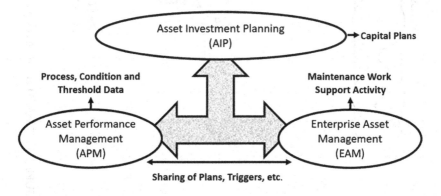

FIGURE 2.4 Asset management tools (EAM, APM, AIP).[2]

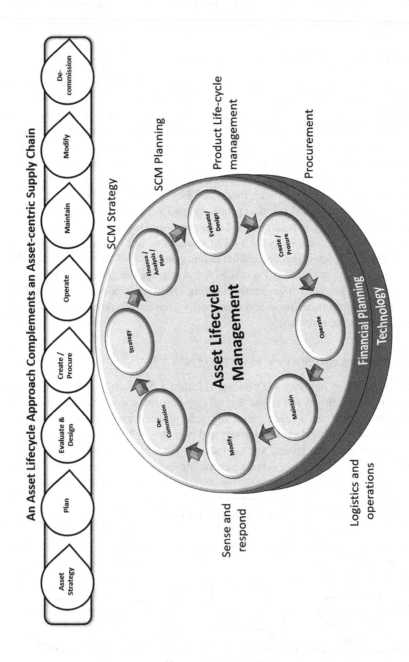

FIGURE 2.5   The ALM approach complements an asset-centric supply chain.

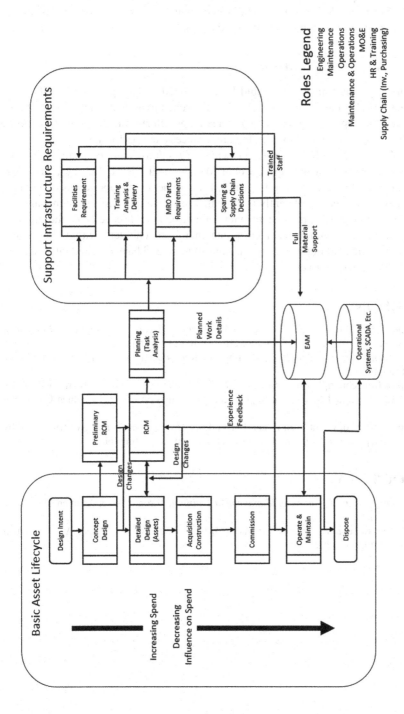

FIGURE 2.6   The ALM framework aligned to maintenance and reliability requirements.

impacts on its operational performance. The measurements can be both internal and external. For example, software can help a company measure and forecast its impact footprint and report within global standards.

2.2.12 ASSET LIFECYCLE MANAGEMENT APPROACH

Operationally, this framework should be formalized and programmatic within the organization. This means applying an ALM approach to asset management systems, integrating the approach into planning and strategy efforts, and using the framework to establish monitoring and metrics to gauge success and performance.

The ALM approach is consistent with asset-centric supply chain elements. Decisions associated with Supply Chain Management (SCM) Strategy, Planning, Product Lifecycle Management, Logistics, Procurement, and Operations of an organization are impacted significantly by how that organization's assets are managed.

Pictured another way, the systems view of a basic asset lifecycle and the supporting infrastructure components incorporate many stakeholders within the supply chain community. All of these elements contribute to and have a vested interest in an effective ALM perspective:

PERMISSIONS

* Chapter adapted from pages 9–19, *Asset Management Excellence: Optimizing Equipment Life-Cycle Decisions*, 2nd Edition by Editor(s), John D. Campbell, Andrew K. S. Jardine, Joel McGlynn Copyright (2011) by Imprint. Reproduced by permission of Taylor & Francis Group.
1 Adapted from copyright from *Physical Asset Management Lecture Notes*, (Part 1 and Part 2), UofT (2019), edited by (Barry). Reproduced by permission of Asset Acumen Consulting Inc.
2 Adapted from copyright from IT and IoT in *Asset Management Training* (2019) edited by (Barry). Reproduced by permission of Asset Acumen Consulting Inc.

BIBLIOGRAPHY

Don M. Barry. *IT and IoT in Asset Management Training*, Toronto, Asset Acumen Consulting Inc., 2019.
Don M. Barry. *Leading Practices in Asset Management*, Toronto, PAM, Asset Acumen Consulting Inc., 2022.

3 A Framework for Asset Management*

Don M. Barry
Previous contributions from Thomas Port,
Joseph Ashun, and Thomas J. Callaghan

3.1 INTRODUCTION

Today's maintenance and physical asset managers face tremendous challenges to increase output, reduce equipment downtime, lower costs, and do it all with less risk to safety and the environment. This chapter addresses the various ways to accomplish these objectives by managing maintenance effectively and efficiently within your organization's unique business environment.

This chapter overviews the multiple aspects of an effective and efficient maintenance management system. Of course, you must make tradeoffs, such as cost versus reliability, to stay profitable in current markets. We show you how to balance the demands of quality, service, output, costs, time, and risk reduction. This chapter examines how maintenance and reliability management can increase profits and add value to the enterprise.

We discuss the levels of competence you must achieve on the road to excellence. There are clear evolutionary development stages. You must ensure the fundamentals are in place to get to the highest levels of expertise. How can you tell if you are ready to advance? We provide a series of charts that will help you decide.

In the final sections of the chapter, we describe the methods used by companies that genuinely strive for continuous improvement and excellence. Therefore, we will briefly touch upon Reliability Centered Maintenance, Root-Cause Failure Analysis, and Decision-Making Optimization. This chapter sets the stage for material presented later in the book.

3.2 ASSET MANAGEMENT: TODAY'S CHALLENGE

Smart organizations know they can no longer afford to see maintenance as just an expense. Used wisely, it supports sustaining productivity and fuel growth while driving down unneeded and unforeseen overall expenses. Effective asset management aims to:

- maximize uptime (productive capacity)
- maximize accuracy (the ability to produce to specified tolerances or quality levels)
- minimize costs per unit produced (the lowest cost practical)

DOI: 10.1201/9781032679600-4

- minimize the risk that productive capacity, quality, or economic production will be lost for unacceptable periods
- prevent safety hazards to employees, and the public, as much as possible
- ensure the lowest possible risk of harming the environment
- conform to national and international regulations on due diligence (i.e., Sarbanes-Oxley, SOX).

In today's competitive environment, all these are strategic necessities to remain in business. The challenge is how best to meet the above demands. In many companies, you have to start at the beginning – put the basics in place – before you attempt to achieve excellence and optimize decisions that will be successful. The ultimate aim is to attain a high degree of control over your maintenance decisions, and in this chapter, we explore what it takes to get there.

3.3   OPTIMIZATION

Optimization is a process that seeks the best solution, given competing priorities. This entails setting priorities and making compromises for what's most important. For example, maximizing profits depends on keeping our assets in working order, yet maintenance sometimes requires downtime away from production capacity. Minimizing downtime is essential to maximize the availability of our plant for production. Optimization will help you find the right balance.

While increasing profit, revenues, asset availability, and reliability while decreasing downtime and cost are all related, you can't consistently achieve them. For example, maximizing revenues can mean producing higher-grade products that command higher prices. But that may require lower production volumes and higher costs per unit produced.

Cost, speed, and quality objectives can compete with each other. An example is improved repair quality by taking additional downtime to correctly do a critical machine alignment. The result will probably be a longer run time before the subsequent failure, but it does cost additional repair downtime in the short term.

The typical tradeoff choices in maintenance arise from trying to provide the maximum value to our "customers."

We want to maximize the following:

- Quality (i.e., repair quality, doing it right the first time, precision techniques)
- Service level (i.e., resolution and prevention of failures)
- Output (i.e., reliability and uptime)

At the same time, we want to minimize the following:

- Time (i.e., response and resolution time and Mean Time to Repair – MTTR)
- Costs (i.e., cost per unit output)
- Risk (i.e., predictability of unavoidable failures)

Management methods seek to balance these factors to deliver the best possible value. You must sometimes educate the affected group about your tradeoff choices to ensure "buy-in" to the solution. For example, a production shift supervisor may not see why you need additional downtime to finish a repair properly. Therefore, you have to convince them of the benefit of extended time before the subsequent failure and downtime.

A maintenance and reliability focus on sustaining the manufacturing or processing assets' productive capacity is needed. By sustaining, we mean maximizing the ability to produce quality output at demonstrated levels. This may require production levels beyond the original design that are realistically sustainable.

3.4 WHERE DOES MAINTENANCE AND RELIABILITY MANAGEMENT FIT IN TODAY'S BUSINESS?

The production assets are merely part of an entire product supply chain that produces a profit for the company. Therefore, it is essential to recognize that maintenance priorities may not be the most critical priorities for the company.

In an entire manufacturing supply chain, materials flow from the source (suppliers) through primary, and sometimes secondary, processing or manufacturing, then outbound to customers through one or more distribution channels. The traditional business focuses on purchasing, materials requirements planning, inventory management, and just-in-time supply concepts. The objective is to minimize work in process and inventory while manufacturing to ship for specific orders (i.e., the pull concept).

To optimize the supply chain, you optimize the flow of information backward, from customers to suppliers, to produce the most output with the least work. Supply chain optimizing strategies:

- Improve profitability by reducing costs
- Improve sales through superior service and tight integration with customer needs
- Improve customer image through quality delivery and products
- Improve competitive position by rapidly introducing and bringing to market new products.

Methods to achieve these include:

- Strategic material sourcing
- Just in time, inbound logistics and raw materials management
- Just-in-time manufacturing management
- Just in time, outbound logistics and distribution management
- Physical infrastructure choices
- Eliminating waste to increase productive capacity
- Using contractors or outsourcing partners
- Inventory management practices

Business processes stakeholders that are involved include:

- Marketing
- Purchasing
- Logistics
- Manufacturing
- Maintenance
- Sales
- Distribution
- Invoicing and collecting

At the plant level, you can improve the manufacturing part of the process by streamlining production processes through just-in-time materials flows. Doing this will eliminate wasted efforts and reduce the production materials and labor needed.

In the past, maintenance received little recognition for contributing to sustaining production capacity. Instead, it was viewed only as a necessary and unavoidable cost.

Even at the department level today, managers typically don't view the entire plant's big picture; they focus only on their departmental issues. Unfortunately, maintenance is often viewed only within the context of keeping costs down. Maintenance shows up as an operating expense in accounting, which should be minimized. After subtracting the cost of raw materials, maintenance is typically only a fraction of manufacturing costs (5–40 % depending on the industry). Those manufacturing costs are a fraction of the products' total selling price.

Reducing maintenance expenses does indeed add to the bottom line directly. Still, since it is a fraction of the total costs, it is often seen as less critical, commanding less management attention. Most budget administrators don't understand maintenance, judging it entirely by historical cost numbers. You could achieve lower maintenance costs by not doing any maintenance! But ignoring maintenance is a poor short-term strategy. When you reduce a maintenance budget, service ultimately declines. Also, the output is usually reduced, and the risk increases when there isn't enough time or money to do the work right the first time.

Of course, the accounting view is one-dimensional because it only looks at costs. When you consider maintenance's value, it becomes much more critical. You generate more revenue and reduce disruptions by sustaining quality production capacity and increasing reliability. This requires the correct application of maintenance and reliability. Of course, doing maintenance right means being proactive and accepting some downtime. Effective maintenance methods are needed to make the best possible use of downtime and the information you collect to deliver the best value to your production customers.

3.5   WHAT MAINTENANCE PROVIDES TO THE BUSINESS

Maintenance enhances production capacity and reduces future capital outlay. It does this by:

- Maximizing uptime
- Maximizing accuracy, producing to specified tolerances or quality levels

- Minimizing costs per unit produced
- Sustaining the lowest practical and affordable risk of loss of production capacity and quality
- Reducing as much as possible the safety risk to employees and the public
- Ensuring the lowest possible risk of harming the environment
- Ensuring a significant contribution to operational excellence, customer experience, and shareholder value

Notice the emphasis on risk reduction. Insurers and classification societies are keenly interested in their clients' maintenance efforts. Your maintenance reduces their risk exposure and helps keep them profitable. Nearly every time a significant accident involves a train, airplane, or ship, there is an in-depth investigation to determine whether improper maintenance was the cause of the disaster.

Maintenance can also provide a strategic advantage. The lowest-cost producer will benefit as companies automate production processes and manufactured goods are treated as commodities. Automation has reduced the size of production crews while increasing the amount and complexity of work for maintenance crews. Maintenance costs will therefore increase relative to direct production costs. Even low-cost producers can expect maintenance costs to rise. That increase must be offset by increased production. The ultimate goal is less downtime, higher production rates, and better quality at low unit cost – that means more effective and efficient maintenance.

Achieving all this requires a concerted effort to manage and control maintenance rather than letting the assets and their random failures control costs. You cannot afford to let that happen in today's highly competitive business environment. Unfortunately, many companies do just that, allowing natural processes to dictate their actions by operating in a "firefighting" mode. Without intervening proactively, these companies can only react after the fact, once failures occur. The consequences are low reliability, availability, and productivity – ingredients for low profitability.

3.6 READY FOR EXCELLENCE? – THE PYRAMID

Optimizing your effectiveness cannot be accomplished in a chaotic and uncontrolled environment. Optimization entails making intelligent and informed decisions. That involves gathering accurate and relevant information to support decisions and act promptly. The saying goes, "When up to the rear in alligators, it's difficult to remember to drain the swamp." You must have your maintenance system and process under control before optimizing effectively. It would be best to tame the alligators with suitable maintenance management methods, followed logically.

Several elements are necessary to achieve asset management (maintenance) excellence.

The pyramid shown below is modified slightly from the original.

Sponsors and leaders need to be aligned and seen as leading. Get aligned and exercise leadership at all times. Without it, change won't be successful.

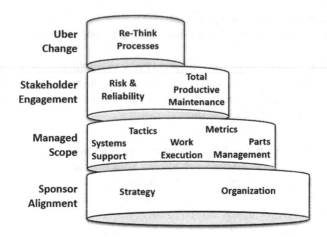

FIGURE 3.1   Maintenance excellence pyramid.[1]

Scope management over the assigned areas in which they have control and execute against their scope across the day-to-day maintenance operation.

Engage the affected cross-functional stakeholder group to facilitate a continuous improvement culture. Once you have this culture in place, you can maintain an industry-leading result.

Promote a tenacious pursuit of rethinking and elevating process effectiveness. This culture will set the stage for accelerated change in asset productivity.

3.6.1   SPONSOR ALIGNMENT

We are frequently advised to have a physical exam by a qualified physician once a year. Most individuals would agree with this advice, but many fail to act upon it. I'm feeling just fine! So, why bother going to the doctor? But, we are more likely to visit the doctor when we are ill and expect to have whatever ails us remedied in quick order. Similarly, a physical asset manager is more likely to inherit a plant struggling to meet safety, production, and cost objectives than they are to inherit one where all systems are in control. This situation requires the manager to start their work from the bottom up, understand how each step contributes to the overall plan, and then move the organization in the intended direction.

Asset management is a journey, not a destination. Sponsor alignment includes displayed leadership. For every journey, leadership is crucial to success.

The essential elements of leadership include the following:

- Before starting in a direction, a maintenance manager must know what they are starting. What resources are available? If we undertake this new work, what will not get done? How do we know where to start, the path forward, and where should this journey lead us?

- A physical asset manager starts with what they have in place and attempts to understand "what is working" and "what is not working." There is no point in expending much time, effort, and resources redesigning a system already delivering the specified outcomes. If a system is in place, is it being implemented as designed? A baseline audit will identify which systems are working and, if they are not working, why they are not working. Is the problem with the system design or how people in the organization are executing the design? Or perhaps both! An audit will measure a range of criteria comparing the system design to the actual execution by various parts of the organization. It will measure overall and unit effectiveness. It will point in the direction the journey must begin.
- Leadership is required at all levels in the organization. Leaders must:
  - Lead from the front
  - Remove barriers
  - Create a path for others to follow
  - Provide and support standards as a guide to operating (i.e., PAS55, ISO55000, or others)
  - Make room for others to contribute
- It is unnecessary to be the chief executive officer (CEO) or the senior manager to lead. A tradesperson, foreperson, planner, or middle manager can also lead. They must do so.

To design and implement a physical asset management process appropriate for the business needs, the manager must understand the following:

- What is the plant meant to do? What are the key performance goals?
- Is the plant safe and reliable? Do employees, management, and the community have confidence that the presence of the plant is a benefit and not an unreasonable or unknown risk to their well-being?
- Does the organization achieve its goals with a common purpose, communication, and teamwork throughout all organizational levels?
- Are people valued throughout the organization? Do they have the opportunity to reach their full potential regardless of where they come from, finance, maintenance, or operations?
- Does the plant have spare capacity, or must it run 24/7 to meet demand?
- Does the maintenance organization add value to the plant? Identifies what needs to be done and executes the work when it needs to be done without wasting human resources, materials, equipment, or process downtime.
- Does the organization optimize its capital resources and infrastructure before replacement?
- What steps are necessary to establish some level of control? For example, can the maintenance manager reduce variation?
- If the organization successfully implemented the intended systems, would we achieve all of the desired results?

3.6.2 Managed Scope

The managed scope highlights the areas that the maintenance team typically has direct influence and control.

- Work execution (i.e., planning and scheduling) to manage work and delivery of the service. Through careful planning, you establish what will be done, using which resources, and provide support for every job performed by the maintainers. You also ensure that resources are available when needed. Scheduling can effectively time jobs to decrease downtime and improve resource utilization. Execution delivers what was planned, when, and how it was planned.

- Parts management practices to support service delivery. A planner's job is to ensure that needed parts and materials are available before work starts. Shutdown schedules cannot wait for the bureaucratic grind of materials delivery. Schedules are set, and material is procured as required by the schedule. You need spare parts and maintenance materials to minimize operations disruptions. Effective materials management has the right parts available, in the right quantities, at the right time and distributes them cost-effectively to the job sites.

- Maintenance tactics for all scenarios – to predict failures that can be predicted, prevent failures that can be prevented, run to failure when safe and economical to do so, and recognize the differences. This is where highly technical practices are deployed, such as vibration analysis, thermography, oil analysis, non-destructive testing, motor current signature analysis, and judicious use of overhauls and shutdowns. These tactics increase the amount of preventive maintenance that can be planned and scheduled and reduce the reactive work needed to clean-up failures.

- Measurements of maintenance inputs, processes, and outputs to help determine what is and isn't working and where changes are needed. By measuring your inputs (labor, materials, services) and outputs (reliability or uptime and costs), you can see whether your management produces desired results. If you also measure the processes, you can control them to align the execution with the system design. Statistical Process Control Charts will tell the maintenance manager if a system is in control and when a process change occurs. Changing a specific "input" such as labor or materials, may not have the desired output. The annals of maintenance history are filled with evidence of throwing human resources at the problem with little to show.

- Data and systems support help manage the control and feedback information flow through these processes. For example, accounting uses computers to keep their books. Purchasing uses computers to track their orders and receipts and control who gets paid for goods received. Likewise, maintenance needs effective systems to deploy the workforce on the many jobs that vary daily and collect feedback to improve management and results. Asset Management systems have grown in type and functionality to include Enterprise Asset Management (EAM), Asset Performance Management (APM), Asset Investment Planning (AIP), solutions leveraging real-time data, and the Internet of Things (IoT). Today, the asset management team will need committed support from their IT teams to achieve and sustain leading practices.

3.6.3 STAKEHOLDER ENGAGEMENT

Engaging all the affected stakeholders is vital to successful continuous improvement. Continuous improvement involves a range of well-known methods and maintenance tactics. The best place to start is with the people closest to work, the operators, and the tradespeople. Many problems can be solved using simple quality management techniques, including on-site data collection, charting, fishbone diagrams, and Pareto charts. There is a time and place for advanced techniques discussed later in this book. But suppose the organization does not have the skills, knowledge, and discipline to execute the fundamental elements of the maintenance system design. In that case, one should not expect more advanced techniques to have meaningful results. Improvements such as Reliability Centered Maintenance (RCM) and Total Productive Maintenance (TPM) methods are described in detail in the chapters ahead.

3.6.4 DRIVING UBER CHANGE

Working environments, processes, and technology are moving at a significant pace. More and more working environments are remote due to a desire to minimize carbon footprints and the impact of cycling pandemics. Technological tools are developing at an accelerated pace as drones, remote monitoring, distributed expert systems, mobile and edge technologies, and the convergence of information technology and operations technology become more focused. The accelerated focus on environmental, social, and (corporate) governance (ESG) and environmental impacts will also influence change and the notion of Industry 4.0 moving to Industry 5.0. The change will not just be constant. The change will accelerate, and competitive organizations must prepare and be ready for "Uber Change."

3.7 ADD VALUE THROUGH PEOPLE

One of the most critical work components for the physical asset manager is to add value by developing people and assigning tasks
The manager adds value by:

- Establishing and communicating the vision for the future
- Establishing the maintenance strategy
- Establishing the fundamental values and standards by which the organization and people interact with each other
- Designing the physical asset management system
- Creating the organizational structure and role descriptions allowing the maintenance system to be executed and for people to reach their full potential
- Assign tasks with consistent clarity and measurable outcomes

- Establishing the accepted standards for the physical plant
- Clarifying issues for the role below

The middle management adds value by:

- Implements the system as designed.
- Acts in a manner consistent with the declared values
- Helps subordinates reach their full potential
- Assigns tasks with consistent clarity and measurable outcomes
- Achieves the established standards for the physical plant
- Removes barriers to successful execution and
- Clarifies issues for the role below

The Foreman and tradespersons add value by:

- Execute the plan and schedule
- Advises management of barriers to successful execution and
- Participate in problem-solving and continuous improvement
- Strive to reach their full potential

The work of the physical asset management leader is to effectively employ the available resources to optimize plant safety and reliability, leading to steady-state capacity. This work is done best when people are valued and there is a systematic approach to designing the required commercial, technical, and social systems in the workplace.

3.8   LEVEL OF ASSET MANAGEMENT MATURITY

The degree to which a company achieves maintenance excellence indicates its level of maturity. A maturity profile is a matrix that describes the organization's characteristic performance in each element. A high-level example of a charted matrix is shown in the following table. Once assessed, the affected organization can visually discuss which areas they need to improve or how they may get, for example, to the "Understanding level" across all elements. This is a good tool for communicating an organization's maturity baseline's current status and starting a discussion for change from this.

A method for determining the spectrum of asset management elements needed for asset management excellence is shown in Figure 3.3. It is presented in a series of profiles, supporting details for a cause-and-effect diagram. The effect we want is asset management (maintenance) excellence.

New elements are added to asset management, and the performance of existing elements improves as excellence is achieved.

Every leg of the diagram comprises several elements, each of which can grow through various levels of excellence:

- Excellence
- Competence

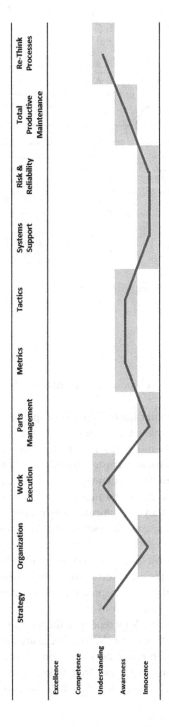

FIGURE 3.2 Example of a high-level asset management maturity matrix.[2]

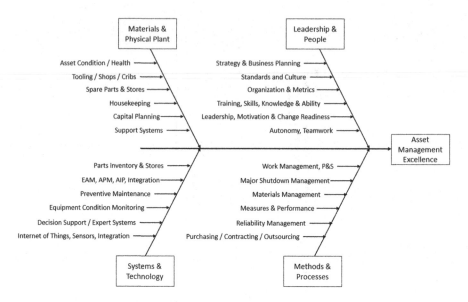

FIGURE 3.3   Elements of asset management excellence.

- Understanding
- Awareness
- Innocence

Tables 3.1–3.4 describe every level for each element from the above cause-and-effect diagram.

The first, Table 3.1, stakeholder leadership and people are the essential elements, although they are not always treated. As you can see, the organization moves from reactive to proactive, depends more heavily on its employees, and shifts from a directed to a more autonomous and trusted workforce. Ultimately it moves to a curated, trusted set of automated systems/processes integrated with operations. Organizations that make these changes often use fewer people to get as much, or more, work done. As a result, they are typically very productive.

The next most essential elements are usually the methods and processes, Table 3.2. These are all about how you manage maintenance. They are the activities that people in the organization do. Methods and processes add structure to the work that gets done. As processes become more effective, people become more productive. Conversely, poor methods and processes produce the much-wasted effort typical of low-performing maintenance organizations.

In Table 3.3, systems and technology represent the tools used by the people implementing the processes and methods you choose. These are the enablers and get most of the attention in maintenance management. With only basic tools and rudimentary technology, some organizations that focus tremendous energy on people and processes still achieve high-performance levels. Other organizations focused on the

TABLE 3.1
Leadership and People

	Strategy & Business Planning	Organization & Numbers	Training, Skills, Knowledge, & Ability	Motivation & Change Readiness	Autonomy & Teamwork
Excellence	Stated strategy with the mission, long-range vision, and goals exist. Goals are specific, measurable, achievable, realistic, and timed (for two or more years). Actions match words. Asset Management strategy linked with corporate goals.	Decentralized teams operate independently of daily maintenance control & may report to production—a constant interaction with production crew members. Maintenance supports teams.	Trades are primarily multi-skilled, with some multi-trade qualified individuals who regularly use their qualifications. Production staff do minor equipment upkeep tasks. Training time is at least two weeks per trade per year.	Trades' compensation has a reward component linked to business results. Competitive forces are accepted as driving the need for beneficial changes. Both management and the workforce initiate changes. Changes are usually successful, and measurable benefits are achieved.	Decentralized teams are self-directed and base decisions on business needs. Excellent cooperation between maintenance and production at all levels. Teamwork is a visible hallmark of the entire organization.
Competence	Strategy (as above) but not linked to corporate goals. Actions close to the words. Improvement assignments and accountability are clear.	Decentralized teams controlled by maintenance have plenty of interaction with production crew members. Leadership has the time and skills to lead.	Trades are primarily multi-skilled and regularly use their skills. Production staff do some minor equipment upkeep tasks. Training time 1 to 2 weeks per trade per year.	A cooperative atmosphere prevails, and trust between management and labor is high. Management always initiates change, and the need for change is explained in advance and widely accepted. Changes are usually successful.	Some self-directed workers and teams. Good cooperation between production and maintenance at all levels. Teamwork may be a feature of the entire organization.

(Continued)

TABLE 3.1 (CONTINUED)
Leadership and People

	Strategy & Business Planning	Organization & Numbers	Training, Skills, Knowledge, & Ability	Motivation & Change Readiness	Autonomy & Teamwork
Understanding	Some goal-setting for long-term and annual plans are used.	A mix of decentralized teams reporting to maintenance and a central shop structure exists.	Trades have some multi-skilling and often use those skills. Production staff do minimal minor equipment upkeep tasks. Training time is less than one week per trade per year. Training needs analysis completed for all trades. Apprenticeship programs are supported.	Some cooperation between management and labor exists, and the level of trust is moderate. The reason for the change is usually explained in advance. Changes sometimes fail.	Directed workforce with some teamwork but little to no team training. Some cooperation between maintenance and production at the operational level.
Awareness	Preventive Maintenance (PM) program in place benefits recognized.	Centralized structure based on trades breakdown. Control through maintenance supervisors/leads in response to production demands.	No multi-skilling is used. Production staff do no equipment upkeep. Training time less than one week per trade per year. Some training needs analysis performed.	Management motivation is explained when questioned. Some distrust but desire to improve exists. Changes often fail.	Directed workforce with no attempt at teamwork outside of shop structure. Good cooperation between production and maintenance leadership.
Innocence	Breakdown maintenance, firefighting, no stated goals.	Centralized structure based on trades breakdown. Action is directed primarily by operations supervisors.	No multi-skilling is used. Production staff do no equipment upkeep. Training is driven by necessity only.	Highly resistant to change. The hourly workforce generally distrusts management motives. No visible desire to improve. Change initiatives usually fail.	Directed workforce with no attempt at teamwork outside of shop structure. The maintenance and production relationship is strained.

TABLE 3.2
Methods and Processes

	Work Management	Major Shutdown Management	Materials Management	Measures & Performance Management	Reliability Management
Excellence	Planning and scheduling are well established. All work is tracked and managed in a system work order. All work except emergencies is planned at least 24 hours in advance. Priorities for work orders are used and respected. Work is usually scheduled at least one week in advance. Very few jobs are "held" because of material problems. "Wrench time" is high. Emergencies are few and far between. The atmosphere is orderly and controlled. Backlogs are managed at 2–4 weeks of work. Long-term planning for capital projects in place.	Shutdowns are planned over six months, with the lockdown of work scope providing sufficient time for long-lead item purchases. Formalized shutdown planning and management process in place. Heavy involvement by production, engineering, and maintenance in the process. Only emergency work arising is added during the shutdown. Shutdowns completed as or better than scheduled and achieved full work scope completion.	Automated inventory control and analysis are used in a fully integrated system. Stock levels are set based on sparing analysis with maintenance input. Automated management features used: stock re-orders, the grouping of purchases by vendor, pick list generation, bar code, or other automated issue and return processes. Maintenance is not involved in obtaining materials beyond the requested/specification stages. Service levels are high and well-managed. Inventory policy manages inventory placement and surplus identification.	Performance measures are a part of everyday life in the plant. All costs are captured and known by the type of cost, area, equipment, and work order. Company, plant, department, and team performance is measured, known, and used to target improvements. Process measures are effective in driving behavior. External benchmarking is used to drive improvement targets.	Plant reliability is high. EAM/APM is an effective tool to identify problem areas for resolution. Data is used from both EAM/APM and Expert/Decision Support systems to optimize decisions in maintenance. These systems are used by engineers, planners, and condition-monitoring teams. Full analysis for condition monitoring and PM tasks completed in all plant areas. Task frequencies and tasks are being refined through work order feedback. Root-cause failure analysis is used on all failures.Reliability is an embedded culture. Staff are all appropriately Risk and Reliability trained. RCM data is being used to automate mitigations to support operational excellence. ML/AI solutions complement Reliability culture where practical.

(Continued)

TABLE 3.2 (CONTINUED)
Methods and Processes

	Work Management	Major Shutdown Management	Materials Management	Measures & Performance Management	Reliability Management
Competence	Materials and resources are planned and usually staged before work is performed. Supervisors may plan maintenance, but trades are now focused on the tools. PM and inspection results are finding problem areas before they become failures. PM and Inspection programs are being changed with more emphasis on inspection. Planners are involved in sourcing materials but not purchasing. Net capacity is used in scheduling. Work order priorities are generally respected. Daily planning meetings handle only a few adjustments to the plan.	Production and stores involved in the shutdown planning process. Production scheduling is under control. There is confidence in the shutdown timing forecast. Shutdowns are usually completed in a scheduled downtime, but not all work is executed successfully. Minimal work is added at the last minute, but some arise during the shutdown. Finished crews usually get "extra" work, which risks extending the shutdown duration.	Stores computer records integrated with a maintenance system. They are using automated pick lists from the planning of work orders. Materials are kitted for work orders/ shutdowns/ maintenance pick-up projects. Service and inventory levels, stock turns known, monitored, and improved. Stores personnel manages inventory fully. All stock items, including capital spares, boneyard spares, and locations, are cataloged. Stock levels are set with maintenance input. Maintenance planning is involved in the sourcing of materials/ parts.	Performance measures are evident on bulletin boards for all maintainers to read. Measures include costs, results, and processes. Areas in general labor, materials, and contract categories break down costs. Some use is made of measures to drive improvement initiatives. Interest exists in comparisons with other plants.	Reliability is improving, with some significant gains known and possibly documented. RCM program training is in place and expanding. RCM programs develop data to be managed dynamically in APM solutions. Targeted improvement programs are in place and generally regarded as successful in achieving their objectives. Some reliability improvement trends are emerging. EAM/APM is being used to help identify problem areas. Formal analysis is being used to target condition monitoring and PM tasks in critical plant areas. Some Root-Cause Failure Analysis is being applied successfully. Risk-based inspections of pressurized containers are fully integrated into RCM.

Understanding	Planner or planning group in place. Technical support is available when needed but remains primarily focused on projects. Scheduling is done weekly with daily adjustments. Weekly and daily "planning" meetings held – lots of adjustments daily. Work priorities changed frequently or were disregarded. Little staging of resources before jobs are started. Resource planning is left to supervisors / lead hands and trades. Some feedback on PMs and Inspections results in changes in the PM program. Planners and supervisors may be doing a lot of purchasing.	Shutdown work arising from PM and inspections are added to the worklist for shutdowns. Some PM work and inspection work are added to shutdowns. Planners support planning and purchasing for shutdown work. Maintenance stages shutdown materials before shutdown. Planning may begin 1 to 2 months in advance if production can promise a downtime window for the shutdown. Shutdowns don't wholly work in scope in allotted downtime or run overtime. Some work was added at the last minute and during the shutdown. Most shutdown works completed indeed require shutdown.	Parts records on the computer. Stock levels set with no maintenance input – depending on vendor recommendations. Lead times and safety stock levels are set but rarely changed. Analysis of inventory levels carried out. Some cataloging of spares in work areas. Some use pre-expanded / ready-use spares in shop areas managed by stores or shops. Establishing stock items is onerous. The heavy involvement of maintenance supervision in managing parts, inventory, and sourcing parts.	Costs are tracked by labor, materials, and contract services but not analyzed. Downtime is measured overall, by area, and by the cause of the downtime. Bickering occurs between production and maintenance about "who's to blame" for downtime. Process performance is not measured, but it is judged in qualitative terms.	Reliability is low. RCM as a discipline is known but not committed. Some pilots have been run, but the program has stalled. Targeted improvement programs are driven by data collected in failure databases by maintainers/engineers. Improvement programs are viewed with skepticism due to their short track record. Improvements are primarily credited to only a few "gifted" individuals who solved the problems. Some Risk-based inspection mindset and disciplines.

(Continued)

TABLE 3.2 (CONTINUED)
Methods and Processes

	Work Management	Major Shutdown Management	Materials Management	Measures & Performance Management	Reliability Management
Awareness	Daily scheduling is attempted but primarily undermined by high-demand maintenance. Some technical help in troubleshooting exists. Inspection work and PM are scheduled.	Annual or other regular shutdown schedules to deal with equipment replacements, significant overhauls, and capital project tie-ins. Maintenance supervisors carry out minimal planning with some input from production. Shutdowns usually start late, finish late, and don't get all the work done. Plenty of work arising is added during the shutdown.	Parts records are kept manually. Stock levels are set but rarely changed – they may be high or low. Zero stock on bin checks triggers re-stock orders. Very difficult to add items to inventory. Maintainers are beginning to accept stores support.	Some downtime records are kept on critical equipment. Costs are known but not under control. Budget overruns are common.	Reliability is low. Little use of downtime records to target problem-solving. Any record keeping is regarded as non-value added.
Innocence	No serious attempt at scheduling or planning daily work.	Annual shutdown with the same work scope each year. Most work is overhauls. Planning is minimal and carried out by maintenance supervisors. Plenty of work was added at the last minute to clear backlogs (regardless of the need for an outage) and during the shutdown.	Sparing is minimal. Plenty of obsolete/unused spares. Some frequently used items are stocked. Large disorganized boneyard with plenty of uncatalogued material from old projects. Maintainers hold several unofficial caches of parts throughout the plant and shops. No or ineffective cataloging. Maintainers are looking out for themselves.	None. Budget overruns in maintenance are commonplace.	None.

TABLE 3.3

Systems and Technology

	Inventory & Stores	EAM/APM	Preventive Maintenance	Equipment Condition Monitoring	Decision Support & Expert Systems
Excellence	Highly integrated stores, maintenance, purchasing, finance, and other corporate systems. Full exploitation of automated features to remove manual effort. Inventory policy has a robust asset lifecycle and initial spare parts culture integrated with local demands.	Fully integrated maintenance, inventory, purchasing, and other systems in place. Maintenance, engineering, and production manage and use information as an asset. APM comprehensively tracks asset health, risks, and related thresholds integrated into the EAM.	PM is used only where analysis dictates – it is targeted at preventable failures that are time or usage-related.	All plant areas are analyzed for condition monitoring and PM needs. Condition monitoring is used extensively – probably by a dedicated crew. Most corrective repair work arises from condition checks.	Analysis/decision support tools for capital planning. Condition monitoring and EAM/APM data are linked for optimum inspection/replacement decision-making. Machine Learning (ML) and Artificial Intelligence (AI) in place and activity leveraged.
Competence	Stores systems linked with maintenance and purchasing. Some automated features and statistical analysis are in use. Some inventory policy data aligned to engineering stocking insights and local demand.	EAM is in place and linked with inventory and purchasing. Integration with other systems planned. Broad access to the system by maintainers in shops. Training completed and users capable.	Some PM is dropped in favor of condition monitoring. Overhauls are infrequent.	Critical plant areas are analyzed to determine condition-monitoring needs. More condition monitoring than PM used. Inspections reveal problem areas for correction.	Expert/analysis systems help manage Asset Health data and read/interpret condition-monitoring data. Business Intelligence tools are actively used. In addition, ML/AI is occasionally used.

(*Continued*)

TABLE 3.3 (CONTINUED)
Systems and Technology

	Inventory & Stores	EAM/APM	Preventive Maintenance	Equipment Condition Monitoring	Decision Support & Expert Systems
Understanding	Computerized stores inventory records. The inventory system may be linked with purchasing but not with maintenance.	EAM in place or being implemented with outside professional help. Although this may be planned, the maintenance system is not linked with inventory or purchasing.	Experience is used to modify manufacturer recommendations for PM frequency.	Time and usage-based inspections. Some condition monitoring (vibration, oil analysis, thermographic) is based on manufacturer recommendations, monitoring equipment vendor recommendations, or experience.	Contracted services interpret some condition-monitoring data (e.g., vibration, oil analysis, or Thermographic). Manual interpretation of results used in-house.
Awareness	Formalized but manual systems only.	Basic (daily) scheduling and work order tracking. No ability to check on parts inventory or other resource needs. EAM may be planned or in place with low functionality.	Reliance on equipment manufacturer recommendations for PM.	Time-based inspections outside of shutdowns.	Experience used to determine actions based on time-based inspections.
Innocence	Informal stock-keeping system.	No system at all. Most work is done on a demand basis. The backlog is managed in supervisor memories or on spreadsheets and notebooks.	Overhauls are used extensively. Annual shutdown for inspections and overhauls only.	None.	None.

TABLE 3.4
Materials and Physical Plant

	Asset Condition & Wellness	Tooling, Shops, & Cribs	Stores & Spares	Housekeeping	Capital Planning
Excellence	Asset health is well-tracked. Plant equipment is considered reliable and looks close to "new," even if old. Technology has been upgraded. The cleanliness of production equipment reveals a sense of ownership.	Purpose-built area centers. Tool crib attendant repairs tooling. Bays for fleet repair work by type of work with dedicated crews. Remote PLC programming. Separate lay down for work in process, materials receiving, and outbound completed work.	Access to stores is controlled but open to maintainers. Staging areas for pre-kitting all parts ordered in and from stock for specific work orders/projects before delivery to shops/maintainers. Large capital spares are located near the point of use or in special storage.	The production plant is well-kept and clean for production, maintenance, and office areas. Area teams take pride in equipment upkeep. Cleaning is seen as an effective tool for keeping equipment in good condition.	Multi-year long-range asset replacement strategy used for capital planning. The annual budget process refines a long-range plan.
Competence	Reliability is good, but improvement is being sought. The plant appears to be in good operating condition. Asset replacement is being done as needs arise with a one-year look ahead at equipment condition as an input to replacement decisions.	Area shops used. Shops adjacent to production areas. Central shops used to support area teams for jobs too large to handle. Fleet shops purpose-designed. Area and central shops all have online access to the maintenance management system. Tools are in a secure area adjacent shop/bay. Tool crib for special tools. Sign out system for tooling.	Access to stores is controlled but open to maintainers. Areas for pre-kitting of parts ordered for specific work orders/projects. Little "dust" indicates a few obsolete items. Boneyard is used only for oversized, weatherproof items.	Production and maintenance areas receive daily clean-up. Cleaning is still a "chore" carried out to medium standards. Dedicated clean-up crews may exist. Unique cleaning routines are used weekly or monthly to keep areas in good condition.	One-year look ahead for capital budget needs supported by equipment condition assessments. Capital expenditure history is used to forecast replacement funding for equipment upgrades. Betterment projects are treated as stand-alone projects.

(Continued)

TABLE 3.4 (CONTINUED)
Systems and Technology

	Asset Condition & Wellness	Tooling, Shops, & Cribs	Stores & Spares	Housekeeping	Capital Planning
Understanding	Reliability is improving quickly in the plant. The plant appears to be a bit run down but generally operable. Asset replacement is considered only if the need is evident and triggered by the annual budget cycle. Repairs carried out to "as good as new" standard.	Formal tool replacement mechanism and tool crib manned on all shifts, controlled and orderly. Tools tracked as stores items. Central shops are designated for each trade or production area. The shop layout follows the material flow. Shops are well-lit and ventilated. Area shops (if any) exist where space is available. Online access to the maintenance management system in shops.	"Ready-use" or "pre-expended" high-use low-value stock available in shop areas. Stores traffic is down, and access is well-controlled. Stores are orderly, and parts can be found quickly. Shops, shipping, and receiving areas are handy for stores. Separate areas exist for quarantine, warranty, receiving and shipping, and repairable awaiting work. The boneyard is cataloged and orderly.	Weekly clean-up of production and maintenance areas used. Standards of cleanliness are good but enforced only in cyclical clean-ups.	Asset replacements are identified only when the budget cycle calls for estimates.
Awareness	Reliability is low. Asset replacements occur only when uneconomic to repair anymore. Repairs are carried out to "as good as before" standard, not "as good as new." The plant is run down / tired. Some plant staff questions why the plant is kept open. Slow leaks allow contamination build-up that could hide other equipment problems.	Tool crib exists with informal control by either maintenance supervision or stores and is manned for one shift only with an arrangement for off-shift access. An informal mechanism is in place to replace tooling. Trade tools are stored in a designated location. Calibration program for tooling/instruments in place and used. The central shop is close to stores, and the layout mimics organizational department clusters, not workflow. The shop may be untidy, lifting equipment, and lay down areas marginal.	Stores appear orderly, but parts can't be found without a store person's help. A location system is in place. Access is controlled but loosely. Lots of maintenance "traffic" in stores. Stores are close to central maintenance shops and maybe close to shipping and receiving areas. Hazardous materials are segregated. Some parts packaging are damaged. Large unkempt boneyard.	Upkeep of plant production and maintenance areas is generally poor and neglected. It's a mess! Any clean-up is done only when repairs or other maintenance is performed.	No budgeting for capital for asset replacements. Each case is handled as it arises.

| Innocence | Reliability is low. The equipment is in disrepair. Numerous "band-aid" type fixes. Technology is out of date. Safety guards may be out of place. Leaking equipment is ignored or is slow to be repaired. | Trades have their own tools/share. No tool specs or standards. No tool crib, or it is uncontrolled. Tools not stored in designated areas. Central shops are generally in disarray. Unofficial production area shops may exist. Shop ventilation, lighting, lifting equipment, and lay-down areas are inadequate. Shop far from stores. Work in process not identified by job/area or work order. | Stores are a mess – disorderly. The location system is known only to stores keeper. Uncontrolled access – maintainers can find and take what they please. Hazardous materials are not segregated from other stores. Parts are not appropriately packaged or are poorly repackaged. Lighting poor. Store location is separate from receiving, shipping, and maintenance shop areas. Large unkempt boneyard. | Dirty, unkept, and uncleared spills may be safety hazards. Minimal clean-up is done before or after repair work. | No budgeting for capital for asset replacements. Each case is handled as it arises. Replacements with the cheapest available alternative are used. |

tools in the following figure haven't. Generally, emphasizing technology without excellence in managing methods, processes, and people will bring limited success. It's like the joke about needing a computer to mess things up. If inefficient or ineffective processes are automated and run with ineffective and de-motivated employees, the result will be disappointing.

Ultimately, how well you manage processes, methods, and the people who use them comes down to the Materials and Physical Plant you maintain. You can use these more detailed descriptions in the Physical Plant and Materials area to best judge your maintenance management effectiveness.

3.9 EFFECTIVE ASSET MANAGEMENT METHODS

Just as you must learn to walk before you run, you must be in firm control of maintenance before successfully beginning continuous improvement. Improvements can include automation as part of executed continuous improvement. You do this by incrementally changing what maintenance is doing to strengthen choices that will optimize business objectives.

Methodologies are systematic methods or procedures used to apply logical principles. The broad category of continuous improvement includes several maintenance methodologies covered in-depth in this book:

- Reliability Centered Maintenance (RCM) focuses on overall equipment reliability
- Root-Cause Failure Analysis (RCFA)
- Total Productive Maintenance (TPM) focuses on achieving high reliability from operators and maintainers
- Optimizing maintenance and materials management decisions

3.9.1 RCM

Reliability Centered Maintenance (RCM) aims to achieve maximum system reliability using maintenance tactics that can effectively apply to specific system failures in the operating environment. The RCM process uses equipment and system knowledge to decide which maintenance interventions to use for each failure mode. That knowledge includes:

- System diagrams and drawings
- Equipment manuals
- Operational and maintenance experience with the system
- Effects of individual failures on the system, the entire operation, and its operating environment

The RCM basic steps are:

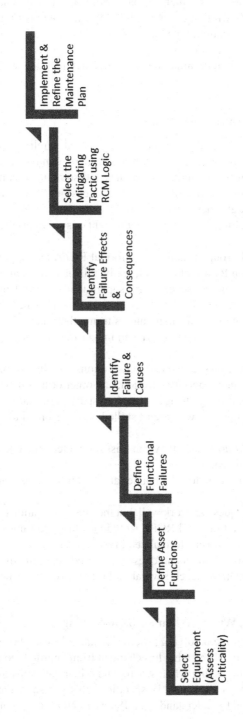

FIGURE 3.4 Reliability-centered Maintenance process steps.

RCM results in a maintenance plan. As part of the overall maintenance workload, the decisions made for each failure are put into logical groupings for detailed planning, scheduling, and execution. Each task states what must be done to prevent, predict, or find specific failures. For each, there is a specified task frequency. You use optimization techniques to determine the best frequency for each task and decide on corrective actions.

3.9.2   ROOT-CAUSE FAILURE ANALYSIS

Root-Cause Failure Analysis (RCFA) is one of the essential reliability enhancement methods. RCFA is relatively easy to perform, and many companies already do it – some using rigorous problem-solving techniques and some informally. Later in the chapter, you'll learn about a formal method based on easy-to-follow cause-and-effect logic, but first, we look at other problem-solving techniques used daily. In all methods, the objective is to eliminate recurring equipment or system problems or substantially diminish them.

Individuals or small groups usually use informal RCFA techniques to determine the best corrective action for a problem. Typically, this involves maintenance tradespeople, technicians, engineers, supervisors, superintendents, and managers. They often have immediate success, drawing heavily on their experience and information from sources like trade periodicals, maintainers from other plants, and contractors.

There are plenty of pitfalls, though, that can impair the informal approach:

- If only tradespeople do the RCFA, their solutions are often limited to repair techniques, parts and materials selection, and other design flaws
- A restrictive engineering change control or spare parts (add to inventory) process can derail people who aren't skilled or accustomed to dealing with bureaucracy
- If only senior staff do RCFA, they can miss out on technical details that the tradespeople would catch
- Some organizations tend to affix blame rather than fix the problem

In short, informal techniques can work well, but they have limitations, making it hard to develop long-term solutions. All RCFA techniques face the same challenges, but they're greater if the process isn't formalized in some way.

More formal problem-solving techniques can be used very effectively. Consulting and educational organizations teach several techniques, two of which we examine here.

3.9.2.1   RCFA-What, Where, When Problem-Solving

The first fundamental problem-solving process is relatively straightforward:

Establish the problem, noting what has changed from "normal" to unacceptable.

Describe the problem, asking what, where, and when questions to determine its extent. Quantify what went wrong and be specific so that you solve the problem only where it exists. You need to understand what IS and IS NOT happening now, as well as where and when.

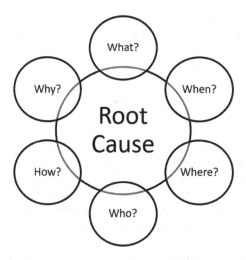

FIGURE 3.5 Cause-and-effect questions.

Identify possible causes.

Identify the most likely cause. Test these possible causes against the "is" and "is not" criteria for the "what, when, and where" of the problem statement.

Verify the cause. Test any assumptions you have made, looking for holes in the argument.

Implement a solution that addresses the cause.

These formal problem-solving techniques are effective and usually performed in a structured group with a facilitator. To use this approach, set up a weekly or monthly meeting to identify and prioritize problems that need solutions. Although day-to-day maintenance and equipment issues will figure prominently, your mandate will likely extend beyond them.

Problem-solving groups should include a cross-section of interested stakeholders such as production, finance, human resources, training, safety representatives, and maintenance. Because it's so broad, the group is often most effectively broken into smaller task teams. Each is assigned a problem to analyze and is usually responsible for solving it completely. These teams report to the problem-solving group on progress and solutions.

Formal problem-solving groups produce credible results because of their broad representation and rigorous analytical processes. This thorough approach ensures solutions that work over the long term.

3.9.2.2 RCFA-Cause-and-Effect

The second RCFA technique we examine is based on cause-and-effect logic. Theoretically, all events result from infinite combinations of pre-existing conditions and triggering events. Events occur in sequence, with each event triggering other events, some being the failures we try to eliminate.

Think of these sequences as "chains" of events, which, like chains, are only as strong as their weakest link. Break that link, and the chain fails – the subsequent events are changed. So while you may eliminate the failure you're targeting, you could also trigger some other chain of events. Remember that the solution to one problem may well be the cause.

To perform a cause-and-effect RCFA, you need to:

- Identify the unacceptable performance.
- Specify what is unacceptable (like the what, when, and where of the previous method).
- Ask, "What is happening?" and "What conditions must exist for this event?".
- Continue to ask this combination of "what" questions until you identify some event that can be controlled. If that event can be changed to prevent the failure from reoccurring, you have a "root cause" that can be addressed.
- Eliminate the "root cause" through an appropriate change in materials, processes, people, systems, or equipment.

By repeating the "what" questions, you usually get a solution within five to seven iterations. A variation of this process asks "why" instead of "what" – both questions work.

The cause-and-effect success rate is also high because it is performed formally, with a cross-section of stakeholders exercising complete control over the solution.

3.10   OPTIMIZING MAINTENANCE DECISIONS – BEYOND RCM

Managing maintenance goes beyond repair and prevention to encompass the entire asset lifecycle, from selection to disposal. Critical asset lifecycle decisions that must be made include:

- Component replacements
- Capital equipment replacements
- Inspection result decisions
- Resource requirements

To make the best choices, you need to consider technical aspects and historical maintenance data, cost information, and sensitivity testing to ensure you meet your objectives in the long run. The chart below describes several situations that can be dealt with effectively:

These methods require accurate maintenance history data that shows what happened and when. Specifically, it would be best to distinguish between repairs due to failures and those that occurred while doing other work. Note that the quality of the information is essential, not quantity. Some of these decisions can be made with relatively little information, so long as it is accurate.

RCM produces a maintenance plan that defines what to do and when. It is based on specific failure modes that are either anticipated or known to occur. Frequency decisions are often made with relatively little data or uncertainty. You can make yours more certain and precise by using the methods described later in this book and Andrew Jardine's perspectives. Along with failure history and cost information, these

TABLE 3.5
Potential Decisions Beyond RCM

Action	Scenario
Asset replacement	• As operating costs increase from usage or time • When equipment degrades from idleness (standby mode)
Capital equipment replacement	• To maximize value • To minimize total costs • To adopt new technology efficiencies over time
Optimize inspection frequencies	• To maximize profit, asset availability • To minimize downtime, total costs
Optimize processes	• Asset Overhaul practices • Group replacement practices
Sensor and monitoring equipment replacement	• Based on inspection results, costs, and history analysis data

decision-making methods show you how to get the most from condition-monitoring inspections to make optimum replacement choices.

These decisions can be made with sufficient data and, in several cases, computerized tools designed for the purpose. These tools are stand-alone and used primarily by engineers and highly trained technicians. When writing, combining them with computerized maintenance management and condition-based monitoring systems is just being explored. Maintenance management systems are growing beyond managing activities and transactions to becoming management and decision support systems. Eventually, this will make the job of monitoring personnel much easier.

PERMISSIONS

* Chapter adapted from pages 23–48, *Asset Management Excellence: Optimizing Equipment Life-Cycle Decisions*, 2nd Edition by Editor(s), John D. Campbell, Andrew K. S. Jardine, Joel McGlynn Copyright (2011) by Imprint. Reproduced by permission of Taylor & Francis Group.

1 Adapted from copyright Figure 13.1, *Asset Management Excellence: Optimizing Equipment Life-Cycle Decisions*, 2nd Edition by Editor(s), John D. Campbell, Andrew K. S. Jardine, Joel McGlynn Copyright (2011) by Imprint. Reproduced by permission of Taylor & Francis Group.

2 Adapted from copyright *Physical Asset Management Lecture Notes* (Part 1 and Part 2), UofT (2019), edited by (Barry). Reproduced by permission of Asset Acumen Consulting Inc.

BIBLIOGRAPHY

A.K.S. Jardine, & A.H.C. Tsang, *Maintenance, Replacement and Reliability*, Boca Raton, CRC Press, Taylor & Francis, 2017.

4 Measurement of Asset Management*

Don M. Barry
Previous contributions from Don M. Barry
and J. Stevens

4.1 INTRODUCTION

Performance management is one of the basic requirements of an effective operation. Effective and automated metrics can provide immediate feedback and influence the asset and maintenance management stakeholder's behavior to action. The key to successful asset management metrics is to have the right metrics serving the stakeholders' needs and driving behavior. Doing it well, though, isn't as straightforward as it may seem. This chapter discusses practical tools to define how to identify which metrics to use – depending on the business dimension and business driver. Getting this matrix of metric elements correct strengthens your cross-functional communications, maintenance performance measurements, and asset and maintenance management stakeholder behavior.

First, we summarize measurement basics with an example showing how numeric measures can be helpful and misleading. Measurement is essential in influencing asset and maintenance management stakeholder behavior, supporting continuous improvement, and identifying and resolving conflicting priorities within your business unit and other internal business units.

We explore metrics within the various maintenance business departments and between maintenance and the rest of the organization's influencers. We look at metrics that primarily serve why: the business (or asset) "**strategically**" exists, then look at metrics that may serve the **operational tactics** of the business (or asset); in more detail, we will look at metrics that may support process **functions** supporting the asset and maintenance activities; and if the focus is on a piece of specific equipment; we will look at **asset-specific** metrics as well. This business dimension approach supports both macro and micro approaches, using a variety of examples. At the macro level, we will explore the notion of a balanced scorecard (BSC) and, at the micro level, a shaft alignment case history.

Maintenance performance measurement business drivers are subdivided into five main components: productivity, organization, work efficiency, cost, and quality, together with some overall measurements of departmental results. You'll learn about consistency and reliability as they apply to measurement. A significant section of this chapter is devoted to individual performance measures, with sample data attached.

 DOI: 10.1201/9781032679600-5

TABLE 4.1
Tangible versus Intangible Benefits of a Focus on Asset Management

Tangible Benefit	Intangible Benefit
• Maintenance Cost Reduction	• Improved Sustainability
• MRO Inventory Reduction	• Improved Financial Control
• Fixed Asset Utilization	• Improved Brand and Customer Service

• Maintenance Cost Reduction	• Improved Sustainability
• MRO Inventory Reduction	• Improved Financial Control
• Fixed Asset Utilization	• Improved Brand & Customer Service

Tangible Benefits		Intangible Benefits

FIGURE 4.1 Example: High-level tangible outcomes versus intangible outcomes from improved maintenance management.[1]

We summarize the data required to complete these measures to decide whether you can use them in your workplace. We also cover the essential tie-in between performance measures and action.

The chapter concludes with a practical look at using the benefits/difficulty matrix to prioritize actions. You will also find a helpful step-by-step guide to implementing performance measures.

4.1.1 MAINTENANCE ANALYSIS: THE WAY INTO THE FUTURE

Some maintenance management outputs are easily recognized and measured – others are harder. As with the inputs, some outputs may be intangible, like the team spirit from completing a difficult task on schedule or business sustainability.

Asset and maintenance management is all about ensuring that an organization can achieve a tangible "return on asset" (ROA) optimally without compromising environmental, safety, or health standards the organization or community it operates.

When assets become more available, and there is a market for the goods or services produced from the asset availability, the benefits from achieving asset and maintenance management success should be considered "tangible."

More effective and efficient maintenance execution can also influence the maintenance cost and the need for parts usage. Therefore, improved asset availability can drive tangible financial output benefits.

"Intangible" outcomes can be attributed to improved asset and maintenance management. They can be high levels of corporate sustainability, better customer service levels (derived from asset availability), and dependable asset and maintenance cost management (derived from the ability to forecast asset risk).

When specifically looking at asset and maintenance performance measurement: there is no mystique to it. The trick is to leverage the results to drive the behaviors that will achieve the needed actions. This requires consistent and reliable data, a good understanding of the influences, high-quality analysis, a clear and persuasive presentation of the information, and a receptive and empowered work environment.

Since asset and maintenance management optimization is targeted at executive management and the boardroom, the results must reflect the basic business notion that asset management and maintenance is a business process turning inputs into usable outputs.

Measuring business drivers that the stakeholders believe they can influence is vital. Otherwise, buy-in to these metrics will be challenged, and the ability to improve outcomes will be undermined with frustration.

While intangibles contribute significantly to maintenance performance, this chapter focuses on tangible measurements.

The following diagram (Figure 4.2) shows the three major elements of the maintenance value equation – the inputs, outputs, and conversion process.

Figure 4.2 shows that many inputs are familiar to the maintenance department and readily measured – such as labor costs, materials, equipment, and contractors. However, some inputs are more challenging to measure accurately – including experience, techniques, teamwork, and work history – yet each can significantly impact results.

Converting the maintenance inputs into the required outputs is typically the core of the maintenance manager's job. Yet rarely is the absolute conversion rate of much interest in itself. Converting labor hours consumed into reliability, for example, makes little or no sense – until it can be used as a comparative measure through time

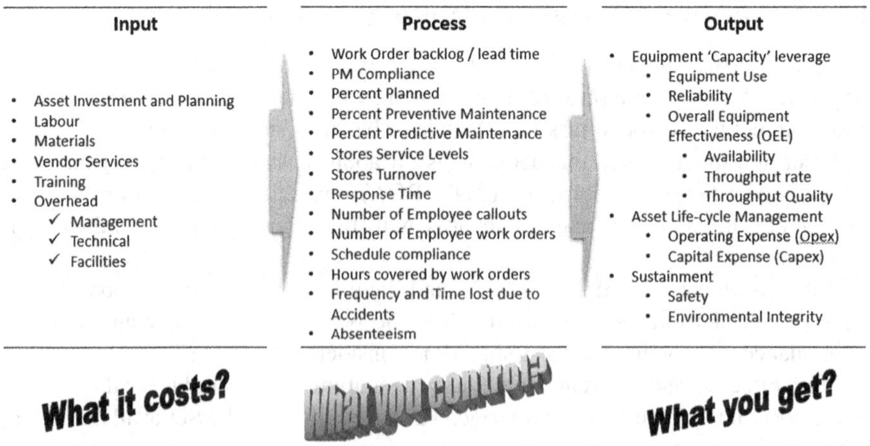

Input	Process	Output

• Asset Investment and Planning
• Labour
• Materials
• Vendor Services
• Training
• Overhead
   ✓ Management
   ✓ Technical
   ✓ Facilities
	• Work Order backlog / lead time
	• PM Compliance
	• Percent Planned
	• Percent Preventive Maintenance
	• Percent Predictive Maintenance
	• Stores Service Levels
	• Stores Turnover
	• Response Time
	• Number of Employee callouts
	• Number of Employee work orders
	• Schedule compliance
	• Hours covered by work orders
	• Frequency and Time lost due to Accidents
	• Absenteeism
		• Equipment 'Capacity' leverage
		   • Equipment Use
		   • Reliability
		   • Overall Equipment Effectiveness (OEE)
		     • Availability
		     • Throughput rate
		     • Throughput Quality
		• Asset Life-cycle Management
		   • Operating Expense (Opex)
		   • Capital Expense (Capex)
		• Sustainment
		   • Safety
		   • Environmental Integrity

What it costs? **What you control?** **What you get?**

FIGURE 4.2   Maintenance as a business process.[1]

or with a similar division or company. Similarly, the average material consumption per work order isn't significant until Press #1 consumes twice as much repair material as Press #2 for the same production throughput. A simple way to reduce materials per work order consumption is to split the jobs, increasing the number of work orders (this doesn't do anything for productivity improvement, of course).

The focus must be on the comparative standing of your company or division or improving maintenance effectiveness from one year to the next. These comparisons highlight another outstanding value of maintenance measurement – it regularly compares progress toward specific goals and targets. Through time, with other divisions or companies, this benchmarking process is increasingly used by senior management as a critical indicator of good asset and maintenance management. Frequently, it discloses surprising discrepancies in performance. For example, a recent benchmarking exercise turned up the following data from the Pulp industry.

The results show some significant discrepancies in the overall cost structure and how Company X does business. For example, it has a heavy management structure and hardly uses outside contractors. You can see from this high-level benchmarking that something needs to be done to preserve Company X's competitiveness in the marketplace. Exactly what, though, isn't apparent. This requires a more detailed analysis.

As you'll see later, the number of potential performance measures far exceeds the maintenance manager's ability to collect, analyze, and act on the data. Therefore, an essential part of any performance measurement implementation is thoroughly understanding the few key performance drivers needed to achieve success. Maximum leverage should always take top priority. So, first, identify the indicators that show results and progress in areas that most critically need improvement. As a place to start, consider 4-3.

If the business could sell more products or services at a lower price, it would be cost-constrained. The maximum payoff will likely come from controlling inputs (i.e., labor, materials, contractor costs, and overheads). If the business can profitably sell all it produces, production is constrained, and it's likely to achieve the most significant payoff from maximizing outputs through asset reliability, availability, and maintainability.

TABLE 4.2
Example of Pulp Industry Maintenance Metrics

Maintenance Metric	Average	Company X
Maintenance costs – $ per ton output	78	98
Maintenance costs – $ per unit of equipment	8900	12,700
Maintenance costs as % of asset value	2.2	2.5
Maintenance management costs as % of total maintenance costs	11.7	14.2
Contractor costs as % of total maintenance Costs	20	4
Materials costs as % of total maintenance costs	45	49
Total number of work orders per year	6600	7100

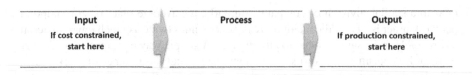

Input	Process	Output
If cost constrained, start here		If production constrained, start here

FIGURE 4.3 Maintenance optimizing – where to start.

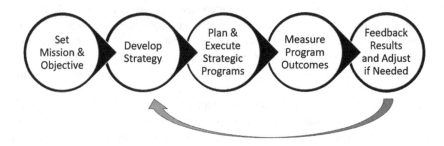

FIGURE 4.4 The maintenance continuous improvement loop.

4.1.1.1 Keeping Maintenance in Context

As an essential part of your organization, maintenance must adhere to the company's overall objectives and direction. Maintenance cannot operate in isolation. The continuous improvement loop (Figure 4.4) is vital to enhancing maintenance that must be driven by and mesh with the corporation's planning, execution, and feedback cycle.

Disconnects frequently occur when the corporate and department levels aren't in synch. For example, if the company places a moratorium on new capital expenditures, this must be fed into the equipment maintenance and replacement strategy. Likewise, if the corporate mission is to produce the highest possible quality product, this probably doesn't correlate with the maintenance department's cost minimization target. This type of disconnect frequently happens inside the maintenance department itself. For example, if your mission is to be the best-performing maintenance department in the business, your strategy must include condition-based maintenance and reliability. Similarly, if the strategy statement calls for a 10% reliability increase, reliable and consistent data must be available to make the comparisons.

4.1.1.2 Understanding the Role of Conflicting Priorities for the Maintenance Manager

In modern industry, all maintenance departments face the same dilemma: which of the many priorities should they put at the top of the list? Should the organization minimize maintenance costs or maximize production throughput? Does it minimize downtime or concentrate on customer satisfaction? Should it spend short-term money on a reliability program to reduce long-term costs?

Corporate priorities are set by the senior executive and ratified by the Board of Directors. These priorities should then flow down to all parts of the organization. When received: the maintenance manager must adopt those priorities, convert them

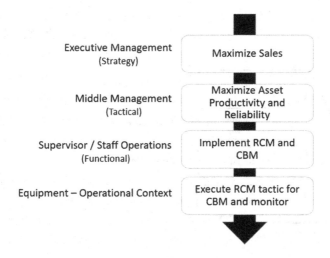

FIGURE 4.5 Inter-relating corporate priorities with maintenance tactics.

into corresponding maintenance priorities, strategies, and tactics to achieve the results, then track and improve them.

Figure 4.5 is an example of how executive management corporate priorities can flow down through the asset and maintenance management priorities and strategies to the tactics and functions that control the everyday work of the maintenance department.

These levels of business dimensions should not be a new or alarming event for the experienced maintenance manager. All enterprises have complementary but often different metrics and influences to satisfy their business dimension level.

A corporation's **executive team** will be looking to apply and measure the **strategies** contributing to ROA, business growth, market success, and improved competitiveness throughout the value stream. At this highest level, they will look for metrics that provide insight into their business integrity focus, including corporate alignment to their annual report, society/sustainment responsibilities, and affirmation of good governance.

Supervisory and middle management will look for BAU (business as usual) business operations/processes/**tactics** throughout the value stream. They may be looking for metrics that provide insight into their capital and asset integrity focus at this level. Supervisory metrics could include return on investment (ROI)/cost-effectiveness, asset lifecycle management, overall equipment effectiveness (OEE), environment, safety, health metrics (ESH), and confirmation of operational alignment.

The **crafts and supporting staff** will be supporting processes that manage **functions** and support operational requirements that contribute to the organization's desired performance and value, often including financial, ESH, operations, and strategy. The staff will look for metrics that provide insight into their process integrity focus at this highest level. Craft metrics may include asset availability, reliability, OEE elements, ESH elements, resource scheduling, operational support, maintenance execution, and supporting functions (i.e., parts, procurement, vendor support).

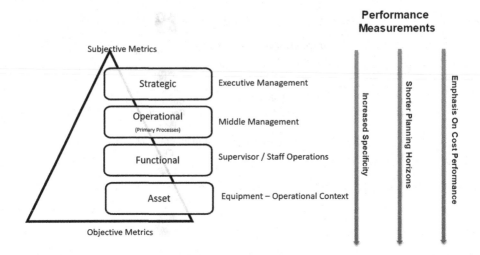

FIGURE 4.6 Business priorities by business dimension.[2]

The team that manages a specific **asset's performance** would review each asset's reliability specific to its' operating context, function, functional failure, and failure mode tracking to a prescribed maintenance tactic, frequency, and execution. At this asset-specific level, the team will be looking for metrics that provide insight into the team's asset integrity focus, including asset availability, reliability, OEE elements, ESH influences, and asset risk, and ensuring that reliability elements are understood and are well executed, providing Operational support, Maintenance execution.

As we look through the "business dimensions" from the executive level to the operational, functional, and asset level for measurements, the data specifics and emphasis on cost performance increase while the planning horizon or timeline shortens.

At the highest corporate level, the BSC will have metrics that include: people, financial, operational, and customer/market business drivers.

The specific "business drivers" a leading enterprise may elect to measure will span many of the elements of a BSC. For example – when specifically focused on asset management, elements of a BSC may see metrics that focus on process and quality, asset productivity, cost management, ESH, and resource utilization.

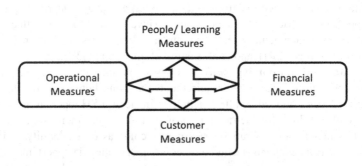

FIGURE 4.7 High-level corporate balanced score card business drivers.[2]

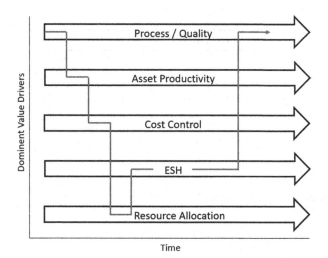

FIGURE 4.8 Dynamic focus change of business drivers depending on the business results or market dynamics.[2]

Like managing cash flow and priorities in a home, the focus may change from year to year or quarter to quarter as the executive team recognizes that a focus must change to keep the enterprise on track. For example, Figure 4.8 suggests that the measurement focus jumped from process/quality to asset productivity, cost control, resources to ESH, and then back to process/quality over a set period.

Maintenance can maximize product sales if the corporate priority is to maximize throughput and equipment reliability. In turn, the maintenance strategies should also reflect this and could include, for example, implementing a formal reliability enhancement program supported by condition-based monitoring (CBM), sensors, or Internet of Things (IoT) calibration. The daily, weekly, and monthly tactics flow out of these strategies. These, in turn, provide lists of individual tasks which then become the jobs that will appear on the work orders from the EAM (Enterprise Asset Management System). Using the work order ensures that the inspections are done across the business. Organizations frequently fail to complete the follow-up analysis and report regularly and timely. The most effective method is to set them up as weekly work order tasks, subject to the same performance tracking as preventive and repair work orders.

Leading practice organizations may elect to create a set of matrixed metrics that show the influence of business dimensions and business domains to display each of the measurements – for example- against a set of maintenance parts management processes. A partial example of a matrixed set of metrics is displayed in Figure 4.9.

You can be confronted with many seemingly conflicting alternatives when trying to improve performance. Numerous review techniques are available to establish how your organization compares to industry standards or leading maintenance practices. The most effective techniques help you map priorities by defining the improvements' benefits. The review techniques tend to be split into macro (covering the entire maintenance department and its relation to the business) and micro (focusing on a specific piece of equipment or a single aspect of the maintenance function).

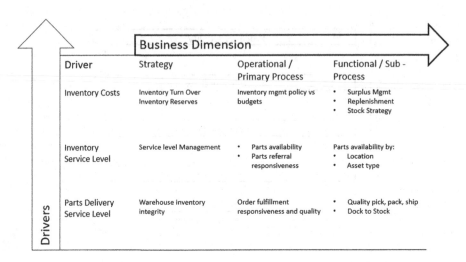

FIGURE 4.9 Example – partial list of business driver measurement points by business dimension (maintenance parts management).[2]

The leading macro techniques are:

- Maintenance Effectiveness Review – involves the overall effectiveness of maintenance and its relationship to its business strategies. These can be conducted internally or externally and typically cover areas such as:
  - Maintenance strategy and communication
  - Maintenance organization
  - Human Resources and employee empowerment
  - Use of maintenance tactics
  - Use of reliability engineering and reliability-based approaches
  - To equipment performance monitoring and improvement
  - Information technology and Management systems
  - Use and effectiveness of planning and scheduling
  - Materials management in support of maintenance operations
- External benchmarks – draw parallels with other organizations to establish how the organization compares to industry standards. Confidentiality is crucial, and results typically show how the organization ranks within performance indicators. Some of the areas covered in benchmarking overlap with the maintenance effectiveness review. Additional topics include:
  - Nature of business operations
  - Current maintenance strategies and practices
  - Planning and scheduling
  - Inventory and stores management practices
  - Budgeting and costing
  - Maintenance performance and measurement
  - Use of EAM and other IS (Information System) tools
  - Maintenance process re-engineering

- Internal comparisons – measure a similar set of parameters as the external benchmark but draw from different departments or plants. They are generally less expensive and illustrate differences in maintenance practices among similar plants if the data is consistent. From this, you can decide which best practices to adopt.
- Leading Practices Review – looks at asset and maintenance management's process and operating standards and compares them against the industry's best. This is generally the starting point for a maintenance process upgrade program, focusing on areas such as:
  - Preventive Maintenance
  - Inventory and Purchasing
  - Maintenance Workflow
  - Operations Involvement
  - Predictive Maintenance
  - Reliability-Based Maintenance
  - Total Productive Maintenance
  - Financial Optimization (Continuous Improvement)
- Overall Equipment Effectiveness (OEE) – measures a plant's overall operating effectiveness after deducting losses due to scheduled and unscheduled downtime, equipment performance, and quality. Each case's sub-components are meticulously defined, providing a few reasonable objectives and widely used equipment performance indicators.

Following is a summary of one company's results. Remember that the individual category results are multiplied to derive the final result. For example, although Company Y achieves 90% or higher in each category, it will only have an OEE of 74% (see Chapter 8 for further details of OEE). By increasing the OEE to 95%, Company Y can increase its production by (95–74)/74 = 28% with minimal capital expenditure. You won't need to build a fourth if you can accomplish this in three plants.

These, then, are some of the high-level indicators of the effectiveness and comparative standing of the maintenance department. They highlight the critical issues at

TABLE 4.3
Example of How OEE Is Measured and Applied to an Asset System

	Target	Company Y
Availability	97%	90%
X		
Utilization rate	97%	92%
X		
Process efficiency	97%	95%
X		
Quality	99%	94%
=		
Overall equipment effectiveness	90%	74%

Macro: External Benchmark		Micro: Internal Benchmark

Macro: External Benchmark	Micro: Internal Benchmark
Maintenance costs per ton are 15% above industry standard	1. Set targets for reduction 2. Implement means of tracking costs to equipment & jobs. 3. Analyse breakdown of costs among equipment, jobs and cost types 4. Examine and compare repair methods 5. Apply "Leading Practice" in all areas

FIGURE 4.10   Relating macro measurements to micro tasks.

the executive level, but a more detailed evaluation is needed to generate specific actions. They also typically require senior management support and corporate funding – not always a given.

Fortunately, many maintenance measures do not require external approval or corporate funding. These are important because they stimulate a climate of improvement and progress. Some of the many indicators at the micro level are:

- Post (systems) implementation review to assess the results of buying and implementing a system (or equipment) against the planned results or initial cost-justification
- Machine reliability analysis/failure rates – targeted at individual machines or production lines
- Labor effectiveness review – measuring staff allocation to jobs or categories of jobs compared to last year
- An analysis of parts usage, equipment availability, utilization, productivity, losses, costs, etc.

These indicators give helpful information about the maintenance business and how well its tasks are being performed. You must select those that directly achieve the maintenance department's goals and the overall business.

Moving from macro or broad-scale measurement and optimization to a micro model can create problems for maintenance managers. You can resolve this by using the macro approach as a project or program and the micro indicators as individual or series of tasks. The example in Figure 4.9 shows how an external benchmark finding can be translated into a series of actions that can be readily implemented.

4.2   MEASURING MAINTENANCE – THE BROAD STROKES

You need measurement capability for all major pain points under review to improve maintenance management. However, as previously mentioned, there usually aren't enough resources to surpass a relatively small number of critical indicators. The following are the major categories that should be considered:

- Maintenance Productivity – measures the effectiveness of resource use
- Maintenance Organization – measures the effectiveness of the organization and planning activities

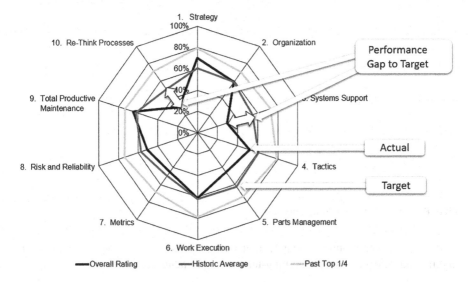

FIGURE 4.11 Spider diagram showing performance gaps.

- Efficiency of Maintenance Work – how well maintenance keeps up with the workload
- Maintenance Costs – overall maintenance cost vs. production cost
- Maintenance Quality – how well the work is performed
- Overall Maintenance Results – measures overall results

Measurements are only as good as the actions they prompt – the results are as important as the numbers themselves. Attractive, well-thought-out graphics will help "sell" the results and stimulate action. Graphic layouts should be informative and easy to interpret, like the spider diagram shown in Figure 4.11. Here, the key indicators are measured on the radial arms in percentage achievements of a given target. Each measurement's goals and targets are also shown, clearly identifying the Performance Gap. Where the gap is most prominent, the shortfall is greatest – and so is the need for immediate action.

The spider diagram in Figure 4.11 shows the situation at a point in time but not its progress through time. For that, you need trend lines. These are best shown as graphs (or series of graphs) with the actual and identified targets, as in Figure 4.12. For a trend line to be effective, the results must be readily quantifiable and reflect the team's efforts.

4.2.1 SELECTING THE BEST MEASUREMENT

Each of the examples above shows only a single measure of performance. The Balanced Scorecard concept broadens the measure beyond the single item. Organizations should develop a BSC to reflect what motivates their business behavior.

Many measurements can be used once the measurable desired outcome is identified. Figure 4.12 (use cross-reference) shows an example that combines the elements

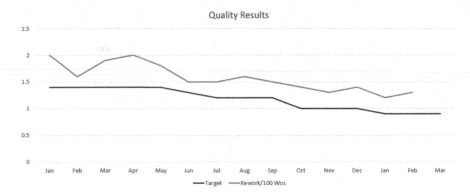

FIGURE 4.12   Trend-line of quality results.

of the input-process-output equation that were referred to earlier with leading and lagging indicators and short and long-term measurements:

- Leading indicators – the change in the measurement (i.e., hours of train-ing) precedes the improvements being sought (i.e., decreases in error rates). Typically, you see these only at a later date.
- Lagging indicators – the change in the measurement (i.e., staff resignations) lags behind the actions that caused it (i.e., overwork or unappreciative boss).

For each of these elements, four representative indicators were developed. While each indicator shows meaningful information, the Matrixed Scorecard provides a good overview of the effectiveness of the entire maintenance organization (see Figure 4.13). In a later section of this chapter, we examine some of the pros and cons of this increasingly popular measurement technique.

The broad performance measures are essential to understand the overall direction and progress of the maintenance function. But within this broad sweep lie multiple opportunities to measure small but significant changes – in equipment operation,

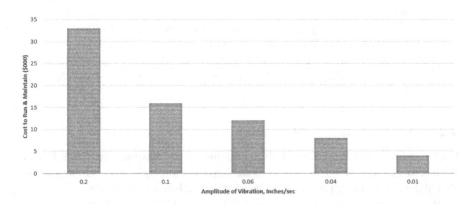

FIGURE 4.13   Internal pump operating and maintenance costs.

TABLE 4.4

Examples of Business Driver Considerations in a Scorecard

Metric Element	Description of an Example
Inputs	• Dollars spent on labor, parts, services, and overhead
Process	• Backlog of work orders • Compliance of actual work with a plan
Outputs	• Critical asset Reliability, Availability, Maintainability • Overall Equipment Effectiveness (OEE)
Short Term	• Number of breakdowns • Number of on-time work order completions
Long Term	• Maintenance costs as a percentage of replacement asset value • Maintenance Parts inventory turnover
Leading indicators	• Number of training hours • Percent of work orders driven by condition-based maintenance (CBM)
Lagging Indicators	• Amount of time lost due to injuries and absences • Total maintenance cost variances from the budget

labor productivity, contractors' performance, material use, and technology and management contribution. The following section examines some of these changes and provides examples that can be used in the workplace.

4.3   MEASURING MAINTENANCE – THE FINE STROKES

To understand individual elements of maintenance functions, you need analysis at a much more detailed level. For example, to evaluate, predict, and improve the performance of a specific machine, you must have its operating condition and repair data, not only for the current period but also historically. Also, it's helpful to compare data from similar machines. Later in this chapter, we examine the sources of this data in more detail. But, for now, you should know that the best data sources are the EAM, APM (Asset Performance Management), CBM (Condition-based Monitoring), SCADA (Supervisory Control and Data Acquisition Systems), and process control systems currently in widespread use.

Reliability, availability, productivity, lifecycle costs, and production losses are ways to track individual equipment performance. Use these techniques to identify problems and their causes to take remedial action. For example, an interesting case study examined a series of high-volume pumps to establish why the running costs (i.e., operating and regular maintenance costs) varied widely among similar models. Shaft alignment proved to be the significant problem, setting off improvements that drove the annual cost per unit down from over $35,000 to under $5000, as shown in Figure 4.13.

Despite this remarkable achievement, the company didn't get the expected overall benefit. After further special analysis, two additional problems were found. First, the "O&M" costs excluded contractor fees and should have been labeled "Internal O&M Costs." Then, the subcontractor's incremental cost to reduce the vibration from the

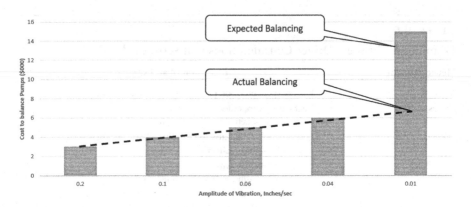

FIGURE 4.14   Subcontractor cost of pump balancing.

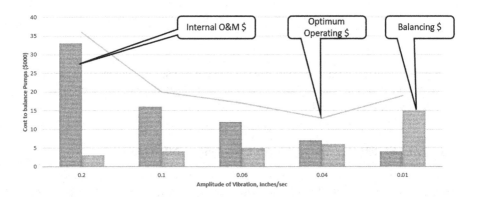

FIGURE 4.15   Total pump O&M costs.

industry standard of 0.04"/sec to the target level of 0.01"/sec (see Figure 4.14) was unexpectedly higher than predicted from earlier improvements.

When the O&M costs were revised to include the subcontractor fee, the extra effort to move from 0.04"/sec to 0.01"/sec was considerably more expensive than the improvement in the operating costs. Therefore, the optimum position for the company was to maintain at 0.04"/sec. (see Figure 4.15).

This type of quasi-forensic analysis can produce dramatic savings. For example, in the case summarized above, the company had fifty pumps operating and reduced costs from $1.6M to $750,000, saving 53%. Yet, in the final step to refine the balancing from 0.04 to 0.01, the company spent $200,000 for no additional tangible benefit.

You can use similar investigative techniques to evaluate and improve labor and materials consumption. Variable labor consumption among similar jobs at several plants can identify different maintenance methodologies and skill levels. Adopting the best practices and adding targeted training can significantly improve and avoid sizeable problems. In one example, a truck motor overheating was traced back to a poorly seated filter through the EAM work order. The maintenance technician responsible for the work had insufficient training for the task. Emergency recalls were issued for six other trucks on the highway that he had fitted the same way, and each one was fixed without damage. The technician received additional training and got to keep his job.

Variations in materials used can be tracked from the work orders and the inventory records, leading to standardized methods. More recently, multi-plant operations have accessed data from multiple databases for analytical and comparative purposes. Parts specification has become standardized, creating huge savings through bulk buying and centralized storage. There are several examples where reduced inventory, reduced supplier base, better-negotiated prices, and removing many hidden purchasing costs through improved productivity have generated savings well beyond 15% in the inbound supply chain.

This chapter introduces various topics to set the scene for a later and more detailed examination. The core issues are the same throughout this book – what should the maintenance manager optimize, and how should it be done? We will return again and again to the elements of this model, which covers the entire maintenance domain, but from several different directions. Its components – machinery, workforce, materials, methods, measures, milieu, and management – create maintenance costs. Controlling these costs and maximizing their return is your primary challenge.

4.4 WHAT'S THE POINT OF MEASURING?

Organizations strive to be successful and considered the best or the industry leader. In an earlier section of this chapter, we reviewed how objectives cascade from the corporate level to the maintenance department and translate into actionable tactics. It's assumed through this process that a common set of consistent and reliable performance measures "proves" that A is better than B. Unfortunately, this is not the case, but common standards are emerging through the works of many.

A major driving force for performance measurement is to achieve excellence. There must be effective measurement methods to withstand the scrutiny of the Board of Directors, shareholders, and senior management. The demand for accurate measurement increases as the Mission Statement cascades down to the maintenance department. This is reason enough to measure maintenance performance.

There are many other reasons, though, to make improvements, including:

- Competitiveness: regardless of whether the goals are price, quality, or service driven, you must compare to establish your competitive advantage.
- Right-sizing, down-sizing, and up-sizing: adjusting the organization's size to deliver products and services while continuing to prosper becomes meaningless if you can't measure performance.
- New processes and technologies) are being introduced rapidly in manufacturing and maintenance (i.e., IoT, machine learning (ML), artificial intelligence (AI)). Therefore, you must keep track of the results to produce the expected improvements.
- Performance measurement is integral when deciding to maintain or replace an item. In another chapter of this book, we included an example of life-cycle costing. We use it again later in this chapter to determine whether to maintain or replace it.

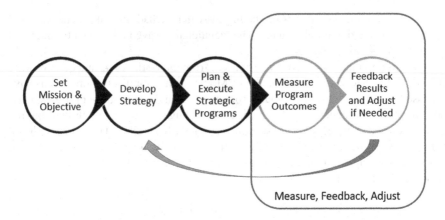

FIGURE 4.16   Measurements as a core part of the performance improvement loop.

- The performance improvement loop (reproduced from earlier in this chapter) is the core process in identifying and implementing progress. Performance measurement and results feedback are essential elements in identifying needed adjustments in this loop.

4.5   WHAT SHOULD WE MEASURE?

The following section looks at various performance measurements with sources for further study. It would help if you concentrated on the need for consistency and reliability. Comparisons over time and between equipment or departments must be consistent to be considered valid. You want the assessment of your operation to be reliably complete without significant omissions.

Maintenance management is a dynamic process, not static. It is inextricably linked to business strategy, not simply a service on demand. Finally, it is an essential part of the business process, not just a functional silo operating in isolation.

Despite this, there are frequent conflicts in setting objectives for an organization. For example, the objectives in Figure 4.17, taken from a strategic review of an enterprise, will be confusing when translated into performance measurement and later action.

From the discussion, you can see no real consensus on the precise source data, what should be measured, and how it should be analyzed. Some suggest a hierarchical approach. Referencing Terry Wireman's point of view in his book, "Developing Performance Indicators for Maintenance Management." Terry offers comprehensive coverage and developed performance measures based on a five-tier hierarchy: Corporate, Financial, Efficiency, Effectiveness, Tactical, and Functional. The following table borrows, expands, overlays, and collates that notion with the strategic, operational, functional, and asset perspectives introduced in Figure 4.6.

The above table shows that you can measure almost anything with unlimited options. However, it is important to emphasize the need to be selective about what is measured.

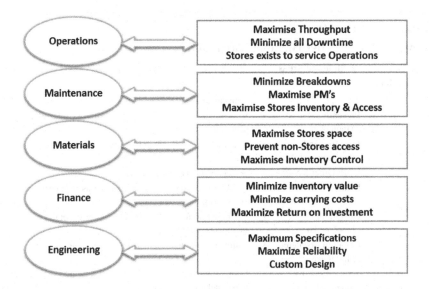

FIGURE 4.17 Conflicting departmental objectives.

TABLE 4.5

Examples of a Five-Tier Hierarchy Approach

Area Covered by Indicators	Functional Areas	Sample Indicator
		Strategic
Corporate	N/A	• Return on assets
		• Total cost to produce
		• Health and Safety
		• Environment, Safety, and Governance (ESG), Green House Gas (GHG)
Financial	N/A	• Maintenance cost per unit produced
		• Repair versus replace values of assets maintained
		• Overall production profit/efficiencies
Efficiency and Effectiveness	Preventive Maintenance	• % of total work that is proactive activities
	Work Order Systems	• % of all work on Work Orders
	EAM / APM	• % of critical assets aligned and integrated into the operational metrics to prescribe production constraint mitigations
	Training	• % of total maintenance rework due to lack of skills
	Operational Involvement	• Year-over-year maintenance-related equipment downtime
	Predictive Maintenance	• % of maintenance costs versus those within a predictive program
	Reliability-centered Maintenance	• % of critical assets supported by an approved reliability analysis
		• Number of repetitive failures versus total failures
	Total Productive Maintenance	• Overall equipment effectiveness combining availability, performance efficiency, quality rate, and applied hours of operation

(Continued)

TABLE 4.5 (CONTINUED)
Examples of a Five-Tier Hierarchy Approach

Area Covered by Indicators	Functional Areas	Sample Indicator
		Operational
Tactical	Preventive Maintenance	• % compliance with the PM (Preventive Maintenance) schedule
	Inventory and Procurements	• % parts lines available versus lines ordered
	Work Order Systems	• % Total WOs versus planned WOs received
	EAM / APM	• Total costs charged to equipment versus total costs from accounting
	Operational Involvement	• PM hours performed by operators as % of total maintenance hours
	Reliability-centered Maintenance	• Number of production disruptions per hour operated
	Total Productive Maintenance	• % of all maintenance work performed by non-maintenance personnel
		Functional
	Preventive Maintenance	• % of total WOs generated from PM Inspections
	Inventory and Procurements	• Inventory integrity metrics (in items and value)
	Work Order Systems	• % of total labor costs from WOs
	Planning and Scheduling	• % of total labor costs that are planned
		• % of maintenance schedule compliance
	EAM / APM	• % of total plant equipment managed in EAM/APM
	Training	• Training hours per employee
	Operational Involvement	• Operator involvement in OEE improvement programs
	Predictive Maintenance	• PdM hours % of Total maintenance hours
	Reliability-centered Maintenance	• % of failures where root cause analysis is performed
	Total Productive Maintenance	• % of critical equipment covered by design studies
	Statistical Financial Optimization	• % of critical equipment where maintenance tasks are audited
		• Bad actors proactively identified
	Continuous Improvement	• Cross-functional improvement programs are encouraged
		Asset
	Operational Involvement	• Reliability commitments by the operator
		• % of total hours worked by operators for equipment improvement
	Reliability-centered Maintenance	• % of critical assets where an RCM analysis program has been applied
	Total Productive Maintenance	• Equipment-specific OEE metrics

A basic approach can focus on needed results, productivity, operational purposeful-ness, and business case. The following approach expands this method by adding extra measures and presenting them as calculations. Check the core data set in section 3.7. We emphasize that new measurement formulae are being continually developed; those presented here are examples only and are not considered a complete or recommended set. What makes the best set of data will vary considerably with each situation.

4.6 MEASURING OVERALL MAINTENANCE PERFORMANCE

The measures are macro-level, showing progress toward achieving the maintenance department's overall goals. Later in this chapter, we cover some micro measurement that applies to specific equipment. Figure 4.18 summarizes some of the Maintenance Business Drivers we use in the examples:

4.6.1 OVERALL MAINTENANCE RESULTS

These Drivers measure whether the maintenance department keeps the equipment productive and produces a quality product. We look at these five measures:

1. Availability is the percentage of time that equipment is available for pro-duction after all scheduled and unscheduled downtime. Note that idle time caused by a lack of product demand isn't deducted from the total time avail-able. The equipment is considered "available" even though no production is demanded.

$$\text{Availability} = \frac{\text{Total Time} - \text{Downtime}}{\text{Total Time}}$$

Downtime includes all scheduled and unscheduled downtime, but not idle time, through lack of demand.

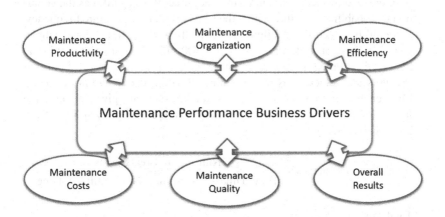

FIGURE 4.18   Maintenance categories for macro analysis.

Total Time = 8760 hours, Downtime = 392 hours

$$\text{Availability} = \frac{8760 - 392}{8760} = 95.5\%$$

2. <u>Mean Time to Failure (MTTF)</u> is a popular measure that will be revisited later in this book. It represents how long a machine can be expected to run before it dies. It measures average uptime and is widely used in production scheduling to determine whether the next batch can likely be produced without interruption. In this example, the equipment is expected to run an average of 647 hours from the previous failure. If the runtime since the last failure has been 250 hours, expect the equipment to have 397 hours remaining working life before the subsequent failure. Note that this assumes the equipment has an equal failure probability throughout its 647 hours. This assumption isn't reasonable since failure rates frequently approximate a normal distribution, which means failure probability increases as you approach the average.

$$\text{MTTF} = \frac{\text{Total Time} - \text{Downtime} - \text{Non-utilized Time}}{\text{Number of breakdowns}}$$

Total Time = 8760 hours,
Downtime = 392 hours,
Non-utilized time = 600 hours,
Number of breakdowns = 12.

$$\text{MTTF} = \frac{8760 - 392 - 600}{12} = 647 \text{ hours}$$

3. <u>Failure frequency or breakdown frequency</u> measures how often the equipment is expected to fail. It is typically used as a comparative measure, not an absolute, and therefore should be trended. It helps to regard it as the conditional probability of failure within the next period. Using Failure Frequency, it should be noted that "failure" is generally a vague term. For example, if a machine that is designed to run at 100 bits per minute is running at 85, is this deemed a "failure"? Similarly, if it produces at 100 bpm, but ten units are defective, is this a "failure"? Adopt the Reliability-centered Maintenance approach – the run and quality rates are given quantifiable failure levels. Anything below has failed, even though it may still be struggling operationally.

$$\text{Breakdown Frequency} = \frac{\text{Number of Breakdowns}}{\text{Total Time} - \text{Downtime} - \text{Non\_utilized Time}}$$

Total Time = 8760 hours
Downtime = 392 hours

Non-utilized time = 600 hours
Number of breakdowns = 12/year
Production rate = 10/hour

$$\text{Breakdown Frequency} = \frac{12}{8760 - 392 - 600} = 0.0015 \left(\text{Failures per hour?} \right)$$

Probability of 0.15% of failure within the next hour
Probability of failure within the next production run of 2500 units

$$= 0.15\% \times 2500/10 = 37.5\%$$

4. Mean Time To Repair (MTTR) is the total time to fix the problem and the equipment operating again. This includes notification, travel, diagnosis, fix, wait times (for parts or cool down), re-assembly, and test times. It reflects how well the organization can respond to a problem, and from the list of the total time components, you can see that it covers areas outside the maintenance department's direct control. MTTR also measures how long Operations will be out of production, broadly indicating maintenance effect on equipment production rate. Note that this can measure all MTTRs averages (i.e., the MTTR for breakdowns or scheduled outages).

$$\text{MTTR} = \frac{\text{Unscheduled Downtime}}{\text{Number of Breakdowns}}$$

Unscheduled downtime = 232 hours
Scheduled outages = 160
Number of Breakdowns = 12
Number of scheduled outages = 6

MTTR for unscheduled downtime = 232 = 19.3 hours 12

MTTR for scheduled outages = 26.7 hours
MTTR for all downtime = 21.8 hours

5. Production Rate Index – maintenance impact on equipment effectiveness. As with the previous indicator, you must interpret results carefully, as operating speeds and conditions impact. To minimize the effect of these variations, it is trended over time as an index and has no value as an absolute number.

$$\text{Production Rate Index} = \frac{\text{Production Rate} \left(\text{Units/Hour} \right)}{\text{Total Time} - \text{Downtime} - \text{Non-utilized Time}}$$

Production Rate = 10 units/hour
Total Time = 8760 hours

Downtime = 392 hours
Non-utilized time = 600 hours

$$\text{Production Rate Index} = \frac{10}{8760 - 392 - 600} = 0.001287$$

4.6.2 Maintenance Productivity

Maintenance Productivity indices measure maintenance's use of resources, including labor, materials, contractors, tools, and equipment. These components also form the cost indicators that will be dealt with later.

1. Workforce utilization is usually called *Wrench Time* because it measures the time consumed by actual maintenance tasks as a percentage of total maintenance time. The calculation includes standby time, wait time, sick time, vacation time, and time set aside for meetings, training, etc. It measures only the time spent on the job. There are frequent problems measuring the results because assigning time to jobs varies within organizations. For example, is travel time assigned to the job? The definitions must be clearly defined, documented, and adhered to for measurements within a single organization. The following example shows a wrench time figure of 69% – a relatively modest standard. High-performance, land-based factory operations will exceed 80%.

$$\text{Manpower Utilization} = \frac{\text{Wrench Time}}{\text{Total Time}}$$

32 staff, total time = 32 × 2088 = 66,816
Wrench Time = 46,100

$$\text{Manpower Utilization} = \frac{46100}{66916} = 69\%$$

2. Workforce efficiency shows how the maintenance jobs completed matched the time allotted during the planning process. Although typically called an "efficiency" measure, it is also a measure of planning accuracy. Many EAM systems can modify the job planning times for repeat jobs based on the average times the job has been completed. The workforce efficiency measure becomes a comparison with this moving average. Many planners reject this measure because they don't want to plan a new job until the teardown and subsequent diagnosis have been completed. They'll apply it only to preventive maintenance or repeat jobs.

$$\text{Manpower Efficiency} = \frac{\text{Time Taken}}{\text{Planned Time}}$$

Time Taken = Wrench Time = 46,100
Planned/Allowed Time = 44,700

$$\text{Manpower Efficiency} = \frac{44,700}{46,100} = 97\%$$

3. Materials usage per work order measures how effectively acquired and used. Again, this is a composite indicator you must refine before taking direct action. The measure shows the average materials consumption per work order. Variations from job to job can occur due to changes in buying practices, pricing or sourcing, inventory costing or accounting practices, how the jobs are specified, parts replacement policy, etc. As noted earlier, subdividing work orders significantly reduces the material cost per work order. Nevertheless, it is a simple trend to plot and, as long as you haven't made significant underlying changes, indicates whether material usage is improving.

$$\text{Material Usage} = \frac{\text{Total Materials \$ Charged to Work Orders}}{\text{Number of Work Orders}}$$

Total materials consumed = $1,400,000
Total WO's = 32,000

$$\text{Material Usage} = \frac{1,400,000}{32,000} = \$44 \text{ per Work Order}$$

4. Total Maintenance Costs as a percentage of total production costs indicate the overall effectiveness of resource use. It suffers from the same variations as the material usage measure shown above, but it will show overall performance improvements or deterioration if you maintain consistent underlying policies and practices.

   Although this indicator is only for total costs, similar indices can be readily created for labor, contractors, special equipment, or any significant cost element.

   Total Maintenance Costs = $4.0 m
   Total Production Costs = $45 m

$$\text{Maintenance Cost Index} = \frac{4.0}{45} = 8.9\%$$

4.6.3 MAINTENANCE ORGANIZATION

Maintenance performance indicators measure the effectiveness of the organization and maintenance planning activities. These metrics are frequently missed when considering the department's overall effectiveness because they predominantly occur at the operational end. Studies have shown that effective planning can significantly impact maintenance's operational effectiveness. One of the early selling features of the EAM products was that allocating 5% of the maintenance department's work effort would increase the overall group's efficiency by about 20%. For example, dedicating one planner 100% from a maintenance team of 20 would increase the operating efficiency of the remaining 19 from 60% to 75%. Also, it would raise the overall weekly wrench turning hours from $20 \times 0.60 \times 40 = 480$ to $19 \times 0.75 \times 40 = 570$ for an increase of 18.75%.

1. Time spent on planned and scheduled tasks as a percentage of total time measures the effectiveness of the organization and maintenance planning activities. This metric focuses on the work-planning phase, as planned work is typically up to ten times as effective as breakdown response. The planning and scheduling index measures the time spent on planned and scheduled tasks as a percentage of total work time. Notice that emphasis is placed on planning **and** scheduling. A job is planned when all the job components are worked out (what is to be done, who is to do it, what materials and equipment, etc.). Scheduling places all of these into a time slot to be available when required. By themselves, planning and scheduling have a positive impact. This impact is greatly multiplied when they are combined.

$$\text{Planning and Scheduling Index} = \frac{\text{Time Planned and Scheduled}}{\text{Total Time}}$$

Time Planned and Scheduled = 26,000 hours
Total Time = 32 employees × 2088 hours each = 66,816 hours

$$\text{Planning and Scheduling Index} = \frac{26,000}{66,816}$$

2. Breakdown time measures the amount of time spent on breakdowns and indicates whether more time is needed to prevent them. Use this index combined with other indices because the numbers will improve as the organization quickly fixes breakdowns. This can lead to a different culture that prides itself on fast recovery rather than initial prevention.
   Breakdown Time = time spent on Breakdowns as % of Total Time
   Total Time = 32 employees × 2088 hours each = 66,816 hours
   Breakdown time = 2200 hours

$$\text{Breakdown Time} = \frac{2200}{66,816} = 3.3\%$$

3. Cost of lost production due to breakdowns is measured because the acid test of the breakdown is how much production capacity was lost. The amount

of time spent by maintenance alone shows only the time charged to the job through the work order, not necessarily how serious the breakdowns are. Wait time – for cooling off, re-start, and materials, for example, is not included but still prevents the equipment from operating. This measure, then, includes run-up time and the costs of lost production due to breakdown and fixing breakdowns.

Breakdown production loss = Cost of Breakdowns as % of total direct cost
Cost of production lost time = $5140 per hour
Maintenance cost of breakdowns = $135 per hr
Total direct cost = $45 m
Breakdown time = 232 hours

$$\text{Breakdown Production Loss} = \frac{232 \times (5140 + 135)}{45m} = 2.72\%$$

4. The number of emergency work orders indicates how well the breakdown problem is controlled. Each "emergency" must be well-defined and consistently applied. In one notable example, a realistic (but cynical) planner defined work priorities regarding how high the requestor's command chain was. You're best to note measures of E-work order numbers in relation to a previous period (last month or last year, for example) and plot them on a graph.

   Number of E-work orders this year = 130
   Number of E-work orders last year = 152

4.6.4 EFFICIENCY OF MAINTENANCE WORK

This set of measures tracks the ability of maintenance to keep up with its workload. It measures three major elements – the number of completions versus new requests, the size of the backlog, and the average response times for a request. Once again, how the measurements are made will significantly affect the results – look for consistency in the data sources and how they are measured.

1. Work order completions versus new requests provides a turnover index. In the following example, an index of 107% shows that more work was completed than demanded. As a result, the backlog shrank, so overall customer service rose. However, no account of the size and complexity of the work requests or work orders is taken.

   WO turnover = Number of tasks completed as a percent of Work Requests

   Work orders completed last month = 3200
   Work Requests last month = 3000

$$\text{WO turnover Index} = \frac{3200}{3000} = 107\%$$

2. Work order backlog shows the relationship between overdue work orders and ones completed. As with most measures, you must formulate and communicate definitions. Maintenance customers will be frustrated if they don't understand how this measure is created. For example: is a work order "overdue" when it passes the requestor's due date or the planner's? If it is the requestor's, how realistic is it?

The example shows 6.8 days – one to two weeks are generally targeted. More than a week, and users will complain that service is lacking. What's needed is better planning and more staff or better screening of work requests. Lower than that suggests over-staffing.

WOs overdue = 720

WOs completed this month = 3200

$$\text{Back log} = \frac{720}{3200} = 22.5 \text{ of one month} = 6.8 \text{ days}$$

3. Job timeliness and response times are measured by the time from receiving the request to when the maintenance technician arrives at the job site. Comparison to a standard is the best method here. The standard should depend on the level of service desired, distance from the dispatch center to the job site, the complexity of the jobs involved, availability of workforce, equipment, and materials, and urgency. There is little value in measuring response times for low-priority jobs, so the trends are usually limited to emergencies and high priorities, with the statistics kept separately. In many cases, an emergency response is prompted by a "work request" that comes over the phone, and the responder is dispatched by mobile phone. No work request will be prepared, and the work order will be completed only after the emergency.

High Priority response time standard = 4 hours

High Priority response time average last month = 3.3 hours

4.6.5 MAINTENANCE COSTS

More attention is probably paid to the maintenance cost indicators than any other measures. This is encouraging, as the link between maintenance and costs (profits) needs to be solid and well-established. As you've seen, though, many factors affect the cost of delivering maintenance services – many of them almost entirely outside the control of the maintenance manager. In many companies, driving down maintenance costs has become a mantra and, in some cases, rightly so. However, you won't necessarily achieve your organization's objectives by reducing costs alone. The company's and maintenance's mission and objectives must be factored in. One way is to relate the maintenance costs to the overall cost of production or, where single or similar product lines are produced, to the number of units produced. For example, maintenance costs per ton of output are widely used in inter-divisional or inter-firm benchmarking.

Within this category, many different measures are used. The examples here show four typical ones:

1. Overall maintenance costs per unit output measure keep track of the overall maintenance cost relative to the product's cost. In a competitive environment, this is very important, particularly if the product lines are more of a commodity than a specialized product. The process industries, for instance, use these measures extensively. You can also subdivide this into the significant maintenance costs components, such as direct versus overheads, materials, workforce, equipment, and contractors.

$$\text{Direct Maintenance Cost Per Unit Output} = \frac{\text{Total Direct Maintenance Cost}}{\text{Total Production Units}}$$

Direct Maintenance costs last month = 285,000
Total units produced last month = 6935

$$\text{DMC} = \frac{285,000}{6935} = \$41/\text{unit}$$

2. Stores turnover measures how effectively you use the inventory to support maintenance. Stores value measures the number of materials retained in stores to service the maintenance work needed. Company policies and practices will directly affect these numbers and limit how much the maintenance department can improve them. Similarly, if manufacturing is far from the source of the material, this will also affect inventory turns. Some organizations argue that the breakdown cost (in financial, environmental, or publicity terms) is so high that the actual store value needed is irrelevant. Inventory turns are best measured against industry standards or last year's results. Inventory values are best measured through time.

$$\text{Inventory Turnover} = \frac{\text{Cost of Issues}}{\text{Inventory Value}}$$

$$\text{Inventory Value Index} = \frac{\text{Inventory Value this Year}}{\text{Inventory Value Last Year}}$$

Materials Issued last year = $1,400,000
Current Inventory Value = $1,800,000
Last year's inventory value = $2,000,000

$$\text{Inventory Turnover} = \frac{1,400,000}{1,800,800} = 0.8 \text{ turns}$$

Inventory value index = $\underline{1,800,000}$ = 0.9 or 90% of the base of 2,000,000

Note that an inventory index of less than 1 indicates a reduced inventory value or a performance improvement. In contrast, an increase in inventory turns shows better use of the organization's inventory investment.

3. Maintenance cost versus the cost of the asset base measures how effectively the maintenance department manages to repair and maintain the overall asset base. The available data uses the asset or replacement value to calculate. Several ways can be cut to select individual cost elements, as with many cost measures. The final measure shows the percentage of the asset's value devoted to repair and maintenance.

$$\text{Direct Maintenance Cost Effectiveness} = \frac{\text{Total Direct Maintenance Cost}}{\text{Asset Value (or) Replacement Cost}}$$

These can have quite significantly different values. i.e., book value compared to replacement cost.

Direct Maintenance Cost = $4,000,000

Asset Value = $40,000,000

$$\text{DMCE} = \frac{4,000,000}{40,000,000} = 10\%$$

4. The overall maintenance effectiveness index shows how maintenance's overall effectiveness is improving or deteriorating. It compares the maintenance costs plus lost production from period to period. The overall index should go down, but increased maintenance costs or production losses due to maintenance will force the index up. This index will measure whether spending more on maintenance has paid off with fewer production losses if a new maintenance program is introduced.

$$\text{Maintenance Improvement Index} = \frac{\text{Total Dir Mtc Cost + Prod Losses Previous Month}}{\text{Total Dir Mtc Cost + Prod Losses Last Month}}$$

$$\text{Maintenance Improvement Index} = \frac{285,000 + 102,800}{270,000 + 128,500}$$

Note that results below 1.0 show an improvement in the index.

4.6.6 MAINTENANCE QUALITY

The pundits frequently ignore maintenance quality when looking at performance measures. Yet, most auto magazines feature this in their "Which car to buy?" columns. Use it to judge how often repeat problems occur and how often the dealer can fix them on the first visit.

There are several ways that the maintenance department can collect and measure this data. At least one EAM has a unique built-in feedback form sent automatically to the requestor when the work is completed.

1. Repeat jobs and repeat breakdowns generally indicate that problems haven't been correctly diagnosed or training or materials aren't up to standard. Many maintenance departments argue that most repeats occur because they can't schedule the equipment for adequate (**maintenance**). The most effective way to get enough maintenance time is to measure and demonstrate the cost of breakdowns. Repeating refers only to corrective and breakdown work, not preventive or predictive tasks.

$$\text{Repeat Jobs Index} = \frac{\text{Number of Repeat Jobs this Year}}{\text{Number of Repeat Jobs Last Year}}$$

Number of repeat jobs this year = 67
Number of repeat jobs last year = 80

$$\text{Repeat Jobs Index} = \frac{67}{80} = 0.84$$

2. Stock-outs are one of the most contentious areas, reflecting the tension between Finance, which wants to minimize inventory, and operations, which, to maintain output, needs spares to support it. You can maintain the balance with good planning – predicting when materials will be needed, knowing the delivery times, and adding a safety margin based on historical predictions. Stock-outs are generally measured against the previous period but are best when tied to the higher priority work. Zero stock-outs aren't necessarily good, indicating overstocking.

$$\text{Stockout Index} = \frac{\text{Stockouts This Year}}{\text{Stockouts Last Year}}$$

Stock-outs this year = 16
Stock-outs last year = 20

$$\text{Stockout Index} = \frac{16}{20} = 0.8$$

3. Work order accuracy measures how closely the planning process from work request to job completion matches reality. The core of the process is applying their workforce and materials to jobs, which is usually the focus of this measure. You can easily measure this through the comment section on the work order. Encourage the maintenance technician to provide feedback on job specification errors, skill requirements, and specified materials to make corrections next time.

$$\text{Work order Accuracy} = 1 - \frac{\text{Number of Work Orders Completed}}{\text{Number of Work Order Errors Identified}}$$

Number of work orders completed = 1300
Number of work order errors identified = 15

$$\text{Work order Accuracy} = 1 - \frac{15}{1300} = 98.85\%$$

4.7   COLLECTING THE DATA

You can select measures to generate the correct information to drive action from the above examples. The data is drawn from CBM, enterprise asset management, engineering, and process control systems. You don't need all of these systems to start the performance evaluation process. Most data is also available from other sources, though computerized systems simplify data collection. Below is a core data set for the examples in section 4.6.

The volume of data available is expanding rapidly. Historically, the maintenance manager's problem did not have enough data to make an informed decision. With today's various computerized systems, the reverse is the problem – too much data. One possible solution to this is the maintenance knowledge database, now available by EAM and APM software solution companies. This concept recognizes that various data sources must be identified. More than a simple point-to-point linkage between these sources is needed. A knowledge base must be constructed to selectively cull the data, analyze it, and use it as a decision-support tool. The real value lies in helping develop actionable management information that attains results. Without this, the data isn't beneficial.

4.8   APPLYING PERFORMANCE MEASUREMENT TO INDIVIDUAL EQUIPMENT

So far, all the performance indicators we have reviewed apply to the broader spectrum of maintenance costs and operations. Many indicators effectively apply at the micro level to individual equipment or jobs. If the organization has numerous examples of the same equipment running, comparative evaluations can show the operating results and costs caused by different running conditions and maintenance methodologies.

The detailed data must be readily available and almost demands you use an EAM and an APM system for forensic maintenance. The data must be accessed in large enough sample sizes to reduce the error probability to acceptable levels. The data must typically be aggregated from individual work orders, pick lists, condition reports, and process control datasheets. Manual collecting from these sources isn't feasible. Two examples will be enough.

Looking at maintenance analysis at this level, you typically seek specific results relating to optimum maintenance intervals, operating parameters, and cost savings.

TABLE 4.6
Core Data Set to Support the Examples Provided for Section 4.6

Item	Value
Total Time – full-time operation	$7 \times 24 \times 365 = 8760$ hours/year
Downtime – scheduled – unscheduled	160 hours /year
	232 hours/year
Non-utilized time	600 hours/year
Number of scheduled outages	6/year
Number of breakdowns	12/year
Production Rate (units per hour)	10/hour
Annual Production (units) – capacity – actual	87,600 units
	77,680 units
Units produced last month	6935
Number of maintenance people	32
Working hours per person per year	2088 hours
Maintenance Hours – capacity	66,816 hours
Total Wrench Time	46,100 hours
Planned Time	44,700 hours
Time Scheduled	26,000 hours
Total Maintenance Costs	$4.0 M
Total Direct Maintenance Costs/year	$3.2 M
Total Direct Maintenance Costs	
• Last month	$270,000
• Previous month	$285,000
Total materials issued per year	$1.4 M
Total workforce costs per year	$1.6 M
Total Work Orders per year	32,000
Total Work Requests last month	3,000
Work Orders completed last month	3,200
Breakdown hours worked	2,200
Work order errors	15
Repeat jobs – this year	67
– last year	80
Overdue work orders	720
Emergency work orders – last year	152
– this year	130
High Priority Response Time – std	4 hours
– last month	3.3 hours
Maintenance cost of breakdowns	$135/hour
Stores Value – last year	$1.8 M
– current	$2.0 M
Stock-outs – this year	16
– last year	20
Total Production Costs	$45 M
Lost Production Time Cost/hour	$5,140
Production losses – last month	20 hours
– previous month	25 hours
Asset Value (Replacement Cost)	$40 M

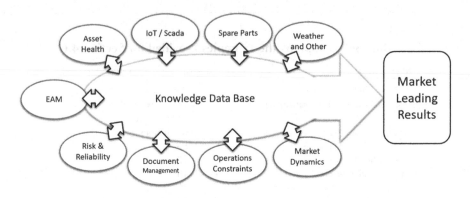

FIGURE 4.19   The knowledge database.

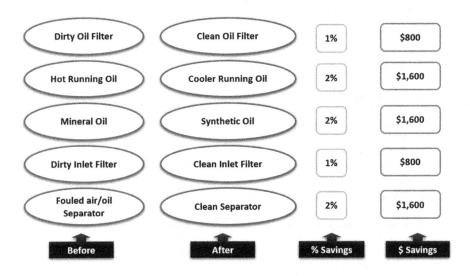

FIGURE 4.20   Energy cost savings generated from maintenance tasks.

For example. Figure 4.20 tracks compressor energy savings as various simple maintenance tasks were done. The improvement in energy costs (i.e., energy cost savings) totaled an annual $6,000, representing a savings of about 7.5% of the total annual "before" cost of $81,000.

Earlier in this chapter, we used an example of lifecycle costing to illustrate the pitfalls you need to be aware of. The following revisits this example but involves a replace versus repair decision. With the repair alternative, the slurry pump can continue to operate for five more years, with a higher annual maintenance cost and the likelihood of a breakdown. To offset this, there will be no purchase or installation costs:

The case for a replacement appears clear-cut, but you need to ask the same qualifying questions as before to be sure you make the "correct" decision. These relate to

TABLE 4.7

A Repair Versus Replace Data Set Example

	Repair	Replace
Purchase cost	0	$20,000
Cost to install	0	$2000
Annual running cost	$3000	$4000
Annual maintenance cost	$9000	$8000
Final Disposal Cost less scrap value	$1000	$500
Pump life (years)	5 years	12 years
Total life cycle cost	$61,000	$167,000
Annual cost	**$12,200**	**$13,900**
Average throughput (gals/hr)	100	175
Lifetime throughput (M gals)	4.38	18.40
Cost (cents/gal)	**1.39**	**0.91**
Average Breakdown Frequency	3 per year	2 per year
Average Breakdown Duration	1 day	1 day
Downtime cost ($/hour)	$1000/hour	$1000/hour
Downtime cost (total lifetime)	$360,000	$576,000
Downtime cost (cents/gallon)	8.22	3.13
Total operating costs (cents/gallon)	**9.61**	**4.04**

capacity, customer satisfaction, and failure data reliability. Also, ask additional questions such as:

- The "repair" case covers a five-year planning period versus twelve years in the "replace" case – is this significant?
- What is the decision-making impact of the zero-purchase cost in the repair case – should a capital cost be included?
- Does the fact that the funds come from two budgets affect the decision – i.e., purchase price from the capital budget and running and maintenance costs from the operating budget?

The base data must be readily available to do any meaningful analysis. Critics frequently and successfully challenge the results based on the integrity of the data. When setting up your data collection process, make it easy to record, be reliable, and consistently analyze.

The other key issue is where to start. Every maintenance manager needs more time and less work, but you know that making time comes from doing things more effectively. The following chart (Figure 4.21) will help determine where to start. Grade the measures and actions high to low, based on how much benefit they create and how difficult they are to implement. Then start with those in the top left quadrant – high payoff, low implementation difficulty.

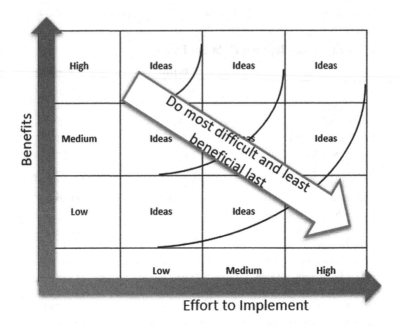

FIGURE 4.21   Benefits and implementation difficulty matrix.

4.9 WHO'S LISTENING? – TURNING MEASUREMENTS INTO INFORMATION

A frequent complaint from the maintenance department is that "they" won't listen. Who are they? Is it management deciding not to release funds for needed improvements? Is it Engineering refusing to accept the maintenance specifications for the new equipment? Is it Finance insisting on return-on-investment calculations at every turn? Is purchasing continuing to buy the same old junk that we know will fail in three months? Or is it production that won't give adequate downtime for maintenance?

It is all of them, and the core reason has usually been the same – maintenance hasn't been able to make its case convincingly. For that, you need facts, figures, and attractive graphs. The cynics claim that graphs were invented because management can't read. This is more than a grain of truth, although it stems from a lack of priority, not time or ability. The trick is to make the results attractive and compelling enough to prioritize and involve the other departments. Buy-in can work both upwards and sideways in your organization and downwards.

For example, Finance has become much more complex and sophisticated than it used to be. Many issues that make Finance question the maintenance department's proposals are technical, requiring inputs from the accounting dept. Co-opt a Finance person onto the team to handle skill-testing questions such as:

- How are inflation and the cost of money handled?
- What are the threshold project ROI levels?
- How is ROI calculated?

- What discount rate is used?
- How far out should you project?
- What backup for the revenue, savings, and cost figures is needed?
- How do you deal with the built-in need for conservatism?
- How do you accommodate risk in the project?

A parallel set of questions will arise with the engineering, systems, and other groups. Involve them in the process where their help is needed to clear barriers.

4.10   SIX STEPS TO IMPLEMENTING A MEASUREMENT SYSTEM

The performance measurement implementation process follows a straightforward and logical pattern. However, because many organizations don't do a full implementation, they get a methodology that can't meet the daily demands of the average maintenance department. The significant steps are highlighted in Figure 4.22:

(1) Define the stakeholder group or area you expect to focus on. Some manufacturing campuses may have different operational areas, such as water or power plant, mechanical group, instrumentation group, or other maintenance focus areas. Once the group is defined, identify related **business dimensions** that will directly interest the success of that focus area (i.e., stakeholders interested in maintenance management success may be slightly different from those interested in the success of maintenance parts management). Typically, there will be:
   - **An executive level that** (in the case of maintenance) will directly interest the organization's strategic success to achieve a ROAs, business growth, gaining market success, and improving competitiveness throughout the value stream.
   - **Operational/maintenance process** level will have responsibility for the maintenance BAU business operations/processes/tactics throughout the value stream.

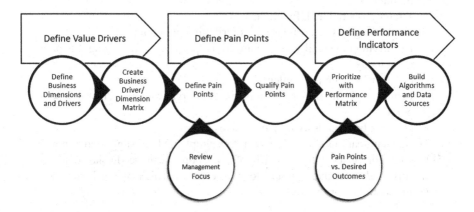

FIGURE 4.22   Steps to implement a performance measurement system.[2]

- **Functional/sub-process** level will support processes that manage functions and support operational requirements that contribute to the organization's desired performance and value, including financial, ESH, operations, and strategy.
- Optionally **an equipment level** that will be focused on crucial asset performance and health set of metrics

Business drivers against each of these business dimensions may be metric elements such as:

- Maintenance results
- Operational or maintenance productivity
- Maintenance organization
- The efficiency of maintenance work
- Maintenance costs
- Maintenance quality
- Environment, safety, health elements (ESH)

You may want to do some research to confirm the current corporate focus, given that focusing on all seven business drivers may be too much to take on in the first cycle.

(2) Step two is to create a blank maintenance measurement matrix with the defined business dimensions and business drivers to facilitate a place for later steps.

You will have to research to develop valid targets related to your organization's current numbers and the hoped-for future improvements. Beware of setting targets that are impossible to reach.

(3) With the business drivers identified: identify the "pain points" the current suggested Business Driver may be experiencing (i.e., high maintenance costs, an asset unavailable for production when needed).

- Define 2-3 pain points for each table section of your blank maintenance measurement matrix (Figure 4.23).

(4) Qualify each identified pain point from step three by:

- Identify which business unit is affected.
- Confirming at what business management level should this pain point be monitored.
- Strategy, operational, functional.
- Confirming this is a primary or contributing pain point.
- Identifying how well your organization "performs" addressing this pain point today?
- (1–10) where ten is excellent
- Identifying how "important" it is that your organization address this pain point?
- (1–10) where ten is extremely

These qualifications can be logged into a simple table as suggested below.

(5) From a completed table: the list of pain points can be displayed in an Importance/Performance graph to help confirm which of the listed pain points should be prioritized.

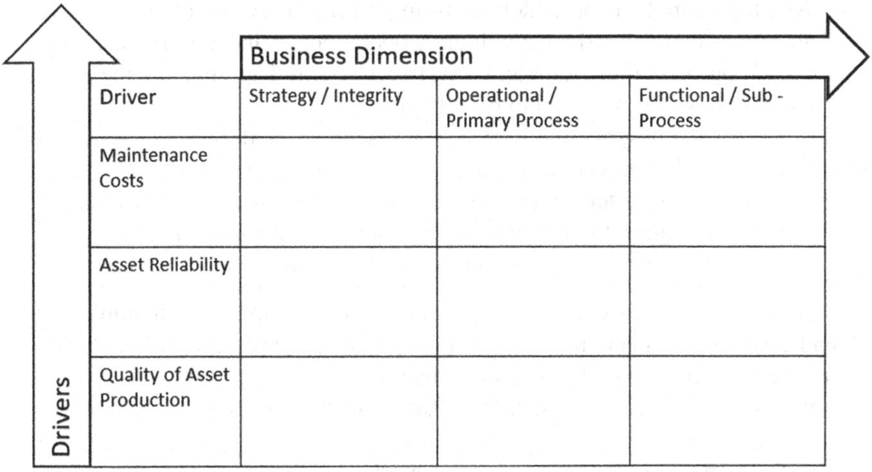

Business Dimension			
Driver	Strategy / Integrity	Operational / Primary Process	Functional / Sub - Process
Maintenance Costs			
Asset Reliability			
Quality of Asset Production			

Drivers

FIGURE 4.23 Example of a blank maintenance measurement matrix (step 2).[2]

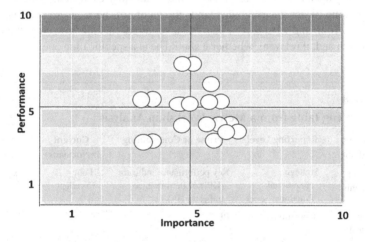

FIGURE 4.24 Example of a blank Importance/Performance graph (Step 5).[2]

In simple terms, an Importance/Performance graph may look like what is shown in Figure 4.24. Typically: the pain points that are most important and have the poorest perceived performance (Lower right quadrant) is the quadrant where the priority focus should start.

Once the pain points are prioritized, redefining the pain into a more positive metric statement is best. For example, a pain point can be a missed production or parts not in stock where the desired outcome could be production percent or parts availability percent).

(6) With the desired outcomes defined from the prioritized pain points, you can now focus on the metric that will drive operational behaviors. You can look at each business dimension and driver section and determine the measurements at this step. You should be able to:
- Define and describe the PI that would address the desired outcome and how the PI would be measured;
- Describe how often (frequency) this desired outcome should be checked for performance to make a difference in your operation; and
- Suggest the potential short and long-term targets.

Document the measures and targets, plus the interpretation of each number or trend. Absolute numbers may not mean anything, so explain it whenever you use a trend or index (i.e., Is Up good or bad?).

In this example, for a Maintenance Parts Operation financial focus:

- the pain point was: ineffective inventory levels and
- the desired outcome was: effective inventory levels.

In setting up the measuring process, make sure that each metric has a process owner, and the owner clearly understands the nature of the measurement, where the data comes from, and what sort of analysis is required. Include how often the reading is to be taken and, if relevant, whether it should be at a specific time or event during the

TABLE 4.8
Suggested Log Table from a Metric Workshop Analysis

Pain Point & Business Unit	Reporting Level	Prime or Contributing	Current Performance	Current Importance
Example and business unit PMs on time	Strategic, operational, functional, asset	Key performance indicator (KPI) or Performance Indicator (PI)	1 Poor – 10 Excellent	1 Minor – 10 Extreme
(Instrument, Mechanical)	(Operational)	PI	(7)	(8)

TABLE 4.9
Example of a PI Metric Table for Parts Inventory Requirements

Performance Indicator Characteristics	Measures	Long-Term Performance Targets	Near-Term Performance Targets:2020	Near-Term Performance Targets:2021
Inventory Turnover	Annual Usage Avg Inventory $	2.0 Turns	1.2 Turns	1.5 Turns
Inventory Reserves	<3% of Avg Inventory	<3% of Avg Inventory	<3% of Avg Inventory	<3% of Avg Inventory

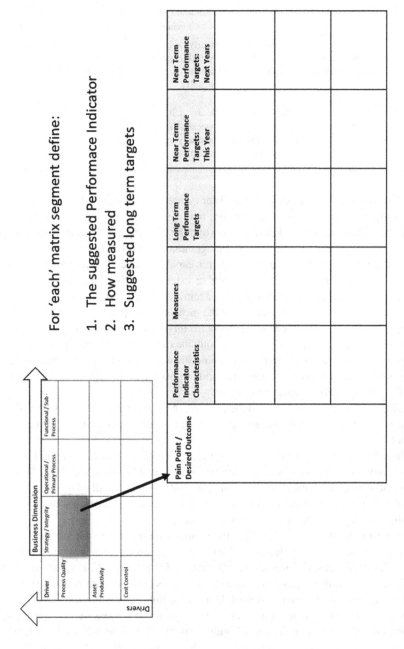

FIGURE 4.25 Example of a blank desired outcome table – with measures identified (Step 6).[2]

TABLE 4.10

Example of a High-Level PI Metric Table Summary for OEE and Parts

Desired Outcome	PI Description	Algorithm	Frequency Should Be Checked	Suggested Dimension	Data Source (System)
High production	Overall equipment effectiveness	% availability % throughput % quality % Production Hours	Hourly, daily, weekly, monthly, annual (weekly)	Strategy, operational, functional	Operational system
High service levels	Parts availability	Parts request versus fulfillment by location	Hourly, daily, weekly, monthly, annual (weekly)		EAM parts system

day (end of the batch, for example). Consistency needs to be stressed. Erratic results will not only sully the measurement, but they may also induce the wrong action.

With the widespread use of EAM systems and the growing use of Asset Investment Planning (AIP) and Asset Performance Management (APM) solutions, available data has grown dramatically. However, the system's capability to collect and maintain the correct data varies dramatically. Once you've established the data set, ensure the computer system makes it easy to collect it accurately. For example, the work order data fields should carry the same labels as the measurement process. They should be mandatory fields so the maintenance technician can understand why they are needed. Build in a value range so that any data entry outside of it will immediately be flagged.

Supplement the macro measures with some measures for critical equipment and bad actors at the micro level. Set standards and targets for each – again, paying attention to the data availability and consistency.

Install the measuring procedure in the EAM/APM system as a weekly work order. This simple step is frequently missed and is a significant reason why measurement systems fail.

Publish the results so all maintenance employees and visitors can see the targets and achievements. This is best done on the departmental notice boards, although some maintenance departments have set up a special war room to display and discuss the results. Set aside time at the weekly meetings to review and discuss the results, especially looking for new ideas to achieve the results.

The measurement point is to target trend implications and remedial actions to improve. A milestone flag on the trend chart effectively shows when a specific action was taken and the subsequent impact. The remedial action should emphasize what has to be done, who does it, when, and what materials and tools are needed. What's needed is a regular work order to record the task for future evaluation.

Most often, measurement projects are seen as just that – measurement projects. But a measurement project can be much more than that. It can be a dynamic program for ongoing change. To make this happen, use the measurement results to re-evaluate the macro and micro standards and targets. Establish a PM-type work order calling for an annual review of each measurement and set new targets to introduce this feedback loop.

4.11 TURNING MEASUREMENTS INTO ACTION

Measurements are only as good as the actions they generate. You miss the whole performance measurement point if you fail to convert them into action. The flip side is the tendency for a good trend to lead to complacency. Some organizations have rejected the basic philosophy of benchmarking, for example, claiming that it forces the organization into a perpetual catch-up mentality. No company ever took the lead in playing catch-up. The breakaway firm needs to think outside the box, and the measurement process helps define the size and shape of the box.

4.12 ROLE OF KEY PERFORMANCE INDICATORS – PROS AND CONS

Management uses performance measurement primarily for monitoring purposes, and many performance indicators have been developed to support operational decisions. These indicators are, at best, descriptive signals that some action needs to be taken. Place decision rules compatible with organizational objectives to make them more useful. You can determine your preferred course of action based on the indicators' values.

You can use indices to measure maintenance performance to clarify trends when the activity level may vary over time or when comparing organizations of different sizes. These can be organized into three categories of commonly used performance measures based on their focus:

(1) Measures of equipment performance – i.e., availability, reliability, and OEE
(2) Measures of cost performance – i.e., operation and maintenance (O&M) labor and material costs
(3) Measures of process performance – i.e., the ratio of planned and unplanned work, schedule compliance

However, the underlying assumptions of these measures are often not considered when interpreting results, so their value can be questionable.

For example, traditional financial measures still tend to encourage managers to focus on short-term results, a definite drawback. This flawed thinking is driven by the investment community's fixation with share prices, mainly by this quarter's earnings. As a result, few managers choose to make (or will receive Board approval for) capital investments and long-term strategic objectives that jeopardize quarterly earnings targets.

Income-based financial figures are lag indicators. They are better at measuring the consequences of yesterday's decisions than indicating tomorrow's performance. Many managers are forced to play this short-term earnings game. For instance, maintenance investment can be cut back to boost quarterly earnings. The detrimental effect of the cutback will only show up as increased operating costs in the future. The manager decided to cut back and may have already been promoted because of the excellent earnings. To compensate for these deficiencies, customer-oriented measures such as response time, service commitments, and satisfaction have become important lead indicators of business success.

Your organization must be financially sound and customer-oriented to assure future rewards. This is possible only with distinctive core competencies enabling you to achieve your business objectives. Furthermore, you must improve and create value continuously by developing your most precious assets – your employees. An organization that excels in only some of these dimensions will be, at best, a mediocre performer. Operational improvements such as faster response, a better quality of service, and reduced waste won't lead to better financial performance unless the spare capacity they create is used or the operation is downsized. Also, maintenance organizations that deliver high-quality services won't remain viable for long if they are slow to develop expertise to meet the emerging needs of the user departments. For example, electromechanical systems are being phased out by electronic and software systems in many automatic facilities. In the face of the new demand, the maintenance service provider has to transform its expertise from primarily electrical and mechanical trades to electronics and information technology.

PERMISSIONS

* Chapter adapted from pages 49–88, *Asset Management Excellence: Optimizing Equipment Life-Cycle Decisions*, 2nd Edition by Editor(s), John D. Campbell, Andrew K. S. Jardine, Joel McGlynn Copyright (2011) by Imprint. Reproduced by permission of Taylor & Francis Group.

1 Adapted from copyright *Physical Asset Management Lecture Notes* (Part 1 and Part 2), UofT (2019), edited by (Barry). Reproduced by permission of Asset Acumen Consulting Inc.

2 Adapted from copyright *Maintenance Parts Management Excellence Program Lecture Notes – 102 – MPE – Key Performance Metrics for Maintenance Parts Management* (2018) edited by (Barry). Reproduced by permission of Asset Acumen Consulting Inc.

BIBLIOGRAPHY

Jasper L. Coetzee, *Maintenance*, Bloomington, IN, Trafford Publishing, 2006.
Terry Wireman, *Developing Performance Indicators for Managing Maintenance*, New York, Industrial Press Inc., 1998.

5 Asset Management Systems and Technology*

Don M. Barry
Previous contributions from Don M. Barry,
Brian Helstrom, Joe Potter, and Ben Stevens

The goals in maintenance often focus on five key business outcomes. They are to:

- improve the quality output of an asset;
- ensure maximum asset functional availability or uptime, or conversely minimal downtime;
- improve the optimal life of the asset;
- ensure a safe operating environment for both the operator/maintenance worker and the environment; and
- integrate with operations to support the effective automation of the company's value proposition.

Many tools have been developed to assist in or accelerate the attainment of these goals within the maintenance process. Fortunately, we live in an information age, in the Industry 4.0 (or now Industry 5.0) era. Decisions are driven by factual information derived from data. In asset management, particularly maintenance, you need readily available data for thorough analysis that produces optimal solutions. The world is growing increasingly intolerant of asset failure, particularly where environment and safety are concerned, so the importance of having good maintenance processes and effectively selecting, planning, executing, and recording maintenance activity is also growing. In this chapter, we discuss the essential aspects of modern computer-based maintenance management systems:

- Maintenance Management Systems (EAM, APM)
- Evolution and Direction of Maintenance Management Systems
- Technology's Impact on Maintenance Systems
- Emerging technical enhancements to Maintenance Management processes

First, we explore Computerized Maintenance Management Systems (CMMS), a basic maintenance management system supporting the asset maintenance of an organization's assets, typically on a single campus. Then we delve into Enterprise Asset

DOI: 10.1201/9781032679600-6

Management (EAM) and Asset Performance Management (APM), how they work, and how they can help you attain maximum results. Often, the terms EAM and APM are used interchangeably. But you will see that an EAM solution facilitates the asset processes and data to manage the actual asset maintenance transaction activity. An APM solution will manage the related asset's health, asset risk, reliability, strategic processes, and data.

Next, we examine how maintenance management systems have evolved to their current state, both functionally and technically. As these systems continue to evolve, so do their potential benefits to you increase.

You want to ensure tangible business improvements once you acquire a maintenance management system. Therefore, we look at the two crucial areas you must get right – selecting an implementing the system to achieve the desired results.

Finally, as many current and evolving EAM/APM solutions are still somewhat limited in what is included in their packaged solution, we will explore some of the many technical enhancements that can be applied to a maintenance management system/process.

5.1 INTRODUCTION: DEFINING MAINTENANCE MANAGEMENT SYSTEMS

First of all, what is a maintenance management system? Nearly everybody is familiar with an accounting system. The CFO or controller would review the accounting data daily. Investment decisions are made keenly on the balance sheet, income statement, and shareholder perception. The accounting system, in effect, monitors the business's health, and there are dozens of standard measures to work with.

Likewise, at an individual level, banks have moved toward "personal" financial systems that ideally record and categorize every expenditure, investment, and income stream. They provide software and internet transactions better to plan expenditures, budget vacations, and track savings. The growth of computer-based financial systems encourages more numeracy.

Similarly, a maintenance management system enables you to monitor the asset base and the activity planned and performed on the asset base, the value of which, especially for resource development and manufacturing enterprises, can add up to billions of dollars. The maintenance management system allows for monitoring the health of the assets and provides standard measures from which to work to optimize the productive life of these assets.

In the late 1990s and early 2000s, maintenance management systems were often called Computerized Maintenance Management Systems (CMMS) and Enterprise Asset Management (EAM) systems. The difference is one of scale, with no clear dividing line. A typical dividing line can be the scope of the asset locations, as early CMMS solutions were server-based and typically managed the maintenance activity of assets on a specific campus. Today, the choice usually comes down to the software supplier's marketing philosophy and strategy. For example, you need sophisticated, high-end asset management applications if your enterprise spans many plants, jurisdictions, and nationalities. At this level, the asset management solution must seamlessly integrate with other business applications (financial, human resources, procurement, security,

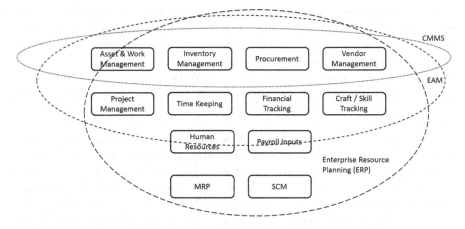

FIGURE 5.1 Comparison between CMMS, EAM, and ERP.

material planning, capital project management, etc.) to produce a total business information solution.

Figure 5.1 provides a simple, functional scope suggestion of the similarities and differences between CMMS and EAM. For example, many EAMs partially support human resources and payroll inputs. And when it comes to asset and work management, some systems, though properly called Enterprise Resource Planning (ERP), do not include this complete function, so the component is partially enclosed.

5.2 EVOLUTION OF MAINTENANCE MANAGEMENT SYSTEMS

Maintenance management systems have developed, over a long time, out of focused business needs. The following table shows how early asset management system functions evolved in the early 2000s:

TABLE 5.1
Functional Developments and Trends in CMMS

From	To
Custom-built	Package Solution
Clear leaders	Many common features
Difficult to interface with other business applications	Integration with ERP systems
Single "site"	Enterprise-wide
Narrow focus (work management)	Total asset management functionality
Difficult to modify	Easily customized and configured
Added functionality	Deployed with embedded/integrated Predictive Maintenance (PdM), Condition-Based Monitoring (CBM), and Reliability-centered Maintenance (RCM) functions; mobile workforce; planning solutions; and document management solutions

The earliest maintenance management applications were custom-built to unique business specifications. Most current packaged applications descend from these pioneer systems, often well-built but costly. There was a considerable difference between these initial systems' functions and implementation, just like the first personal computer applications. At first, there were clear leaders, but each product's best features became standard over time. Vendors consolidated, and the selection process became more complex. Today, you need to make a detailed comparison to detect the differences.

The earlier systems were built to suit each customer's functional and technical needs. As a result, they could not easily be deployed elsewhere. Maintenance management software customization changed with the advent of the many "de facto" standards: Microsoft Windows, server and web platforms, relational databases, Structured Query Language (SQL), etc. Today, it is mandatory for all EAM applications to share their data with other business applications, except in the smallest EAM implementations.

As organizations grow and evolve, so should their core business systems. Consolidation, mainly, has driven the need for business systems to be proper enterprise applications, dealing with multiple physical plants, sites, currencies, time zones, and even languages. World-class ERP systems offer this kind of location transparency, and the leading maintenance management applications are now comparably complex.

The scope of leading maintenance management applications is extensive, incorporating modules for:

- Asset hierarchies with parent/child and asset class relationships
- Maintenance (work order management, scheduling, estimating, workflow, preventive maintenance, equipment hierarchies, equipment tracking, capital project management)
- Inventory Management (parts lists, repairable items management, catalogs, warehouse management, inventory replenishment algorithms, parts kitting, parts reservations)
- Procurement (Purchase Order processes, vendor agreements, contractor management and administration)
- Human Resources (health and safety, time control, skills management, payroll, benefits, recruitment, training)
- Financials (General Ledger Accounts, Payable Accounts, Receivable, fixed assets, Activity Based Costing, budgeting)

Also, the leading maintenance management vendors offer sophisticated performance measurement and reporting capabilities.

This wide range contrasts with earlier maintenance management applications, which focused solely on asset logs and tactical work management, usually involving work order initiation, resources required, processing, and closing.

Finally, modern asset management is turning to optimization methods to improve maintenance effectiveness. Asset-intensive organizations now employ EAM and APM solutions as parts of their asset management technology solution. They now

TABLE 5.2
Technical Developments in CMMS Applications

From	To
Mainframe	Micro/Mini
Data files	Relational database
Terminal/file server	Client/server/Cloud-based platforms (in-house or outsourced), Software as a Service (SaaS)
Proprietary	Open standards
Dedicated infrastructure	TCP/IP/internet-enabled
Paper help	Context-sensitive/online help
Classroom training	Computer-based training

need to manage: CBM techniques (where maintenance is condition-driven rather than time interval-driven), Reliability-centered Maintenance (RCM; described in Chapter 7), and optimal repair/replacement analysis decisions, drawing on rich stores of historical information in the maintenance management system database. These techniques often require additional software modules embedded in the leading maintenance management products.

Technically, there have been several significant transitions:

From the list above, you can see a gradual standardization process developing. Adopting the relational database "standard" has made inter-system communication accessible. As well as a new architecture for distributed systems, the internet provided Transmission Connection Protocol/Internet Protocol (TCP/IP), the current communication standard for networks. These evolving protocols, in turn, have allowed systems to inter-connect as never before.

It used to be that racks of paper information for large business systems would sit in the system administrator's office, virtually inaccessible to users. As a result, the paper has disappeared, and current applications (i.e., mobile solutions) now have extensive online help systems. Nevertheless, as application complexity increases, training continues to be a problem.

You cannot consider current business systems without including the internet and e-enabled initiatives (in this chapter, the word "internet" is used for all related network terms such as intranet and extranet). Today, finding a maintenance management system Request for Proposal (RFP) that did not specify the need for web-enabled, Cloud, or Software as a Service (SaaS) infrastructure/functionality would be challenging today. But one of the most significant difficulties facing maintenance management system specifiers is understanding what that means and the associated trade-offs. There are many opportunities for web-enabling a maintenance management system, notably:

- Complete application delivery, including application leasing
- Internet-enabled workflow (e-mail driven)
- Management reporting
- Supplier management/procurement

- Standardized human-to-computer interaction
- Facilities management

The following sections provide a brief description of each initiative.

5.2.1 APPLICATION DELIVERY

As long as your hardware infrastructure is sufficiently robust (bandwidth, reliability, performance, etc.), you can deploy some EAM products ultimately through a standard web browser. The advantages of this include version control, potentially smaller (less powerful, therefore cheaper) desktop machines, and ease of expansion. A recent twist is to lease, not buy, the EAM application. This is akin to mainframe "time-sharing," which is still used for functions like payroll processing.

5.2.2 INTERNET-ENABLED WORKFLOW

One of the most popular maintenance management system features is comprehensive workflow support so that business rules can be written and changed at will. By including internet technology, you can expand the scope of your business rules to cover the globe and, if you can imagine, perhaps even further. Ongoing research is into expanding the internet to include spacecraft and extra-terrestrial sites!

5.2.3 MANAGEMENT REPORTING

Closer to home, the internet is ideal for disseminating management reports and is a growing part of asset management systems (i.e., EAM/APM). You need current data to make better business decisions. In some cases, data even a week old is out of date. You need to be accessing an up-to-the-minute corporate database from any internet connection.

5.2.4 SUPPLIER MANAGEMENT AND E-PROCUREMENT

Supplier management has received much attention, primarily because it is internet-driven. The basic procurement cycle is essentially the same whether you buy capital items or office supplies. E-procurement automates time and labor-intensive processes and enforces a single enterprise-wide policy through a buying interface.

The process of E-procurement includes online supplier catalogs (several new companies have sprung up to provide them), requisitions sent over the internet, purchase orders, receipt, and billing confirmation (essentially electronic data interchange (EDI), with a new twist), and guaranteed security for financial transactions. Of course, we are only skimming the surface here as entire books have been written on these topics.

5.2.5 STANDARDIZED HUMAN-TO-COMPUTER INTERACTIONS

Ease of use has always been a fundamental issue. The people who manage the system and its data must interact with a workstation to gather the information. Browser-based interaction is intuitive and supports ubiquitous ease of use and implementation. The features of all and any application can be managed easily through this comfortable interaction. This reduces the training time of users and implementation strategies for technology changes. Further, with the advent of newer handheld (mobile) devices, this browser-style interaction has become necessary to support these smaller ubiquitous devices.

5.2.6 FACILITIES MANAGEMENT

It has been around for some time and was considered a different arena. However, facilities management is not that different from managing other assets. For example, facilities have inventory and equipment that require work management. What is extending now is recognizing that this is no different from what is managed within EAM systems. Locations are vital data points in EAM systems, but these become fundamental items for proper facility management. Consolidating facilities management software products aligns with the EAM solutions to offer a completely integrated solution to facilities and maintenance of the facilities, the associated equipment, and inventory. The integrated workplace management systems (IWMS)'s market introduction has a functional asset management component in its scope. So the market has confirmed the need for asset management software functionality to support facilities management.

5.3 TECHNOLOGY-ASSISTED MAINTENANCE

Utilizing technology to assist maintenance is not new in the 21st century; however, it is definitely in its infancy when considering what can be applied and how few organizations have fully applied it. Most organizations that manage assets do have an implemented EAM solution in place. However, most do not fully utilize the software solution they have implemented, and most also are working with dated or "old versions" of the processes and applications.

One industry that has demonstrated that it is ahead of the maturity curve when applying technology to asset and maintenance management is, not surprisingly, the "computer industry." Collecting data to help the asset manager/maintainer understand mean-time-to-repair (MTTR)/mean-time-between-failure (MTBF) data has been around throughout the latter half of the 20th century in the computer industry. Having a system or the asset itself collect and analyze its failure data or its "potential to fail" data was well in place in the late 1960s and 1970s. Converting that data to flag or even initiate a request for service through an external telephone system was well-exercised in the late 1970s. For example, a multi-national computer services provider implemented a maintenance management and response system to support the 1984 Olympic installation in Los Angeles. They took advantage of their existing analysis process capabilities and introduced a radio frequency (RF) communications process to ensure that the maintenance craftsmen could service the 1984 Olympics to minimize MTTR. This RF terminal technology effectively automated the asset "call-in process" to a central dispatch system. In addition, it facilitated two-way communications between the technical craftsman and the supported work management processes.

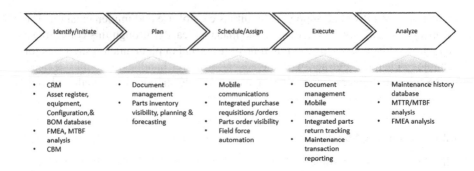

Identify/Initiate	Plan	Schedule/Assign	Execute	Analyze

• CRM • Asset register, equipment, Configuration,& BOM database • FMEA, MTBF analysis • CBM	• Document management • Parts inventory visibility, planning & forecasting	• Mobile communications • Integrated purchase requisitions /orders • Parts order visibility • Field force automation	• Document management • Mobile management • Integrated parts return tracking • Maintenance transaction reporting	• Maintenance history database • MTTR/MTBF analysis • FMEA analysis

FIGURE 5.2   Technology solutions known to support the 1984 LA Olympics.

This work management process worked in an automated fashion as listed:

1. The machine asset determines it has a problem and signals the pre-defined fault code to the supporting service organization;
2. The service organization receives the call and registers it as a specific supported asset with a specific fault code and determines which available craftsman with the appropriate skills can support this fault code, and sends the appropriate craftsman a notification through the radiofrequency system;
3. The assigned service craftsman receives the notification through the personal portable RF terminal, acknowledges the call to the dispatch system, interprets the fault code, and the recommended action received. They then decide if an order should be placed for the recommended parts associated with the fault code;
4. The assigned technician orders parts via the same RF portable terminal informs his dispatch system of his estimated arrival time (ETA) and proceeds to the machine location to support the service call;
5. The parts organization receives the parts order, fills it, and ships it to arrive at the machine location within one hour from when the order was sent.

While so many tactics and strategies support the availability of a production asset, the Los Angeles Olympic example proves that technology was significantly leveraged even almost 40 years ago to minimize response time to the process components of a work management process (a maintenance call).

The Los Angeles Olympic example provided suggests that technology has effectively been used to:

1. Track potential failure conditions and initiate a call or work request to the service organization based on predetermined criteria established in the assets" expert system;
2. Accept the call in a work management dispatch process that has predetermined who the best resource is to accept the maintenance call based on skill and availability;

3. Provide two-way communications via a RF technology for the call management process as well as the parts ordering and second-level maintenance support;

4. Provide a predetermined course of action, including a recommended list of repair parts that could solve the issue, based on the asset fault code.

This set of call-response and workflow accelerators was subsequently established across all locations that this service provider supported in North America and many other countries. Over time, the tools were upgraded to support more effective and updated technologies. As a result, this set of initiatives created significant service enhancements for their clients while ensuring they could deliver this service with optimal delivery and support resources. In other words, the users of the assets got a proactive asset management service and world-class emergency call support. In addition, the delivery costs for the service provider were significantly reduced through this integrated set of technology solutions.

5.3.1 Asset Management Evolving Technology

Asset Management has been around for centuries. The advancement of tools, processes, and standards has recently come to the forefront with asset strategy initiatives over the past decade (i.e., PAS55, ISO 55000, GFMAM) and the acceleration of data/systems advances to align with Industry 4.0. The spotlight on the asset management strategy structure (via ISO 55000) has promoted the notion of a list of things to do to improve the approach to asset management and productivity. The asset management strategy allows the organization to align its asset management priorities across its enterprise globally. The strategy would include a strategic asset management plan (SAMP) reflecting asset-related and agreed-upon organization goals and objectives.

Before the practical application of computers, asset and maintenance management would have been managed manually. As previously discussed, the CMMS solutions primarily provide basic asset tracking and maintenance execution support with foundational parts and procurement support. Additionally, the CMMS is typically a location-specific solution. Still, with this solution offered via a Cloud-based platform, many organizations can apply this basic asset management functionality across multiple locations.

Enterprise Asset Management (EAM) systems offered the basic functionality of the CMMS with many more features and functions that could be applied enterprise-wide. An EAM solution would typically support asset management throughout its lifecycle. EAM, with some process automation and data analytics functionality, is typically a solution that manages the asset and the efficiency of maintenance execution. EAM solutions typically do not focus on the process and data functionality to help the organization manage "what" maintenance should be done. A new asset management software market emerged called Asset Performance Management (APM) that supported this niche's process and data needs. APM set out to optimize the availability of operational assets by holding the data that helped predict when a maintenance task should be performed. APM essentially supports managing the asset risk

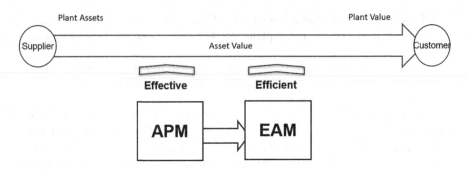

FIGURE 5.3 APM and EAM together (what maintenance to do efficiency).[1]

and reliability to suboptimal operations. As these markets grow and both the EAM and APM vendors work to grow their share, an overlap of functionality emerges. Some EAM vendors now promote themselves as providing APM support, and some APM vendors either provide more asset forecast insights or operational excellence alignment and insights. EAM focuses on asset management execution "efficiency," and APM solutions focus on asset management "effectiveness."

At a high level, the functionality of these two market segments provides functionality, as shown in the figure below.

APM solutions support what maintenance to do "effectively" while EAM solutions support doing that maintenance activity "efficiently" against a specific asset in a specific location and operating context.

TABLE 5.3
APM and EAM High-Level Functionality List

Enterprise Asset Management (EAM) Features	Asset Performance Management (APM) Features
Asset management,	Asset criticality and prioritization management
Work management (including planning and scheduling)	Risk and Reliability-Centered Maintenance (RCM) management
Parts inventory management	Hazardous Operations and Fault Tree Analysis (FTA) support
Procurement and vendor management	
Financial management	Strategy decision process management
Workforce management	Asset Safety Integrity Level (SIL) management
Analytics and reporting functionality	Asset health and policy workflow management
User support features	Asset performance tracker management
Interfaces with other systems	Digital twinning support
Administration support	Asset performance recommendation management
CRM functionality (optional)	Alerts and work order generation
	Reporting, Analytics, and User support features
	Interfaces with other systems

Sources: Adapted from copyright from *IT and IoT in Asset Management Training* (2019) edited by (Barry). Reproduced by permission of Asset Acumen Consulting Inc.

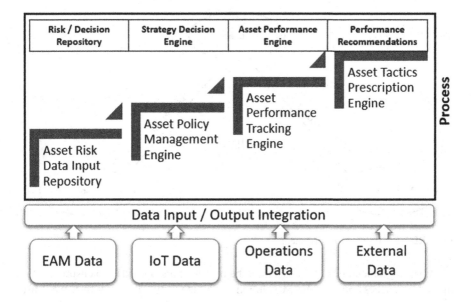

FIGURE 5.4 APM processes and Interfaces.[1]

The scope of an APM functionality will interface with many datasets, including EAM, IoT, operations, and external data.

Given the overlap, this chapter will combine the discussion of asset management solutions and band them together (i.e., EAM/APM).

The full spectrum of Industry 4.0 and Industry 5.0 impacts the "Art of the Possible" when IT commits to support asset management is further expanded in Chapter 20.

5.3.2 ASSET MANAGEMENT IMPLEMENTATION SUCCESS FACTORS

Implementing an asset management solution has many considerations that must be managed concurrently if you desire the assured success of any technological implementation solution. Looking at any implementation strategy's typical business transformation triangle, we influence the process, technology, and people. However, these are not alone; we must also include the essential issues of sound leadership, training, and security to support them. Once all these are in place and managed, the last concern that can significantly impact the success is the corporation's culture and people adopting the new processes and technology.

We need to engage through multiple streams of effort to effect the implementation through all these concerns and effectively drive the implementation to success through the four stages of delivering a solution. This includes working through initiation and stakeholder engagement, solution design, actual implementation, rollout, and finally finishing with the assessment for managed sustainability (which includes assessing the effectiveness of the implementation by looking at the expected delivery rewards).

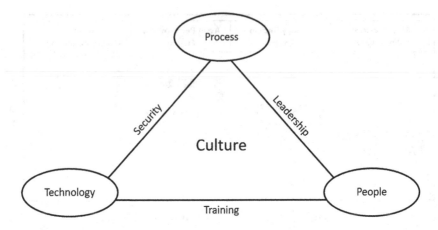

FIGURE 5.5   Points of involvement.

What implementation streams should be managed and included to ensure implementation success and the EAM solution's sustainability? To know this, we must again look back at the business goals and issues discovered through the highlights of the same concerns and issues. As we begin any project, we can recognize the need for project management. In addition, we must also look at the streams for managing the technology and its integration into the business. With the understanding that EAM cannot stand alone as a technological implementation if it is to be successful, we must consider the stream of the business process that delivers to the people their actions to align with the enabling technology to perform. This further implies that we are educating and training people on how to use technology and business processes to affect change in doing business. However, such change cannot happen without understanding the culture and the solution's impact on the culture. For this, you require Change Management. This Change Management stream needs to be well managed to ensure the success of the implementation. The technology, process, and people must be supported to recognize the planned benefit. All this can be summarized in a nice picture, as shown below.

5.3.3   MEASURING SUCCESS

So how do we know we are successful and have obtained the targeted realization of value for all the effort to implement an EAM solution? The important thing is to know where you are starting from and assess the critical performance metrics extracted from the daily performance that will show the return on investment to the business. We must select our measure earlier and perform the measurements in the old system to establish a baseline from which we started. Hence in the early stages of the project, we need to select what indicators will best measure success. Engaging the business and the technology stream to ensure that the metrics can identify the success of the new maintenance efforts and be measured and trended. This can sometimes be the greatest challenge of implementing the measures for success, so careful

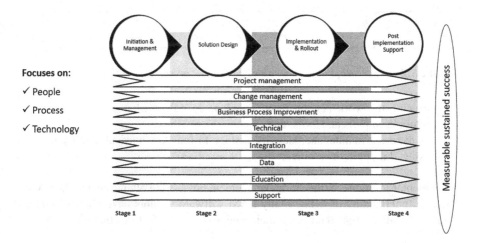

FIGURE 5.6   EAM project, seven swim-lanes of work, performed over four stages.

consideration is required in selecting these metrics. This, of course, is followed up by prudent reporting and review.

5.4   EMERGING TECHNICAL ENHANCEMENTS TO THE MAINTENANCE MANAGEMENT PROCESSES

Many of the leading EAM solutions today provide the basic requirements to support the basic maintenance management processes and some form of integration to Parts, Procurement, Human Resources, and Financials. However, the need to meet the goals listed at the beginning of the chapter and to optimize the "mean time between failure" (MTBF) and "mean time to repair" (MTTR) in the pursuit of these goals has driven many organizations to search for and implement enhancements to their packaged EAM solution.

To capture some of the technology approaches successfully applied across many industries, we will discuss some of the critical applications, features, and benefits of applying technology to contribute to the goals listed at the beginning of the chapter. We will also briefly explore how some of these technologies have evolved. Finally, we will describe what variations of the technical solutions exist in the market today, what is expected in the near future, and their known dependencies.

The maintenance execution process typically takes five high-level steps. They are to:

- identify or initiate the asset and work that needs to be completed;
- have a plan pre-developed that clearly defines how to do the required work with an understanding of the required skills, tools, and parts;
- schedule and assign the work with the required skills, tools, parts, operational alignment, and management approvals;

FIGURE 5.7 Simple steps of a maintenance work order process.

- execute the work per the provided plan, document what was done, and return used parts and tools; and
- analyze the maintenance activity history to look for unusual activity per asset, skill, part, or resource.

Fundamental solutions for the work management process are often found in the EAM software functionality of today. Many EAM software vendors have improved their products to differentiate them from their competitors with features that promote:

Listing the supported industries and asset types
Efficient work management processes (i.e., work order planning and scheduling tools),
Control through enhanced asset, resource, parts, and tools management, and
Superior analysis of the asset, resource, parts, and tool activity can improve asset availability.

Because of this, new and exciting technologies are being deployed to better assist the operations and maintenance solutions by aligning with existing EAM software solutions and enhancing the capabilities of the end-users and maintenance staff.

5.4.1 Mobile Technologies

5.4.1.1 History of Paper-Based Versus Mobile Technologies

Initially, Field Operations tended to be paper-based, with little or no data analysis performed on field data. Operators collect information for different assets through readings gathered and logged using log sheets and notations made about required maintenance items obtained through observation of operational deficiencies. These may be addressed by issuing work orders or scheduled routine checks. Following this effort, the information is assessed, resulting in maintenance work requests being generated, and these will likely be entered into some form of record-keeping or EAM/APM system. These work requests would then be planned and scheduled by the planner to be assigned to maintenance personnel or maintenance activities for issuance.

Alternatively, Field Maintenance receives a paper copy of the work order (whether from some form of EAM or not) using a manual data entry. They then would assign these tasks for action. The work orders may correspond to work requests created by operations personnel or another maintenance person. Once the work has been assigned and completed by the maintenance personnel, it is reported as work performed, likely directly on the maintenance system (i.e., workstation entry). Finally, the completed work orders are manually updated or filed for future analysis and management reporting (preferable to the EAM/APM system).

5.4.1.2 Limitations and Issues of a Paper-Based Process

1. Critical operational data is not immediately available to the field personnel where it is needed most, thus impacting equipment reliability/efficiency and, ultimately, equipment or plant availability.
2. Field operations are entirely paper-based and full of hand-written readings filed in a cabinet with little or no analysis being performed for preventative maintenance (PM) purposes. Most analysis is performed after the fact or as needed for investigations.
3. Requesting operational readings requires sifting through log sheets to obtain readings, which takes time away from regular duties.
4. Work requests may not be inputted into EAM/APM for days, primarily due to the busy work schedules (sometimes, they may even forget to input work requests).
5. Paper-based processes lead to erroneous data or the loss of capturing critical information into EAM/APM solutions.
6. Maintenance personnel frequently perform work orders never documented in the EAM/APM because of workload or loss of paper records.
7. Critical maintenance data (e.g., repair history) is not readily available to the maintenance personnel in the field where it is needed most. This may impact the equipment's reliability/efficiency, repair timing, and plant availability.
8. Manual data entry of completed work orders is a time-consuming task. It, therefore, has the potential for a backlog of work orders to be inputted into the EAM/APM solution.
9. Not all work requests get inputted into EAM/APM promptly. Thus, delays in this work being assigned and completed on time are inevitable. This may result in emergency work rather than routine work being performed.

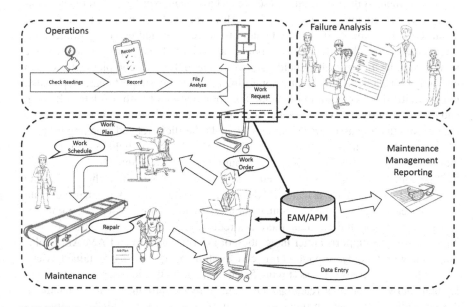

FIGURE 5.8 Field operations/maintenance process flow.

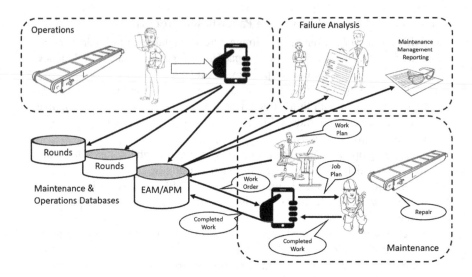

FIGURE 5.9   Field operations/maintenance of tomorrow.

10. Analysis and assessment of maintenance history for RCM and management reporting are slow and time-consuming. Much of the information may be difficult to find or even lost. This results in poor or inaccurate reporting and analysis, leading to poor decision-making regarding an asset.

5.4.1.3   What Is Expected in a Mobile-Based Solution

If we look at the preceding figure, it is probably not much different from what has been implemented in your plant. These issues have been experienced in many maintenance environments in many industries. However, with the current technological advancements, some of these problematic issues can be addressed by implementing handheld or mobile devices to support the in-the-field personnel. Using mobile system architecture with handheld computing technology to complement the data currently collected via remote site monitoring methods and support makes it possible to have critical information in the field. The combination of both the data collection and the availability of crucial maintenance information will provide a more accurate picture for enabling asset preservation, PM, and other field-related planning tasks.

Handheld computing devices enable the streamlining of field operations and maintenance processes by gathering and putting information into the hands of the field person. This simplifies and ensures the information is in the right place at the right time. In addition, these handheld devices remove the heavy dependence on data entry processes by bringing such entry out to the field person (closer to the point of occurrence). Through doing this, there is a reduction in loss of information through human error and supports better integration of information into an EAM/APM solution. This will, thereby, provide improved overall management of maintained assets.

The field operations process (Figure 5.9 above) allows the field personnel to transfer their daily operational readings and routes directly into and from the handheld device. This can be done through some form of docking or, if available, through a wireless network. This allows personnel to use their handheld device to collect their current operational readings, observe equipment, and create necessary work requests while in the field. This information would then be transferred back into the EAM/APM system for action without additional data entry or another personnel handling.

Similarly, Maintenance personnel will perform similar steps to get their work orders and information about the asset. The maintenance personnel could then use their handheld device to document and complete the assigned work orders and create new ones. Once they complete their work on the asset, they will update the work order and other information directly into the handheld device. This can be transmitted directly into the EAM/APM system through wireless or be held and transferred directly into EAM/APM at the end of the day through some docking station. This provides a more current view of an asset's state and further readiness for the maintenance personnel to perform new tasks.

Subsequent reporting and analysis of critical operational and maintenance data will be managed and fully collected by using the information within EAM/APM. Report analysis and performance characteristics are readily stored and available for better analysis and generation of failure reporting and management reports.

5.4.1.4 Business Benefits

In general, the application of handheld computing to field operations/maintenance can offer numerous advantages that are aligned with key business drivers such as:

1. **Improving Operational Availability** (i.e., Equipment reliability/efficiency, planned maintenance outage, information availability)
   - The EAM/APM has accurate and timely data on all work orders and equipment status, enabling more accurate and timely reporting and organizing of maintenance schedules.
   - There is a reduction of erroneous data input, which will increase the quality of data in the EAM/APM and thereby provide the opportunity for increased maintenance efficiency.
   - Field personnel can access accurate and timely data where and when they need it most – in the field, which could reduce the average repair time.
2. **Managing Operational Costs** (i.e., Workforce cost, Planned maintenance schedule)
   - This reduces paperwork and the need for manual data input into the EAM/APM solution.
   - Improved accuracy and time needed to collect data and subsequently report on it.
   - The ability to immediately generate work requests/orders in the field as needed enhances the performance of the asset by reducing "missed" opportunities for repair and potentially reducing lost productivity resulting from failures.
3. Other benefits from the introduction of Mobile technology may include:
   - Replacement of obsolete technology and infrastructure (i.e., computer infrastructure and hardware)

- Opportunity to integrate warehouse material management functions onto a standard handheld device
- Extending the use of bar-code infrastructure to other functional areas (i.e., bar-coding of equipment, RFIF tagging, etc.).

5.4.2 DATA MANAGEMENT

Within any system or process that underlies maintenance management is the data. The data is the heart of any EAM/APM solution. It is the foundation by which we can evaluate the impact and measure the successes of achieving better maintenance processes and improving the lifetime value of the maintained assets. Good data allows for the ability to extract information to make good decisions. Good maintenance data comes from data that follows the business processes of sound maintenance solutions. Therefore, good data management will significantly benefit managing through an effective maintenance solution.

So let's look at good data management within any EAM/APM. The first thing to ensure is a clear definition for each piece of data stored, used, or entered into the system. Data must represent assets or attributes of the asset or its associated

---

TABLE 5.4
Sample Mobility Solution Business Benefits Review

Business Drivers	Issues With Manual Process	Benefits of Mobile Integration
Plant availability: Availability of Information	Limited operational or maintenance available in the field	Maintenance and operational data made available on handheld devices
Plant availability: Analysis/reporting of critical operational data	Minimal analysis performed on critical operational data	Reports can provide up-to-date operational data quickly, and relevant parties can perform subsequent analysis (i.e., engineers, site management)
Plant availability: Accurate and timely input of critical data	Paper-based processed lead to: Erroneous data entry in the system Loss or delay of the capture of field information (work requests/orders relevant database or EAM/APM) Backlog of work orders to be input	Work requests/orders can be: Captured and completed in-the-field iteration Uploaded via a device with no need for manual entry
Plant costs: Time savings	Operator foreman spends time: Analyzing paper log sheets for weekly replenishment ordering Creating work requests for operators or maintenance Gathering operational reading for engineering	The operator supervisor can use the system reports determining weekly replenishment orders Engineers can view operation reading reports via the internet Field operations can upload field-generated work requests Maintenance personnel can now complete most work orders

transactional information against these assets. Thereby the data in an EAM/APM system must be of value to the management of the assets.

So how do we know we are managing the data within the EAM/APM solution with some assurance? Some tools can help us evaluate and ascertain that our data is essential. But they are only tools, and accurate analysis has to come from the people who use the data. Who assures that the information entered is correct and valuable. Data used regularly will tend to be good data as it is representative of the assets in use.

So how do we make these assurances? The best method is usually the simplest: Implement security on who can update, enter, or modify data items to only those who need to modify these data elements. Many EAM/APM tools have such internal systems to control data change, and some time should be spent to ensure these security features align with the business processes.

In addition to security to ensure data, we can institute limitations on the contents entered into data fields. For example, making the number of choices that can be entered into a data field supports keeping the data useful and allows us to analyze based on these features.

Open text is tough to manage; therefore, limited choices and restricted access can improve the data values you enter. However, there are times when such open text is all we can use to handle recording information. You might wish to establish how this open text can be entered in these cases. Then have periodic data samplings to ensure the standards are being practiced.

Many things are involved in managing data within our EAM/APM solution. But we have described only managing the value of the data we have in the system. Although this is one of the most critical factors in managing data, we also have issues that fall into the technology realm for data management. These include storage management, archiving, backup, and data recoverability. So as computer technology changes, too do these technology issues.

Historically, maintenance data was managed on paper, and then over time, this data has been moved onto large computer systems or mainframes. Then, over time, we migrated our data solutions from these monolithic data files onto more versatile storage solutions and implemented our data into relational databases on servers throughout the organization. As we move to a more internet-based solution, data access methods are being changed to accommodate this.

Maintenance management solutions operate on many platforms, including a relational database engine in the background. Some solutions run on a single personal computer, others are now provided as a Cloud solution, and some run on a client-server configuration. All of these offer differing opportunities for managing the data of the EAM/APM solution.

So what can you expect in the future? As technology changes, so will the data management capabilities and the means of accessing that data. We have come far in technology, and we continue to evolve. The wireless capabilities of the future may offer us ways to connect to the data stores more instantly so that data is available everywhere and instantly. As bandwidth continues to improve and the performance of the handheld devices and workstations continues to expand, we can expect to find a closer association of the data available to the source of use. Further, as assets become more evolved and computerized, they may become data entry points and be able to update the data directly into your asset management software. In any case, some disciplines to check in your data culture could include:

- Getting the data elements right
- Understanding the interrelated data dependencies
- Confirming the source and quality of the data
- Getting comfortable that decisions can be made from data insights
- Building the process execution into the system to automate process triggers, alerts, and steps
- Driving value for the data dynamics

5.4.3 Planning and Scheduling Tools

The concept of planning and scheduling has been around for a long time. It is not new when it comes to managing the completion of any undertaking efficiently and timely, especially when there are many intricacies of aligning and coordinating the delivery of material, labor, tools, and services. However, being a good planner and scheduler used to be quite the art. Maintenance planning took many hours of organizing and coordinating with a complete understanding of what would be done and how it would be accomplished. But as more sophisticated methods and tools evolve, the art of planning and scheduling becomes more of a science. The skill and experience needed to build and manage the schedule can now be supported within the tools, thus taking the complexities of planning and scheduling and simplifying it into a process of routines.

We are all familiar with Gantt charts and organizing our tasks visibly for easy analysis. Many technology tools exist that assist us in building these for simplified planning analysis. But which tools and how we use them is always the question that must be considered. Planning and scheduling take time, even with the correct tool. So how do we identify which tool to choose?

The EAM portion of the EAM/APM combination traditionally could create and manage a job plan tied to a work order. The ability to schedule the written plan against other plans and resource dynamics is now offered in many EAM solutions. They are fully integrated with the resource management dynamics (i.e., craft, skillset, parts, and tools availability). If practical, this schedule should be automatically aligned with the operations schedule.

Independent planning tools available on the market have been developed over many years to support planning and scheduling efforts and work with a clear start and finish. The effort involved in such planning and scheduling can be simple to complex, depending on the nature and issues of the work. It is important to remember that it is not the tool that perfects the planning but the people using it, as it can only be as good as the information given. Desktop tools like Microsoft Project or Primavera tools have been around for many years and can aid in developing and managing these planned undertakings. These tools currently support the planning of tasks through sophisticated algorithms used to balance and manage tasks within the undertaking. Still, they now include the ability to manage time, resources, skills, and costs. This provides the

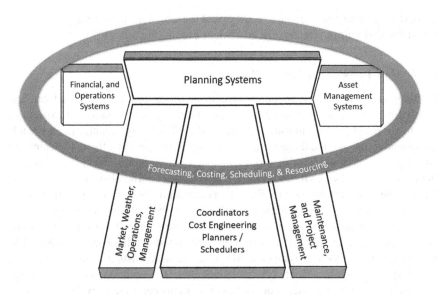

FIGURE 5.10 Asset management, finance, and planning integration.

planner and scheduler with information that can be applied and reviewed to fine-tune and improve upon repeating work.

Synchronizing your team performance through such tools goes far beyond planning and scheduling: These tools can provide a standard, comprehensive platform for maintenance and project managers. Using these advanced collaboration tools, with role-based features and action alerts to support centralized information that aids in keeping efforts on track with cross-functional activities in sync, is done through integrating other critical systems where data only has to be entered once and reused. This integration provides trusted forecasts and reports for maintenance turnaround solutions. This is what the focus of a planning and scheduling tool should be.

Planning systems are evolving to support integrating financial, market, weather, operations, and asset management systems (EAM/APM) (see Figure 5.10) to produce a completed solution for delivering managed and maintained assets to the organization. Since the tools for planning and scheduling are much more entrenched and have been developed around such common practice principles, it behooves the EAM/APM tools to utilize these capable process systems to develop the schedule for maintenance activities. Forecasting, costing, resource management, and scheduling encapsulate all finance, planning, and asset management needs. Therefore, we must integrate these processes in a bi-directional focus to take the work requests from the EAM/APM solution into the planning and scheduling tool to return the successful solution to the EAM/APM system for tracking and gate approval management. The practical solution for such planning and scheduling involves several groups within the organization. The planning and scheduling tools assist us with communicating between management, the coordinators, and the planners, through to the technicians and fore-people of field operations, in ensuring the right resources are aligned to complete the undertaking in the best cost-controlled manner. Why would we not want to use these in an integrated fashion within any asset management execution strategy?

5.4.4 The Internet of Things (IoT)

The Internet of Things (IoT) has enabled connectivity across the globe. IoT can connect IoT-connected sensors and & analysis, visual inspection data, and expert systems to the EAM/APM solutions with the needed IT security. Expert systems can be augmented to the asset and maintenance management arena by effectively using statistical process controls or other programmed automation forms. Meters are installed on the assets to measure vibration, wear, and viscosity continuously, or simply a time wherein these sensors provide feedback to a central processor where calculations are done to assess the asset's condition. Depending on the state it encounters, this can automatically request service in a part swap or maintenance. The example provided early is a rudimentary type of expert system. With today's Internet of Things (IoT) connectivity technology, we can endeavor to get more sophisticated assessments and track history so that the expert system can diagnose the problem, attempt to correct it by itself, and issue a service request. Expert systems continue to evolve as the ability to understand the asset increases, IT security and the power of the infrastructure processors improve.

5.4.5 Document Management Tools

Document management has come a long way since its inception. The origins of document management first fell in the hands of administrators and librarians who managed filing cabinets and shelves full of such information. The Dewey-decimal classification was developed to assist in handling large volumes of information, but with the advent of the computer, we have now moved to a new way of storing and retrieving such information. Yes, much information is still contained in books and such, and we will still need the means to find information in them, but more and more information is now being stored electronically. Document management tools were devised to manage this electronic storehouse of information. These tools have improved store, search, and retrieve capabilities to bring this documentation to the forefront of managing an asset.

It would be amiss not to discuss the impact a document management system can have on any maintenance program, especially implementing an EAM/APM solution. As with any plant or organization where maintenance is involved, understanding the asset, what the manufacturer used to build the asset, or what repairable parts exist within an assembly requires viewing engineering drawings or specifications. These and the reliability specifications are now typically stored in an electronic document. These electronically stored pieces can now be managed through some tool that understands the document and the need to access it by content.

Technology can now leverage Machine Learning (ML) and Artificial Intelligence (AI) to discern the content of many unstructured data sets. This analysis can be as simple as interpreting what a technician reported in their work order close-out comments. On the other hand, it can be as complex as having a system review thousands of documents to discern a potential course of action against a specific set of problem statements. For example, many medical issues are being addressed by what has been called cognitive computing to help to narrow down a course of action for a specific patient. Leveraging ML and AI could be installed to do a similar assessment against a prioritized set of critical assets.

Asset management systems are typically set up to formulate alerts based on structured data. However, by definition, documents are comprised of unstructured data, and today's technology can start to include this unstructured data in a real-time analysis to complement the risk and reliability strategies and the asset health prioritizations.

5.4.6 Integration

Many books and papers have been written about integration. Therefore, we will not attempt to cover all the facets of such a topic. Instead, we will view some of the directions that integration takes in applying EAM/APM with other application areas.

The design of integration has been in a state of constant change. As technology improves, more efficient means are being developed to support the heavy demand for an integrated business. Within EAM/APM, we find integration demand for financials, operations, external, and IoT data, documentation, planning, scheduling, mobility tools, etc. These pose not only technical issues but also process-related issues. Therefore, integration is not only about technology but must also include the issues associated with the business processes that are being integrated.

Originally integration meant large amounts of data exchange, extracting data from one application and importing it into another. Since many applications used a proprietary structure, this was a sophisticated task best handled only by well-experienced programmers. But along came the relational databases that started the path forward to better application integration. We could now generate SQL queries to extra data, and since the data was already in a standard database format, it could be easily parsed and imported into another database. This enabled us to generate backend processes that moved information from one system to another. Although we eliminated manually triggering these activities, it still left us with data duplication across multiple systems, even though it may not have appeared as such.

Along comes the newest architectures that leverage the capabilities of the relational database. This new architecture focuses the data content to be related to the object of its service. As developers build their applications to suit this new architecture, we see the growth of services within the application and the proper marriage of information across the enterprise. The service-oriented architecture supports a service approach to designing and building an application. This "Web Services" approach makes the integration layer more complex with advantageous features and control. The clear separation of its components has granted greater capability in the integration space. Integration is now assembling many parts and components to deliver an end-to-end solution.

Integrating the EAM/APM solution is just one component of managing the organization's assets through technology.

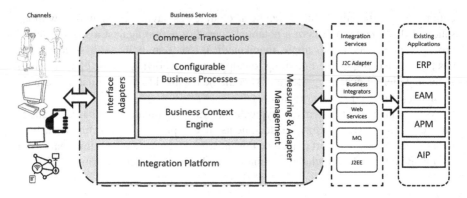

FIGURE 5.11   Asset management systems integration landscape.

5.4.7 Systems Infrastructure Supporting Asset Management

Technology supporting asset management infrastructure and pursuing automated operations has developed rapidly over the past decade. Trends suggest that the campus-only CMMS systems have evolved to enterprise-wide EAM solutions, and those EAM solutions are now supported by APM systems with integrated and automated feeds between each other.

The evolution of Information Technology and Operations Technology (IT/OT) emergence drives the need for broader and more dynamic IT solutions. Other trends supporting asset management include IoT, mobile, big data, visual recognition, ML, and AI support solutions.

The system platforms supporting asset management that have been in place have evolved from campus-only server hardware support to on-premise internet or intranet-connected support through a worldwide web. As the need grows to automate sensing at the source of an asset or system and respond intelligently: IT platforms have provided centralized platform agility and distributed intelligent solutions closer to the asset/system being tracked. In addition, the central site platform support requirements have grown and optionally been outsourced to a Cloud provider. The Cloud provides access to greater storage, computing, and data analysis power and agility as a business grows or the requirements change (i.e., daily, seasonally, or year over year.)

Central Cloud support exists at each end of the IT platform story, and edge computing is closer to the asset sensor.

Cloud computing supports heavy IT workload processing in a remote Cloud IT platform. Clouds are places where data can be stored or applications can run. They are software-defined environments created by data centers or, often, server farms.

Clouds are computing platform resources, typically virtual machines (VMs) & containers, but some offer physical servers. They will offer network connectivity, data storage resources, and many application tools on compatible operating systems. Generally, Cloud computing can be cost-effective over an on-premise solution when the computing needs dynamically change. However, this can sometimes be a more

TABLE 5.5
Cloud versus Edge Computing Considerations

Consider Cloud When You Need To	Consider Edge Solutions When You Need To
• Consolidate data	• Low latency
• Supporting many users	• Ability to compact data for transmission
• Use a scalable resource platform	• Control functionality at the site
• Heavy computing capabilities	
• Functions such as ML/AI, reporting tools, etc.	

Sources: Adapted from copyright from *IT and IoT in Asset Management Training* (2019) edited by (Barry). Reproduced by permission of Asset Acumen Consulting Inc.

expensive solution if the Cloud resource management is not managed and adjusted as resource needs dynamically change. Also, Cloud computing may be an issue for some organizations if they are concerned about government access to their data in some geographic scenarios.

Edge computing (or edges) are often remote locations where data is collected. They are physical environments made up of hardware outside a data center. Often the edge platform will have limited capability to collect the sensed information, discern a situation based on thresholds, and alert the central platform of an event or trigger.

The hardware device constraints edge computing applied. A mobile phone can be an edge device, but that would likely be a costly application versus a fit-for-use device. Once installed, the edge device is typically seldom upgraded, as many have been in the field for decades. They provide efficient (at the source) decision-making and communications. Remote control of a device, performing recovery restarts, and security must be thought through before applying the edge device. Often, remote sensing devices that have been in place for years do not have the security controls to defend against the modern hacker.

The table below provides high-level thoughts about Cloud and edge computing solutions.

5.5   ASSET MANAGEMENT AND ITIL

Information technology management (IT) has been included in an international standard under ISA20000. The Information Technology Infrastructure Library (ITIL) is a significant component in defining processes and a framework to operate IT. Chapter 2 suggests that IT assets are one of five asset classes. Of course, we all recognize that IT assets must be maintained similarly to all other enterprise assets.

So what exactly is ITIL? ITIL is a collection of books identifying established processes for managing and maintaining any IT-based organization's IT assets and services. These books include many concepts and processes that can be adapted for asset maintenance. The only thing in comparing the two is the semantics of the processes and tasks. But once you get past this, you will discover that ITIL and

EAM/APM have much in common from the process and goal purposes. EAM/APM, as does ITIL, aims to drive better service and uptime assets to be more productive. Because of this, there has been a crossover of the tools that manage IT assets with those that manage other maintainable assets. Thus, you will likely see more and more that the tools we are using for EAM or IT are being cross-purposed to perform many of the same functions in managing the enterprise's assets.

As the tool developers refine their tools and processes for asset management and ITIL tools, they will likely merge and overlap (or become the same). So get ready for the infusion of IT assets into your maintenance space as we move to a more integrated world within the enterprise. This truly will become an EAM solution.

5.6 CONCLUSION

It must be clear that an EAM/APM is an indispensable tool for today's asset manager. Organizations look at asset management as a core competency, and, in many cases, the asset manager is on par with the Chief Financial Officer (CFO).

The available asset management systems solutions cover all organizations – from small single-plant operations to multi-plant, multi-national companies. These modern systems use the latest technological advances, such as the internet, IoT, mobile, data, ML, AI, etc. Functionally, they are vibrant and provide features for asset management's strategic aspects (long-range budgeting, capital project management) and maintenance's tactical and operational aspects. The latest EAM/APM advances include in-built "intelligence" to customize maintenance for a particular piece of the plant (Predictive Maintenance). The potential cost savings are significant without affecting availability. It, too, is usually improved.

No wonder we are heading into a universal notion, as it is not just maintenance management but asset management supporting operational excellence. Further, as the worlds of different asset classes in an organization converge into a single solution of both processes and framework, it becomes even more evident that it will become a single enterprise set of tools. So selecting the right tool for the job becomes even more critical, as are the additional features you add. Spending the time and doing it right is the best advice.

Knowing what you are asking for and WHY is most important during the selection process. In other words, you must understand the requirements and have a good business case for buying an EAM/APM. It is not unusual for organizations to spend huge sums on asset management systems and not see any improvement in equipment availability or maintenance costs!

Not everything you are looking for in an EAM/APM solution may be found in a single package. Many emerging leading processes have been in limited use for several years but are just coming to be available in an EAM/APM package or as a "bolt-on." Care should be taken to consider whether implementing a "bolt-on" enhancement to your selected asset management solution is worth the additional process and systems management time. With large ERP solutions growing in their asset management capability and "Best of Breed" asset management solutions continuing to lead in functionality, there continues to be a gap between integrated EAM/APM and functionality or the "user-friendliness" of a "Best of Breed" package.

The implementation process is where "the rubber meets the road." Even the best EAM/APM can be crippled by poor implementation decisions, training, and support. Operating an EAM/APM is a process-driven exercise like any extensive system. So it should be no surprise that the results will be unsatisfactory if users don't follow the process. The appendix for this chapter supports asset management solution selection and implementation.

Business processes will constantly change due to industry conditions, personnel issues, etc. Recognize that the EAM/APM is, at heart, just a computer system. Like our automobiles, it needs to be regularly tuned up, with process changes, to operate at maximum efficiency.

Chapter 20 delves further into the emergence of Industry 4.0 and how asset management in information technology must partner to help their organization succeed. Every organization must change with the times and adapt to newly available technologies to stay competitive.

CHAPTER 5 APPENDIX: ASSET MANAGEMENT SYSTEM SELECTION AND IMPLEMENTATION

SYSTEM SELECTION

If your organization acquires a new asset management system and makes you responsible for selecting and deploying it, this could be a career opportunity. However, you need to ask some critical questions before you proceed further. Once you receive satisfactory answers, the following requirement is a robust system selection methodology roadmap.

PRELIMINARY CONSIDERATIONS

For the most part, you use computer-based business solutions to increase your organization's effectiveness. The solution has to enhance profitability. Typically, you make

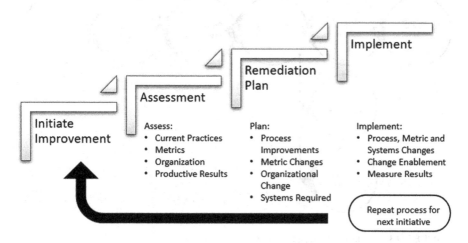

FIGURE 5.12 Business improvement cycle.

a business case to describe, justify, and financially estimate the expected benefits. Figure 5.10 shows that acquiring a business system like an EAM/APM is usually part of an improvement cycle. The business case is part of this cycle's remediation plan and other future success measures.

For example, management may expect the EAM/APM to improve productivity and Overall Equipment Effectiveness (OEE) by a certain percentage, increase maintenance parts inventory turns, etc. The team responsible for selecting the EAM/APM needs to understand these specific goals and the overall strategic context for acquiring the EAM/APM.

SYSTEM SELECTION PROCESS

Most packaged systems are selected similarly. However, unlike custom systems, when you procure a packaged system, you decide what you need and evaluate various vendor offerings to see which fits best. The general approach is:

In the sections that follow, we describe each step in the process. In practice, most organizations use an outside consultant to help execute the process or take it over completely. The advantage is that the consultant can fast-track many tasks, minimizing the cost of disrupting the organization.

ESTABLISH TEAMS

Your goal is establishing working teams to set requirements and validate vendor offerings against them. Teams usually include users, maintenance managers, and others who have a significant stake in the success of the EAM/APM.

If the planned EAM/APM scope is enterprise-wide, you may need several teams representing different plants, sites, or locations. If the EAM/APM is complex,

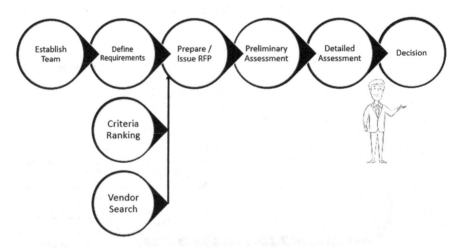

FIGURE 5.13 System selection roadmap.

consider forming teams with particular domain expertise. There could be teams from maintenance, inventory management, procurement, etc. As teams grow, they coordinate their outputs into cohesive requirements. You need to form each team carefully, clearly defining what it is expected to deliver.

The key deliverables are:

- Project charter defining what teams are expected to deliver and how much detail. The charter should also define each team's specific responsibilities.
- Task schedule defining the timeline for each team. Do not expect results after a few meetings since capturing requirements can be demanding.

The overall manager of this task should recognize that team-building skills will be needed, and they will frequently have to adjudicate in situations where responsibilities aren't clear.

VENDOR SEARCH

Assemble an initial list of pre-qualified vendors you are inviting to bid. Typically, this is done with a Request for Information (RFI), including an EAM/APM scope outline to move the process and obtain consensus. At this stage, you likely cannot decide based on functionality alone since most prime systems provide more functions than even the most advanced user needs. But if you have precise functional requirements that point to specific vendors, consider that.

An RFI should also ask about the vendors' commercial and financial viability, track record of comparable installations (particularly if the proposed installation sites are far from the vendors' home offices), product support capability, and other "due diligence" considerations.

You should issue, return, and analyze the RFI in time to meet the RFP's planned issue date. Usually, you won't need all the teams to accomplish this, probably just team leaders.

To re-cap, typical deliverables from this task are:

Request for Information (RFI)
Initial vendor list

DEFINE REQUIREMENTS

The quality of vendor proposals broadly will reflect how complete and precise your requirements document is. Also known as the system specification, it is the core document against which the EAM/APM application is acquired, implemented, and tested. So it needs to be assembled with care.

At the highest level, group requirements into major categories. Then break them down into sub-categories and, if necessary, into specific requirement criteria. Here is an example of what a requirements hierarchy could look like:

TABLE 5.6
High-Level Example of CMMS/EAM Base Requires

Category	Sub-Category	Sub-Sub-Category/Issue
Operations	Equipment	Asset hierarchies
Data analysis	Drill down, graphical, history	
	Work management	Blanket WOs, approvals, resources, scheduling, safety, crew certification, contractors, condition reporting
		Can labor hours charged to a Work order be broken down into regular and overtime hours?
	Preventive maintenance	*Can the system trigger an alarm when equipment inspection measurements trend outside a user-defined criterion?*
	Inventory	*Reordering, vendor catalogs, multiple warehouses, repairable spares, multiple part numbers, ABC support, service level costs*
		Can the system support multiple warehouses?
		Is a warehouse hierarchy supported?
	Procurement	
	Resources	
Financial	Electronic data collection	*How would bar-coding support issues and receipts?*
	Reporting	*How does the system use/generate a cycle count report?*
	Accounting methods	
Technical	Concurrent users, # licenses, "Power" users vs. casual users, architecture, scalability, performance, security/audit logs, databases supported, integration with other systems, data import/export, workflow solution, application architecture, database management, client configurations, development tools, interfaces, capacity performance	*Does the system use constraints (also cascading)?*
Human	Documentation	
	Training	
	User interface	*Can a user have multiple-screen access? If so, how?*
	Services	

Note that where detailed requirements are stated above as questions, they should be presented in an RFP in the form "the system shall…"; otherwise, if you are uncertain about specific needs, pose the requirement as a question, inviting vendor comment.

Requirements are best gathered using business process maps as the context. You should have generated these as part of the assessment and desired future state work shown in Figure 5.10 above. For example, consider work order processes – raising, approving, executing, closing, reporting, etc. The team(s) responsible for gathering requirements should conduct workshop interviews with user groups with these maps. A workshop format, bringing together different perspectives, stimulates maximum input.

DEFINE RANKING CRITERIA

At the end of the exercise, the requirements hierarchy will be extensive, with potentially hundreds of detailed system requirements across all of the major categories. You will need a quantitative approach to compare all the vendor responses.

Two numeric scores are relevant to each specification item – degree of need and degree of compliance. The degree of need represents how badly you must have the specification item, from mandatory to "would be nice." A mandatory requirement should be scored higher than one that is optional. In the sample above, the requirement for the system to support multiple warehouses would probably be mandatory and scored, say, 5 in a degree of need. In contrast, if supporting a warehouse hierarchy is optional, it could be scored 3, representing a "highly desirable" requirement. An unimportant specification item would be scored at only 1. To simplify the rating, you can use a binary score – 5 for mandatory and 0 for non-mandatory. Choose your approach by how detailed your evaluation needs to be.

The other score to set up is the degree of compliance. In the example given above: Rate vendor compliance scores on the following criteria:

- 5 fully compliant with the current system, no customization required
- 3 compliant with the current system, customization included by the vendor
- 1 not compliant without 3rd party customization, the requirement cannot be met

Here, using a range of scores rather than the binary approach is better. Why? Vendors can often supply the needed requirement with minor customization. In this case, they would score 3, and you must distinguish between minor, vendor-provided customization and powerful third-party add-ons. When you issue the RFP, give vendors the above list and the degree of need scores so that completed bid sheets have two scores for each requirement. Then, multiply both scores to get a "raw" score for the requirement when evaluating the bids.

TABLE 5.7
High-Level Example of CMMS/EAM Decision Weighting

	Relative Importance (Need)	Vendor Compliance
Multiple warehouses	5	
Warehouse hierarchy	3	

What about non-functional issues, such as vendor track record, support, financial stability, non-fixed-price arrangements, or implementation partners? These are as important, sometimes even more than functional requirements. You need a scoring scheme to compare vendor offerings in these areas. Check with your organization's procurement department. It should have guidelines to follow and standard scoring rules.

PREPARE/ISSUE THE REQUEST FOR PROPOSAL

The RFP is a system specification of the EAM/APM's functional requirements. Depending on your organization's procurement practices, there are standard terms and conditions, forms of tender, bid bonds, guarantees, and so on, which you must assemble, check, and issue. Public organizations' RFP issue and management process must conform to that jurisdiction's procurement rules. For example, if one vendor raises a query during the bidding period, you may have to issue clarifications to them formally. Bid opening can be public, with formal processes to manage appeals.

Once you've issued the RFP, the selection teams should develop demonstration scripts for vendors selected for detailed assessment. Demonstration scripts provide an agreed-to basis to judge how the candidate systems and vendors operate and perform. Without them, vendors, without any constraining requirements placed upon them, will naturally showcase the best features of their system, and you inevitably wind up with an "apples to oranges" comparison. They should focus on critical functional issues and reference business process maps. Typically, the scripts should be comprehensive enough to cover two to three days of detailed product demonstrations, a reasonable time for modern EAM/APM applications.

To summarize, task deliverables are:

- Request for Proposal, reviewed and approved by all involved teams
- Clarifications issued during the proposal preparation period
- Communications from vendors
- Detailed demonstration scripts used during the assessment

INITIAL (1ST CUT) ASSESSMENT

The initial assessment is reasonably mechanical. As we mentioned, the scores for each functional requirement are multiplied, and the result is used as the raw score from each vendor. Next, calculate the ideal scores (degree of need score times the fully compliant score) to calibrate all bids. If the bid results are significantly lower than the ideal, it does not necessarily mean a poor response. Perhaps the requirements list was highly detailed in areas outside the EAM/APM market. For this reason, using an expert consultant to build the system specification is a good idea. They should be highly familiar with each vendor's product and know whether specific requirements can be easily met.

Tabulate the non-functional responses (commercial, financial, etc.) and apply initial scores if a scoring scheme has been set up. Often, organizations visually inspect the results, leading to interpretation problems. For example, which of the following responses to the track record question is "better?"

- We have ten installations in your industry sector, three matching your user count.
- We have six completed installations in your industry sector, each matching your user count.

Although trivial, this example illustrates the potential for making decisions based on qualitative assessments.

Now, distribute the initial results to the selection teams and seek opinions. This is not always easy. Ideally, the rules for joint decision-making should be defined upfront as part of the project charter. Does a majority decision carry? Is a majority defined as 50% plus one, or should it be a significant majority? This critical question should be addressed early on before the decision needs to be made.

The output of this task includes:

- A documented initial assessment was reviewed and signed off by each team lead
- A shortlist of vendors (we suggest a maximum of four) who will be evaluated in detail, with supporting documentation for their inclusion

DETAILED ASSESSMENT

We suggest a two-stage approach. First, a presentation by each shortlisted vendor concentrates on the product overview, corporate background, financial stability, and ability to deliver high-quality services. Follow this with a detailed scripted demonstration/presentation by the two best vendors, emphasizing product software and services. Of course, you do not have to limit the demonstrations to two vendors. It isn't unusual for three vendors to be involved. The process is time-consuming, though, and expensive. Weigh the benefits of having more than two vendors involved at this stage against the cost of the extra effort. As we describe the detailed demonstration steps, assume that only two vendors are involved for clarity.

Invite the shortlisted vendors (again, we suggest no more than four) to present their credentials in a three-hour presentation. This ensures the vendor's philosophy and business are consistent with yours in crucial areas such as services, support, and company background. Firmly steer the vendor away from detailed software demonstrations at this stage to concentrate instead on their approach to implementation, experience in the industry sector (manufacturing, resource development, utilities, etc.), training methods, etc. A typical schedule would include:

- General introduction (15 mins)
- Company overview (45 mins)

- Questions from the selection team (60 mins)
- Software demonstration (30 mins)
- Wind-up and remaining questions (30 mins)

Use a scoring scheme for each topic above to help you decide on the two best finalists. In particular, team questions and a response rating system should be decided beforehand. This isn't very easy because the answers will be delivered interactively. Also, ensure that each selection team has the exact expectations from the brief software demonstration so that they're looking for the same thing. This isn't an exact process and will require a lot of discussions to work out.

Select the two best vendors based on the presentation and evaluation criteria and prepare written justifications. Also, notify the losing bidders, clearly spelling out why they were eliminated. Everyone involved should be advised about who made the final selection and why.

From here on, you begin a detailed assessment in earnest. Specific steps include:

- Invite each of the two vendors to a site visit. You want them to understand your operational needs better, collect data for the final detailed demonstration, and reflect on your processes in the final software review.
- Undertake initial reference checks simultaneously with the site visit if you wish. This can include conference calls/visits to each reference. You want to ensure the vendors' information is consistent with the reference user's experience. References must be chosen carefully, as their operating environments must be relevant to yours. You should advise the reference in advance about the nature and length of your call so they can adequately prepare. Naturally, the vendor isn't included in the reference call or visit.
- Invite each finalist for a detailed presentation and software demonstration, following scripts prepared and supplied in advance. There are two primary objectives: to ensure that the software is genuinely suitable and that the vendor can provide a high-quality and effective implementation. Vendors, naturally, will demonstrate software attributes that show their system in the best light. So that they address your needs, predetermine that the demonstrations must reflect how the system will be used in your application. Similarly, you want to know how the vendor or a business partner would implement the system.

Again, this interactive process is best served by preparing in advance. Selection team members need a shared understanding of what they are looking for, and some agreed pass/fail criteria for the scripted demonstrations.

A critical part of the evaluation is to analyze the implementation and post-implementation services required and the vendor's (or implementation partner's) ability to supply them. Include customer support, system upgrades, training quality, post-implementation training, user group meetings/conferences/websites, location, and support quality. The selected vendor should provide a sample implementation plan as part of the final presentation/demonstration, followed by a detailed plan to be approved by the selection teams before awarding a final contract.

At this point, the winning vendor will, most likely, be apparent. Carefully document your justification and present it to senior management for ratification. Next, notify the successful vendor and the second-place candidate. The second-placed candidate should also be advised that they may be invited to continue the evaluation process if a final agreement can not be reached with the preferred vendor.

What if there is not a clear winner? You could do a detailed functionality test of the two finalists to filter out the best solution. Then, you develop a test instruction set based on each criterion. Because the functionality depth in most leading EAM/APM applications is a significant task, taking several weeks of detailed analysis. We recommend that only mandatory functions be included to keep it within reasonable bounds. Consider this optional analysis only if you're still deadlocked over the final choice after the detailed demonstrations, reference checks, site visits, and selection team discussions.

CONTRACT AWARD

Before proceeding to the contract, hold final discussions with the successful vendor to clarify all aspects of the proposed scope, pricing, resources, and schedule. While it is unlikely that anything significant will be uncovered at this stage, remember that up to now, the primary focus has been on functionality. This is your opportunity to deal with other vital aspects of the vendor's proposal, demanding your full attention. Once completed, the next step is usually to issue a purchase order. The selection team should also prepare a careful selection and justification record.

SYSTEM IMPLEMENTATION

To effectively implement packaged computer systems, three elements must work together:

- People
  - Willingness to change
  - Role changes (i.e., planners, schedulers)
  - Organization change => reporting line change
  - Training Effectiveness
- Processes
  - How business is done now
  - How business should be done
- Technology
  - Hardware, IT platforms, and operating systems
  - Application software
  - Connectivity (network)
  - Interfaces
  - Data

A well-designed implementation project addresses each element so that users will effectively and accept the system. You can apply the outline steps that follow most packaged business systems. However, the detail applies specifically to EAM/APM implementations.

READINESS ASSESSMENT

There are some preparatory steps before the implementation teams arrive on-site with software and hardware. The first should be to conduct what we term a Readiness Assessment, covering:

- Organization and culture issues: This review asks questions: Is this company ready for system and process change? Is there a consistent sense of excitement, or is there tangible resistance? Is senior management supportive of the initiative and prepared to act as change agents throughout the implementation?
- Business processes: Are they documented, practiced, and understood? Is process change necessary?
- Technology: Is there a need for remedial work before the system is deployed (network, communications, staffing, etc.)?
- Business case: Is the conclusion understood, and are appropriate key performance indicators agreed upon? How can we be sure the EAM/APM delivers the expected benefits?
- Project team: Has it been formed? Do the members understand their roles and responsibilities? For example, if they are drawn from operational staff, do they commit to dealing with project, not operational, issues?

Some of these topics will have been (or should have been) addressed before the RFP was issued. However, it is good practice for the implementation project manager to review them again. More change enablement insights are described in Chapter 18.

IMPLEMENTATION PROJECT ORGANIZATION

Several user groups are needed to implement an enterprise-level EAM/APM successfully. If you think about where the EAM/APM sits functionally in your organization, this should not be a surprise. Maintenance certainly is front and center, but other skills and staff need to be included, from warehouse and inventory, procurement and purchasing, accounting, engineering and project, and IT support are also needed to configure and manage the system. And that is just for a "routine" EAM/APM solution set!

The following diagram shows the relationship between the project teams:

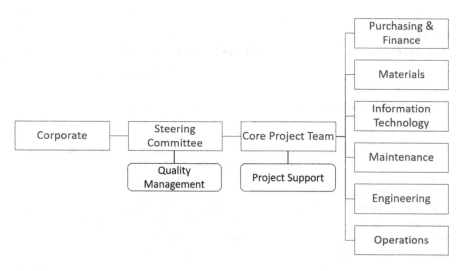

FIGURE 5.14   Typical project organization.

IMPLEMENTATION PLAN

An EAM/APM implementation generally proceeds with the following high-level timeline:

Now, we will deal briefly with each of the above project stages.

PROJECT INITIATION AND MANAGEMENT

This is an ongoing task lasting for the project's duration. The key activities and deliverables are:

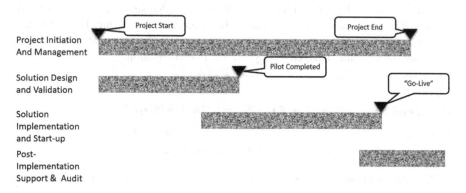

FIGURE 5.15   Typical EAM/APM project plan.

TABLE 5.8
High-Level Description of an EAM Project "Initiate Phase"

Activity	Deliverable
Prepare Services Contract	
Confirm objectives, expectations, and Critical Success Factors (CSFs)	Services Contract document
Finalize project budget	Project cost report
Prepare project schedule and develop for initial work	Project schedule – expanded to detail level for system configuration and validation
Define and document project procedures (reporting, change control, etc.)	Project procedures document
Detailed activity planning	Updated schedules (rolling wave)
Budget and Change Control	Change Reports and Budget/Actual Reports
Project kick-off meeting	Working project plan

DESIGN AND VALIDATION

You may find that developed and documented business processes from previous work can be excellent in helping you select the new system. However, if the implementation team is unfamiliar, conduct a review to ensure they can be configured into the system. With fit problems, you must make some process changes (hopefully minor – underlining the advantage of having reasonable process maps for the selection teams). Of course, these will be developed in conjunction with user groups.

The conference room pilot (CRP), also referred to as a proof-of-concept configuration, is where all business processes and user procedures defined earlier are tested and validated. It is here that you will implement most configuration changes. The CRP environment is ideal because the data volume is low, there are knowledgeable users, and there are minimal configuration changes.

CRP sign-off, which documents that the system adequately meets the defined functional requirements, typically follows. However, it is also essential to document shortcomings that can be addressed in subsequent phases to ensure that Client-raised issues are tracked and managed throughout the implementation.

TABLE 5.9
High-Level Description of an EAM Project "Design Phase"

Activity	Deliverable
Define & document business processes and user procedures	Business processes document
Configure the system with sample data	Conference Room Pilot (CRP) system ready for validation
Provide infrastructure support	Database sizing sheets, network recommendations
Train implementation team members	Training services/materials
Conduct the CRP	Modified system configuration and associated documentation
Complete the CRP and sign-off	CRP completion document signed off by the core team members

TABLE 5.10
High-Level Description of an EAM Project "Implement Phase"

Activity	Deliverable
Detailed activity/task planning	Updated project schedule
Defined data conversion requirements	Data mapping documents
Define interface requirements	Technical design documents, test plans
Develop user training materials	Training package
Develop and implement system testing	System test plan, test results
Deliver user training	Trained users
The final conversion of production data	Converted data on the target application
Verify interfaces	Interface sign-offs
Final readiness checks/define resources, etc., disaster planning	Go-live check-lists, resource lists, contingency plan
Go live	Production system

IMPLEMENTATION AND STARTUP

During this stage, you assemble the production system, having previously validated the base configuration, and undertake support activities not directly shown above, including infrastructure changes (usually network, hardware, and database-related).

The importance of thorough system testing cannot be overstated. Unfortunately, it is often inadequate, causing frustration among the user community after go-live. One reason is that implementation team members do not have the time, inclination, or training to develop and conduct detailed test cases. Depending on staff availability, there is a good argument for bringing in fresh minds to focus on testing. While this can be expensive, it's often the cheapest alternative in the long run, mainly where there are several integration paths between the EAM, APM, or other systems.

POST-IMPLEMENTATION AUDIT

After the initial production operation (go-live), put a rigorous monitoring process to ensure that the system is technically stable (i.e., performance, availability), being used correctly, and producing business benefits after an operation. Although this is

TABLE 5.11
High-Level Description of an EAM Project "Execute Phase"

Activity	Deliverable
Monitor system	Performance reports, database tuning changes, etc.
Obtain user feedback	
Measure achievements against critical success factors (CSFs) and key performance indicators (KPIs)	Analysis of results, where available (maybe time-dependent)
Implement required changes where possible	Configuration changes

why the EAM/APM was procured in the first place, it is often given scant attention. However, if you set up business measures at the project's outset, they can easily be measured after an appropriate time.

PERMISSIONS

6 Maintenance Parts Management Optimization*

Don M. Barry
Previous contributions from Don M. Barry,
Eric Olsen, and Monique Petit

6.1 INTRODUCTION

Maintenance Parts Management (also known as Maintenance, Repair, and Operations or Overhaul – MRO) is not always available when and where required for many organizations, despite significant expenditures to stock them locally and heroic efforts by inventory managers and procurement to meet unpredictable demands.

When looking at maintenance parts management that supports a maintenance organization, the complexities can range from simple to complex. For example, are we supporting the functionality for assets on one campus or across many and perhaps many in diverse regions or countries? Other considerations and complexities include whether we expect to manage rotating parts (identified parts to be repaired and returned to inventory once repaired), parts warranties, or vendor-owned inventories.

Maintenance Parts Management is as complex as asset management. Unfortunately, many techs often go to their parts counter to request their parts without understanding the role they and others in their enterprise community need to perform to ensure that maintenance parts management is successful. Although Maintenance Parts is just one component of the Asset Management Excellence pyramid, it commands its unique ten unique focus elements to succeed.

The maintenance parts focus elements include strategy, organization, logistics (including stores warehousing), metrics, inventory planning, parts sourcing and refurbishment, systems support, asset lifecycle alignment (with maintenance parts management), inventory optimization, and re-thinking existing processes.

The Maintenance Parts Excellence Pyramid can be leveraged to reflect how it supports maintenance excellence.

When a new asset is introduced into the organization's asset lifecycle, multiple stakeholders need to work together and communicate, among other things, the parts that may be reasonably used in the assets' lifecycle. In addition, the initial spare parts (ISPs) also needed to be identified and stocked to support the asset's lifecycle.

DOI: 10.1201/9781032679600-7

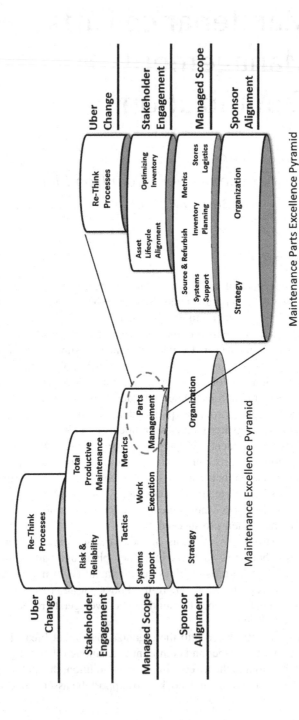

FIGURE 6.1 Maintenance parts management excellence is as complex as asset management excellence.[1]

The list of stakeholders includes the maintenance parts inventory planning and warehouse teams, maintenance, operations, engineering, procurement, vendors, finance, legal, HR, and IT.

Figure 2.5 demonstrates the diverse stakeholders needed to support a new asset early in its lifecycle and when some data transfers should happen. In that example, the stakeholder group to support that set of processes includes design engineers, vendors, finance, procurement, inventory planners, maintenance, and operations. Getting the initial asset data correct at the birth of a new asset can go a long way to getting the related parts support for the asset lifecycle.

To organize some of the areas we expect to focus on in this chapter, we will use a simple, high-level model shown in Figure 6.2. This model shows an optimized macro view of maintenance parts management that we should have, at a minimum, to focus on:

- Key performance indicators (KPIs) for material management to track how this infrastructure supports the maintenance of the asset functionality effectively;
- Inventory management policy that supports all of the maintenance parts management areas and can ensure that the critical performance metrics are tracked and tracking to plan;
- Physical maintenance parts management set of activities that manage the receiving, physical storing, and distributing of parts; and lastly
- Procurement of parts for inventory and to support work orders. Procurement of the repair of rotating parts designated as serialized repairable parts that go to inventory once repaired.

Availability of maintenance parts (or spare/service parts) on a timely basis is critical to successfully executing a maintenance plan. They are sourced, placed, managed, and utilized to support the sustainability and lifecycle of the expected functions of valued assets. **Overall Equipment Effectiveness** (OEE, defined in Chapter 4) will suffer if critical spare parts are unavailable for planned or unplanned maintenance. In a successful maintenance operation, spare parts are available as and where required

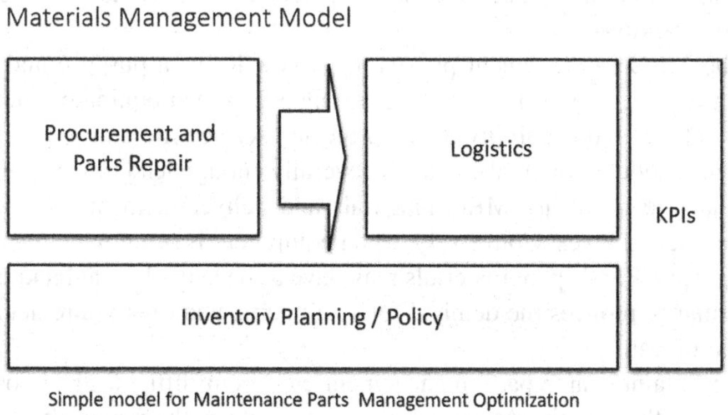

Materials Management Model

Simple model for Maintenance Parts Management Optimization

FIGURE 6.2   Maintenance parts management model.

to maintain the asset's function (i.e., equipment used in a critical manufacturing process or a power generator in a plant).

This chapter will discuss fundamentals and some of the complexities of sourcing and delivering maintenance parts (or MRO materials). Before that, we introduce the maintenance parts management lifecycle dynamics and how it supports the asset lifecycle. Figure 6.1 displays the high-level macro view of a maintenance parts management set of processes. Then, working from right to left on this diagram, we will introduce some of the metrics successfully leveraged in an optimized leading maintenance parts management organization.

Maintenance parts management begins with understanding the demand for the materials required for MRO for the planning period. Once the specifications, quantities, and timing are known, procurement can find the best suppliers based on multiple factors, including service, quality, and total cost. If suppliers could bundle all the necessary MRO materials for the specific maintenance task and deliver them from their shelves direct to the job when needed, there would be no need for local inventory. The diversity of materials required and irregular demand usually justify local inventory in central or satellite warehouses, depots near specific equipment, or service vehicles. The number of materials, multiple stocking locations, and irregular demand present significant complexity for inventory management. Improvement in MRO management is a balancing act between the availability of parts (service) and the cost of making them available.

Success in maintenance parts management optimization requires data, including material specifications, Bills of Materials (BOM) for specific maintenance tasks, historical usage, inventory integrity counts by location, lead times, order quantities, logistics costs, and more. The consequences of material not being available when needed must also be known and are equally important. A rational process to manage and mitigate business risks related to spare parts availability must involve operations, maintenance, and finance. For many organizations, assembling the relevant data and making sound risk-based decisions for thousands of materials each year, one material at a time can be daunting. Fortunately, information management systems can make these decisions easier by grouping materials for similar treatment and judicious application of the Pareto principle. Getting procurement and inventory control for the limited number of items representing the most value is the obvious way to get started for these organizations.

While the MRO improvement path is a journey, it has a purpose and becomes easier when it merges with maintenance excellence. When equipment delivers the expected functions and capability for its projected service life and requires maintenance only at scheduled intervals, there is generally enough lead time to have suppliers bundle all the necessary MRO materials and deliver them directly to the job needed from their shelves. Conversely, when equipment is often breaking and without warning, spares and spare materials may have to be kept close at hand. Excellent asset performance reduces the demand for spare parts, and the requirement for substantial local inventory.

Improving maintenance parts management can seem difficult or impossible for organizations with extensive MRO inventories to support the maintenance of assets regularly failing in service. All elements of effective change management, the people side of things, are required to supplement the improvement of processes for strategic procurement and inventory management in support of maintenance excellence, including a robust and well-communicated "case for action," participation in analysis and solution design by all affected stakeholders, performance measurement and intentional performance management, and visible and steadfast leadership from executives and senior managers.

Many organizations spend a lot of money on maintenance parts management, and it's often poorly controlled. Usually, there isn't enough consideration given to how a part will be procured or replenished, the best inventory placement, managing surplus, or delivering parts so that the technicians can be confident of reliable parts delivery. Inventory reserves for scrap and inventory surplus management are often misunderstood and underfunded. As a result, inventory is often not optimized, and suppliers' service levels and inventory are low.

The fundamental rationale for storing tens of thousands of part numbers or stock-keeping units (SKUs) on-site is to reduce the mean-time-to-repair (MTTR) for critical equipment. The farther an operation is from parts and supplies distribution centers, the more safety stock required, and the more critical inventory optimization becomes. The position of an organization on the "innocence to excellence" scale of Figure 6.3 is a visual representation of how much the inventory is rationalized, inventory control is optimized, and stores purchases are strategically sourced. The lower the position on the scale, the more opportunity there is to reduce spending on MRO and improve OEE and service to users – the main job of maintenance, after all.

While manufacturing inventory is often kept at locations determined by use in manufacturing, service parts inventory is more likely to be stocked for proximity to the supporting asset base. Maintenance parts inventory can be kept in multiple stock rooms across the company, in service vehicles, depots, and asset locations. Inventory

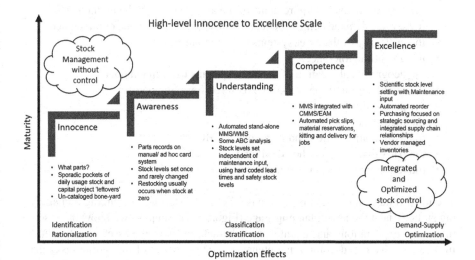

FIGURE 6.3 Innocence to excellence scale for spare parts delivery.

placement in these locations is an optimization issue, balancing proximity (service) with cost.

Inventory planning is based on sales forecasts and orders in a manufacturing environment. One tactic to optimize costs in a manufacturing environment is to use these forecasts to manage a **"just-in-time"** product inventory. For example, maintenance parts are stocked for unpredictable equipment failures driven by risk-averse management approaches. Stocking for OEE in a supply chain is often called the **just-in-case supply chain**.

6.2 ASSET MANAGEMENT LIFE CYCLE

In maintenance, the availability of a spare part alone does not entirely fulfill a customer's request. A craft/tradesperson must perform the maintenance action and install the part. The spare part, the technician with the right qualifications, and the necessary technical documentation must be brought together ("rendezvous") to satisfy the maintenance demand efficiently.

Understanding the asset criticality, configuration, component makeup (BOM), and planned service strategies contributes to a leading parts inventory planning strategy.

To accomplish this effectively, the maintenance parts management organization should understand all the attributes of the maintenance service strategy and expected maintenance tasks (planned and unplanned). This planning approach applies before the asset is installed and starts operating and includes understanding when it is decommissioned to sell surplus spare parts or scrapped as appropriate.

Compared to a typical manufacturer's supply chain, the (spare parts) maintenance parts management challenges are unique. Effective inventory strategy and placement require an aligned understanding of the assets they support. A simple list of some of the areas of focus includes:

- the service lifecycle requirements of the asset/systems to be maintained;
- the service demand being scheduled or unscheduled (planned or unplanned);
- the criticality of the asset/systems to be maintained;
- the downtime tolerance;
- the geographical location of the asset/systems to be maintained;
- the qualification of the workforce required for maintenance;
- the possibility of repair;
- the location of repair;
- the asset configuration and its associated BOMs;
- the parts that make up the BOM; and
- the requirement for reverse logistics.

The ideal time in the asset management lifecycle to develop a sourcing and stocking strategy is at the asset planning stage (Figure 6.4). Suppose we know the asset's functional expectations and maintenance characteristics of the asset's components. In that case, we can develop a maintenance strategy or set of mitigating tasks that will have parts associated with them. This can be done through a basic reliability centered maintenance (RCM) initiative during the asset planning phase of its lifecycle

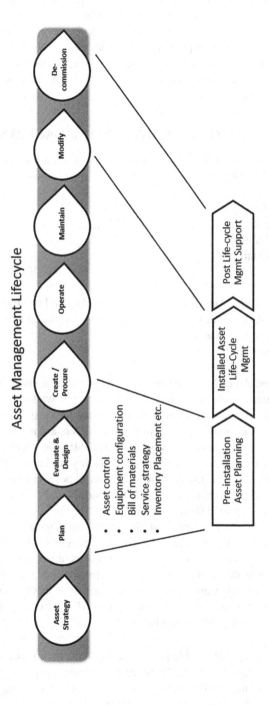

FIGURE 6.4 Maintenance parts management lifecycle.

(see Chapter 8 on reliability). With this information, we can determine a basic parts requirements list, understand how the part may be needed (i.e., corrective maintenance vs. preventive maintenance), and determine where we should stock parts and the quantity to fulfill the scheduled maintenance tasks. At the other end of the lifecycle, we can dispose of their spare parts if we know specific assets will be decommissioned. For maintenance organizations that have existing assets, an RCM initiative can provide a similar result; however, the best time to do this assessment is during the asset planning phase.

6.3 PERFORMANCE MEASUREMENT AND MANAGEMENT

At a high level, KPIs that would serve as valid indicators of a leading maintenance parts operation could include:

- Parts Availability
- Parts Acquisition Time
- Systems Availability
- Distribution Quality
- Parts Quality
- Parts Costs
- Inventory Turnover (Annual Usage vs. Average Annual Inventory)
- Inventory vs. Asset Value (Value supported Assets produced in one year vs. Average Annual Inventory)
- Inventory Reserves

Parts availability would ideally be measured for every craft or planner's parts request. Leading organizations would measure how long the craft waited for parts or the time the asset availability was impacted due to waiting for parts. Systems must be "available" for a craft to order a part, confirm prior orders are reserved, picked, or staged, or confirm that the parts are in stock and available. The warehouse would be measured on inventory integrity and distribution quality (right part, right quantity, right place, and within the committed time). The overall process could be measured against the cost of the part versus "new" or "street value" and the confirmation that the part performed as expected (parts quality).

These metrics refer to the health of the overall maintenance parts process. Comprehensive performance measurement typically requires sub-metrics for each supporting process to ensure the overall process meets its objectives.

Many metrics can be developed to manage the typical maintenance parts process. Figure 6.5 provides a small example of additional measurements to support the ultimate leading metrics provided above. An example of KPIs for the repair of returned defective parts and rotating parts would be:

- Confirmation that all repairable parts are captured in their process;
- Monitoring the repair yields of the received defective part;
- "Out of box failure" (OBF) would monitor the new defective rate reported by a craft for a refurbished part or rotating part;

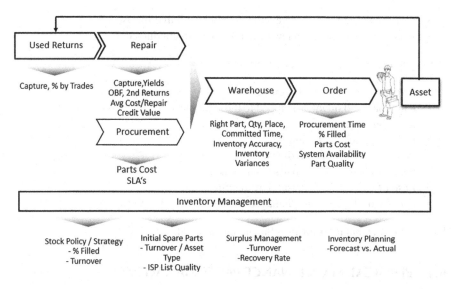

FIGURE 6.5   Maintenance parts management sub-KPIs.

- Logging parts that have been returned for repair for the same symptom a second time in the past year; and
- Cost of repair versus the cost of a new purchase.

6.3.1   EXAMPLE OF SOME KPIs FOR MAINTENANCE PARTS MANAGEMENT

Initial spare parts:

- Inventory turnover/asset type
- Service level/asset type
- Parts inventory value/asset production value
- Initial spare parts (ISP) list quality

Additional characteristics of successful material management and procurement organization include:

- Stockouts representing less than 3% of orders placed at the storeroom.
- A central tool crib for special tools.
- Control procedures are followed for all company-owned tools and supplies, such as drills, special saws, ladders, etc.
- Inventory cycle counts are conducted.
- Inventory is reviewed regularly to delete obsolete or very infrequently used items.
- Purchasing/stores can source and acquire rush emergency parts that are not stocked quickly in time to avoid plant downtime.
- Blanket and system contracts and orders minimize redundant paperwork and administrative effort.

- Stores catalog is up-to-date and is readily available for use.
- Vendor performance reviews and analyses are conducted.

Some key metrics that support delivery include:

- Inventory variance and accuracy;
- Right part and quantity shipped to the right place within the committed time (distribution quality);
- Dock-to-stock cycle times;
- Order to delivery cycle times;
- Percent of lines filled (parts availability);
- Ease of Use to order (system availability); and
- Quality of the part (not damaged by handlers or shipping).

Please refer to Chapter 4 for insights on how to create new KPIs.

6.4   PHYSICAL MAINTENANCE PARTS MANAGEMENT

The maintenance parts need to be ordered by the tradesperson, they need to be purchased, and the inventory needs to be managed. This section discusses some critical points about maintenance parts management and handling physical logistics.

Traditionally, tradespeople have often turned to stores to collect their parts and search for the "right" part. This has been mainly due to a poorly maintained (or completely lacking) catalog or parts book. The stockroom has often been the first interface between the trades and the process interface with the asset management process.

The idea of service levels from a stock room is simple from the trade's point of view.

When they come to the parts counter to order and receive their parts, the quality service criteria for maintenance parts management are easy to understand. "They want the part when they want it."

Six Rs define complete optimization of the sourcing and delivery of spare parts and materials to maintain, repair, and operate equipment assets:

- Right parts
- Right quality
- Right quantity
- Right place
- Right time
- Right price

Distribution quality cannot be achieved without a well-defined and executed inventory management and procurement set of processes; however, the physical logistics process also needs to be well-defined and executed. The parts have to be physically handled to complete the Six Rs.

For example, a critical step to delivering a part when requested is to ensure that the part is receipted correctly into the inventory system when shipped to and received at the dock. Receiving accuracy is critical to inventory accuracy and control. Inventory integrity will never be correct if the part receipted is incorrectly acknowledged. There are several key events in receiving:

- Invoices are paid by matching item quantities and attributes to the purchase order, then confirming with the accounting systems
- The item as ordered is confirmed
- The item is marshaled for end-use (inspection, storage, delivery)
- All variances or inconsistencies are recorded and monitored and used to track vendor performance.

A significant consideration is how much to centralize receiving. You must determine whether trained personnel should handle all receipts at a designated location or end-users should be responsible for receipts of non-inventoried purchases. While either decision has merits, you must review internal factors:

- What is the training/motivation level of the staff? Any employee who receives goods from the organization must implicitly agree to be rigorous, prompt, and accurate.
- Who will perform the task? Should the engineer at the site receive the goods or the warehouseman? Who should be fiscally responsible for the task?
- Does centralized receiving of all goods increase the lead time to the end-user? Is this a receiving function or an internal communication or systems limitation?

Another example: Best Practice warehouses or storerooms must have appropriate security for the type of item stored in terms of its deterioration or risk of theft. They are intended to provide maintenance personnel with high levels of service. For example, they may have procedures to alert maintenance personnel regarding the receipt of their materials.

Maintaining inventory involves identification, storage, and auditing. Inventory auditing is the process of confirming absolute inventory accuracy. Proving inventory accuracy is the equivalent of an operations quality control program. You must instill and maintain user confidence that whatever the system says is in the warehouse.

Service level (the frequency of tolerated fulfilled orders) measures inventory placement, procurement service levels, and inventory integrity/control performance. Reduced lead times, inventory accuracy, and the ability to find the right part contribute to parts availability service levels. Establishing different service levels for each commodity reduces costs while ensuring the availability of necessary items. Fulfillment misses can be broken down into two categories:

- Not Stocked: parts that were not intended to be stocked at the local stock room)
- Out of Stock: parts that are normally stocked in the local stock room but are currently not available in the stock room

For inventory control service levels, traditional organizations annually "count everything in a weekend" and then, as a result, adjust stock balances and value. Unfortunately, this method is inefficient, often inaccurate, and usually done under strict time and resource constraints.

Perpetual Stock Count or Cycle Counting is a better alternative. You set up numerous counts, typically less than 200 items, to be counted regularly. Over a year, all high-value inventory (at least A and B class) is counted at least once. The benefits of Cycle Counting are:

- More accurate counts, as the number of items is relatively tiny, and variances can be easily researched.
- Stock identity validated against the description, and changes noted, reducing duplicates, increasing parts recognition efficiency, and helping standardize the catalog.
- Confirmed stock location.
- Accuracy levels (dollar value and quantity variances) that can be used as a performance measure.

6.5 UNDERSTANDING INVENTORY MANAGEMENT DYNAMICS

An effective plan to ensure maintenance, repair, and operations (MRO) parts will be available when needed includes a holistic understanding of each asset's parts service requirement dynamic and the maintenance parts management infrastructure you have to support this need. The simple maintenance parts management model introduced earlier in this chapter (Figure 6.1) ensures that this infrastructure will support the business needs falls into the inventory management and policy area.

The process typically begins with assessing what maintenance tasks will be done during the asset management planning period and what parts will be required. For example, for a single part, i.e., a gasket, maintenance parts management would like answers to the following questions (and others):

- What is the associated task? i.e., overhaul a pump
- When will the task be done? Is the timing predictable with confidence and precision?
- Where should the part best be stored?
- What is the lead time for delivery to the job site?
- What is the reliability of delivery within the required timeframe?
- What is the designated stock room's economic order quantity (EOQ)?
- Is the cost of shipping significant?
- Have we established a supplier of choice for the material (gasket)?
- Can the supplier bundle the material with some or all of the other parts required to complete the task?

With that information, procurement decisions, including when to order and inventory decisions, such as stocking location and quantities, can be made rationally with lead time to optimize service and cost. In practice, this information is usually not available on a comprehensive basis. However, using the information listed on high-cost items, the total expenditure for MRO parts can be managed effectively by taking a systematic approach.

6.5.1 Inventory Planning

Demand for many MRO materials is random. Except for parts used for preventative maintenance (notably for time-based replacement and overhaul), most MRO stores' items are used irregularly. In contrast, the consumption rate of tires per automobile manufactured or catalyst used per barrel of oil is highly predictable. For MRO goods, it can be more costly than effective to utilize automated reorder points and a high EOQ according to traditional materials analysis. An exception would be for a large install base and, therefore, high potential for the same randomly failing component; using EOQ theory for these parts would be a practical and leading exercise. For MRO materials with random or "lumpy" demand, manual intervention in stocking quantities and location decisions is appropriate for those with high value or high potential for causing downtime or other unacceptable risks.

The role of an inventory planner is critical for both the inventory dollars managed and the service levels derived from the inventory placement. The inventory planner manages the dynamics of inventory inputs and disbursements. The planner forecasts and defends the inventory (at an enterprise level) and lobbies and facilitates the scrapping of extraneous parts to asset risk requirements. The elements that could make up this "inventory policy" (see Figure 6.9) are the control points to the inputs and disbursements.

When defining the final inventory policy, location, delivery, asset maintenance, and stocking strategies are considered.

Typically the inventory planner would declare the inventory and its expected influences monthly. This declaration helps the executive teams understand their risks to MRO service levels and financial expectations in a balanced view.

6.5.2 Location and Delivery

Deciding what to stock where can become very complicated for many maintenance organizations. Figure 6.7 suggests that in some cases, the organization's many stock

---

TABLE 6.1
Inventory Considerations that Could Influence Inventory Policy

Inventory Inputs	Inventory Disbursements
• Initial spare parts/recommended spare parts	• Technician parts usage on a work order
• Surplus asset commissioning parts	• Parts warranty or quality Issue
• New parts returns from customer/technician	• Parts sale to an external entity
• Parts replenishment (from inventory minimum stock policy thresholds)	• Surplus sale
• Unique parts orders or asset enhancements	• Surplus scrap
• Warranty replacements	

Maintenance Parts Inventory

Maintenance Parts Inventory Additions	Maintenance Parts Disbursements
Initial Spare Parts / Recommended Spare Parts	
Surplus Asset Commissioning Parts	Technician Parts Usage on a Work Order
New Parts Returns from Customer or Technician	Parts Sale to External entity
Parts Replenishment	Parts Warranty or Quality Issue
Unique Parts Usage or Asset Enhancement	Surplus Sale
Warranty Replacements	Parts Scrap

Service Levels · Inventory $

FIGURE 6.6   Example of inventory inputs and disbursement dynamics.

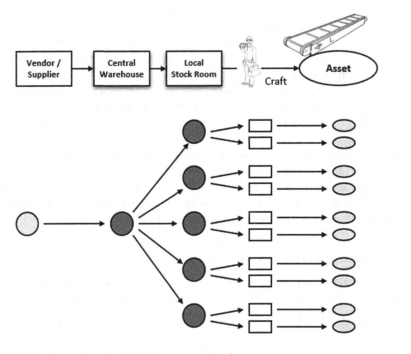

Vendor / Supplier → Central Warehouse → Local Stock Room → Craft → Asset

FIGURE 6.7   Example of an MRO inventory network.

room locations may offer the opportunity and challenge of multiple choices for where to stock the part best. Figure 6.8 shows that the choices range from the suppliers' shelves to the equipment or asset itself.

One option: leaving materials on suppliers' shelves (and on their books) until required, which is the effect of ordering direct-to-job, offers several significant advantages. If an order is placed to supply parts for a scheduled maintenance task, the

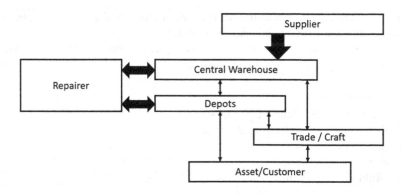

FIGURE 6.8 Maintenance parts flowthrough of an extensive parts network.

specifications and quantity are generally well known, so there is less tendency to order too much, potentially building inventory with a surplus and less requirement for returns and substitutions. If the order can be kitted, meaning all the parts required are assembled by the supplier or in a reserved receiving area on-site, placing them physically in bins and later retrieving them can be averted. Kitting usually makes maintenance execution more efficient, as little or no time is lost looking or waiting for the "one last (necessary) part." Just-in-time ordering reduces the average value of inventory, a key indicator of maintenance parts management effectiveness, provided service levels meet or exceed requirements.

Most organizations decide to maintain inventory on-site, often in a central warehouse. Spare parts inventory is usually expensive to purchase initially and maintain – inventory carrying costs typically range from 20% to 30% of book value per year. In addition, many organizations, especially capital-intensive industries such as mining or chemicals manufacturing, have stocked items that are seldom required – items that have sat unused in the warehouse for years. Obsolescence, including electronic components past their "use before" date, and spare parts for equipment removed from service, can further reduce the effectiveness of inventory spend. Despite these cost factors, having locally available inventory may be justified if demand is unpredictable and the significant consequences of a stockout. Some parts and materials are used frequently to repair or maintain several assets, for which significant efficiencies in ordering and handling are possible. When these items are low unit value (i.e., gloves or studs and nuts), they should be ordered in economic quantities and handled efficiently.

For similar reasons, satellite warehouses, tool cribs, service vehicles, or a craftsman's workstation may be the location of choice. Gains in the efficiency of maintenance work execution can outweigh the incremental costs of placing and replenishing spare parts inventory in these places.

Generally, adding locations to the MRO parts network will add cost. This cost must be justified by improved service and, ultimately, enhanced maintenance efficiency. Failure to account for potential downtime, waiting time, travel time, and inconvenience may lead an organization to over-rely on just-in-time ordering or a central warehouse. Off-system or "squirrel inventories" will proliferate if needed parts cannot be obtained quickly and conveniently.

One of the critical areas in the maintenance parts supply chain focus shown in Figure 6.6 is inventory management. Essential elements of inventory management include:

- Location-specific demand management
- Service-specific forecasting algorithms
- Part criticality
- Inventory effectiveness
- (Multiple replenishment) inventory echelon management
- Automated replenishment plans
- Inventory surplus management
- Inventory surplus scrap programs

The best organizations determine demand characteristics and develop a way to leverage automated replenishment to ensure they have the parts where they need them. For example, leading organizations will work with the maintenance planners to confirm when planned maintenance is scheduled and leverage their suppliers' inventory rather than their own to keep inventory carrying costs down and service levels up for random demand and typically urgent needs. This may result in an inventory stock room network and tailored inventory placement within the network. Leading MRO operations will have service-level agreements with suppliers to ship planned parts requests directly to the work order or craft. The stockrooms quite often carry only minimal inventory for planned work. Figure 6.8 provides an example of an inventory network diagram for a large organization with multiple tiers of maintenance parts sourcing options.

System solutions exist to provide inventory planning based on the actual cost of stocking each part, including what it costs to stock a part and the costs associated with not having a part when it is needed (stockout). Such planning would optimize the placement of warehouses and field stocking locations locally, by country, or globally. It could define where sites must support critical asset availability and where existing sites are no longer needed. This planning would optimize the target stock level for each part number and site and provide intelligent logic to maintain the target stock levels through replenishment and inventory rebalancing.

Advanced scenario modeling tools exist to provide multiple "what-if" responses, allowing maintenance organizations to fine-tune their service levels on a company-wide scale.

Inventory planning system solutions can include real-time, web-based (cloud) interfaces that provide an up-to-the-minute snapshot of parts availability and materials requirements.

Simple ABC Inventory categorization can help manage inventory placement and stocking strategies within these tools. The figure below shows how some organizations have interpreted MRO parts within an ABC or "3 levels" context to work for them. Once defined, these strategies can be automated in a system-managed replenishment program.

TABLE 6.2
MRP Strategies for Maintenance Parts Inventory Management

Stocking Strategies	ABC	Active Versus Insurance Stock	Criticality Stocking	Optimization
Level 1	• Top 80% used items by $ value • Less than 10% of stocked items	• Top used parts (typically 2 usages in 3 months) • Less than 10% of stocked items	• Level 1 is supported through either ABC or active Level 1 process • Less than 10% of stocked items	• Low-value parts that are deemed to be used in the lifecycle • Balanced with stocking and expediting costs
Level 2	• Next 15% used items by $ value • Less than 20% of stocked items	• New part in the past year or at least one usage in the past year • Less than 20% of stocked items	• High criticality, high usage parts not covered in level 1 • Can be 20% of parts stocked	• Cost-effective stocking of parts expected to be used once a cycle (i.e., year) • Can be 60% of stocked items
Level 3	• The bottom 5% used items by $ value • Can be than 70% of stocked items	• Parts deemed to be stocked "just in case" • Can be 70% of stocked items	• Lower criticality parts and lower usage parts • Can be 70% of stocked items	• High-value low-usage parts are stocked, often at a consolidation center
Comments	• An often manual initial stock process with Min/max support	• Often scientific initial stock process with Min/max support	• It can have many levels of criticality and echelon support depending on the support network	• Considers all costs/impacts in stocking optimization calculations by network location

Notes: Parts deemed as "critical" or required for "insurance" are typically part of level 3.

6.5.3 INVENTORY OPTIMIZATION ASSESSMENT PROCESS

The objective is to optimize your stocking decision by balancing two conflicting cost drivers: stocking materials to minimize stockout costs versus reducing inventory ownership costs.

Using this basic method, you:

- identify all MRO (procurement, repair, returns, etc.) sources
- identify goods that need to be stocked
- develop new and efficient ways of dealing with goods that should not be stocked

The three steps of inventory optimization are analysis, evaluation, and optimization.

6.5.4 ANALYSIS: IDENTIFICATION AND RATIONALIZATION

The first step in optimizing inventories is to analyze current inventory sources and MRO practices. This will help you develop fundamental processes to develop a strategy to reduce costs and optimize stocking decisions.

"Inventory" also includes items that are not held in a warehouse. Often, materials are purchased directly and stored on the shop floor or in designated end-user lay-downs (local convenient storage areas in the plant). These inventories are kept for numerous reasons, most often distrust inventory management, but it's an inefficient practice. It can mean poor stock visibility, inappropriate charges, no assured adherence to specifications or loss protection, and excessive on-hand quantities. All of this is costly to the organization.

You want to consolidate this inventory with all other types into a centralized inventory information source and manage it accordingly. This will help reduce the MRO spend and ensure the material is available.

The basic tasks performed at this point are:

Task 1
Identify the entire inventory within the organization, including items "off the books."

- List each item's supply and usage date by referring to the inventory information sources.
- Identify all inventory outside the warehouse, including satellite shops, scrap yards, lay-down areas, lockers, squirreled inventory, etc.
- Construct matrices that segment the inventory according to such criteria as value, transaction frequency, criticality, and likelihood to be stolen, using tools that perform ABC analysis (see Figure 6.9).

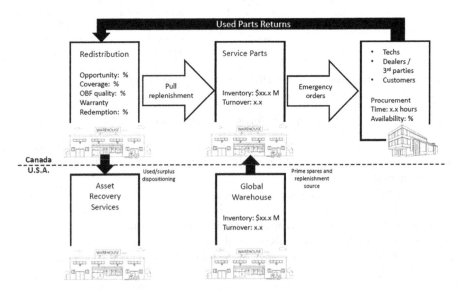

FIGURE 6.9   High-level parts flow diagram.

TABLE 6.3
Inventory Matrix for MRO

Inventory Classifications (A, B, C)	M	R	Ov	Op
A	• Contracts • Large quantities of highly active parts • Short planning horizon • Low impact	• Small purchase/contracts • Small quantities, highly managed • Short/medium PH • Medium – critical impact	• One-off purchases • Minimal quantities, highly managed • Very long specific PH • Highly critical impact	• Contracts • Med-large quantities, medium managed • Not in PH • Low impact on maintenance – may be critical to the operation
B	• Contracts/VMS • Large quantities, highly managed • Short planning horizon • Low impact on downtime	• VMS • Small quantities, medium managed • Low impact	• One-off/VMS • Rarely purchased, highly managed • Specialized PH • Medium to critical	• VMS/contracts • Large quantities, not managed • Not in Planning horizon • No impact
C	• VMS • Large quantities, not managed • No planning horizon • Little impact	• Not often found	• Not often found	• One-off purchases • Small quantities, not managed • Low planning horizon • Low impact

• Purchasing strategies (VMS – vendor-managed solutions)
• Inventory strategies
• Planning Horizon (PH)
• Impact to downtime

MRO – Maintenance Repair and Operations Inventory
Traditionally, the "O" in MRO meant overhaul.
The above matrix refers to both options.

- Ensure that all inventory items are uniquely identified and adequately described and that all components are currently in use.
- Assess standard functions and uses across operations and identify and eliminate duplicate part numbers and sources.

Task 2
Develop a strategy for rationalizing inventories.

- Identify the volume and location of material held as inventory. Establish a suitable stock control system based on inventory size and distribution, transaction volumes, and the integration you need for procurement and work management systems.

6.5.5 Evaluation: Classification and Stratification

Make stocking decisions first on a commodity level and then, where warranted, at the individual item level. This ensures that the right amount of inventory is available and the suitable types of inventory are stocked and controlled.

To make the appropriate stocking decision, partition the inventory into segments, and apply stocking and sourcing strategies to each unit (see Table 6.2). You can estimate cost savings using performance measures or industry KPIs, as described in Chapter 3. Also:

- Establish transactional stock control and auditing processes to get a clear stock position and set a control framework for each item.
- Determine stock level requirements (using statistical stock models such as EOQ if demand is predictable or input from the end-user if demand is highly variable, seasonal, or varies other than with time) to set optimum levels for individual items.
- Determine how service levels will be calculated and applied.

From Table 6.2, several factors apply to each cell on the matrix, representing activity level (usage volume and frequency) and type (why the item was purchased) for each commodity.

If the item you are evaluating is "A" class (the 10–20% of the inventory that accounts for 80% of the value or spend) and was purchased for Operations, you could use the following strategies:

Purchasing Strategies	Contracts
Inventory strategies:	medium to large quantities, medium amount of stock management (if stocked)
Planning horizon:	none
Impact on downtime:	none to maintenance; may be critical to the daily operation of an organization

6.5.6 INVENTORY CATEGORIZATION AND OPTIMIZATION

You now have an optimal stocking decision for each commodity and maximum availability (service levels) at minimum cost.

The basic tasks at this step are:

Task 1

Determine the stocking decision for each commodity. There are several stocking choices, depending on whether the item is:

- Regular inventory
- Unique (highly managed) inventory
- Vendor-managed inventory (consignment, vendor-managed at the site)
- Vendor-held inventory (the vendor becomes a remote "warehouse" of the organization)
- Stock-less (the item is cataloged and its source identified but not stocked)
- No stock (purchase as required)

Task 2

Increase maintenance's ability to find the right part for the job. This has become easier now that you have a standard user catalog. The next step is to link each part with the equipment used by assigning parts in the Bills of Materials (BOM), simplifying identifying material to complete tasks. Accurate and complete BOMs reduce duplicates and are a straightforward, reliable tool for requisitioning matercials. You can use them as well to determine stock levels accurately. There are two types of BOM:

- "Where-used BOM": lists all of the repair parts and installed quantities for a piece of equipment or one of its components
- "Job BOM": lists all the parts and consumables required for a particular repair job.

Task 3

Develop management processes for difficult-to-manage or unique stocks.

- Develop an inventory recovery (surplus stock) management system, including a decision matrix to help you retain or dispose of surplus stock. This is essential to ensure that surplus stock is managed as a capital value on the company's financial books. Include a divestment model in this matrix

to appraise the immediate disposal return against the probability of future repurchases. Include a plan to identify and value all non-stocked inventories.
- Formalize the repairable stock management process. Establish accounting practices and policies to control repairable components effectively and stocked materials (i.e., motors, mechanical seals, pump pullout units, and transformers).

Automating inventory management is a leading practice. If we understand the influences of why we stock the part, this can be set into an algorithm and run daily, weekly, or monthly. Some enterprise asset management (EAM) and enterprise resource planning (ERP) systems do this in a small way. Most do not yet do it, so the organization is comfortable allowing the algorithm to run without an inventory analyst reviewing each suggested inventory addition. A leading practice would have this set to optimize inventory and service levels and run without line-by-line analysis. Products are now available to manage this specifically for maintenance parts. Material requirements planning (MRP) can be used for the highly active or predictable activity, but tailored applications are required for managing less active or insurance parts.

One way to approach this inventory management auto-replenishment challenge is to outsource the process to organizations that provide "black box" inventory management services, analyze the recommendations, and execute inventory adds as appropriate. However, this can be labor-intensive and assumes the third party understands your business.

A second way would be to run an application within your organization, monitor the recommendations, and adjust the algorithm. This focuses on the line-by-line analysis to the algorithm analysis and directly places control and responsibility on inventory management. Once proven, less resource oversight should be needed.

Strategic financial and customer service modeler solutions now exist for inventory management. They can be tailored to provide automated analysis to identify short-term optimized inventory and service strategies while modeling longer-term, continuous improvement benefits. They can sense and quickly respond with adjustments to inventory parameters throughout the product's life. This can allow users to control lifecycle decisions by updating the SKU service targets. See Chapter 20 on Technology Trends in Asset Management for more detail.

6.6 PROCUREMENT AND PARTS REPAIR

Today parts procurement has multiple source categories. For example, at a high level, it can be sourced from the original equipment manufacturer/supplier or a parts repair/refurbishment source.

The parts repair source is often a mix of vendor and in-house services that repair the defective part and then return it to stock as if it were as good as a new one. The

MRO supply chain is a "two-way" supply chain in these cases. As parts are replaced in the field, many are returned for repair or refurbishment. In many industries, such as aircraft maintenance, the field service is the physical swap of a field replacement unit (FRU), and the used part is a rotating part (or "rotatable") that will be returned to the bench for repair or overhaul and then held in inventory. To support the management of used parts, a "reverse logistics" process must be established to ensure that the used defective part is tracked so that it can be repaired and the opportunity to establish a repaired part back to inventory is maintained. One significant benefit of doing this is that the repair cost is typically significantly less than the cost of new, allowing lower-cost quality parts into the inventory.

So managing used parts returns and repairing used parts is a growing part of a maintenance parts procurement strategy. For more traditional parts sourcing, procurement can be summarized as three main activities:

- Strategic commodity sourcing;
- Supplier relationship management; and
- Transaction management.

The Inventory Management group typically helps us understand what commodities we want to procure and when but procurement will help us refine these requirements. For example, **Strategic commodity sourcing** will require a better understanding of the commodities we need and analyzing our past and planned spending in these areas. In this area, procurement would assess the market dynamics for these commodities, coordinate the request for information (RFI) and request for proposal (RFP) activities and develop a suitable commodities strategy. **Supplier relationship management** would empower procurement to engage and manage selected suppliers, contract specifications, pricing, and compliance and assess supplier performance to establish service-level agreements. With the internalization of the internet to most companies and the personnel that use it, **transaction management** can execute in an automated interface between the two businesses, leveraging business-to-business (B to B) transactions and allowing procurement to focus more on the first two activities of procurement. The internet now helps manage parts catalogs, purchase orders, order requisitions, order confirmations, and receipt confirmations. The procurement group can have a less hands-on role, and the physical logistics folks can complete the cycle with receipt validation.

6.7 MAINTENANCE PARTS ELEMENT INSIGHTS

The maintenance parts management elements are described in detail in the *Maintenance Parts Management Excellence: A Holistic Anatomy* book. Like this book, the parts-focused book has questions an organization can use to provoke thought on areas they wish to improve.

A summary of the elements and the common areas of interest are revealed in the following table:

TABLE 6.4
Maintenance Parts Management Elements and Examples of Leading Practices

	Element	Leading Practice: Pain Point Example
1	Strategy	• Have a single person with the mandate (responsibility + accountability + authority) to manage plant-wide or company-wide stores. • A complete asset lifecycle support understanding of the value and costs that are influenced by a solid parts management infrastructure and execution.
2	Organization	• The roles between design engineering, maintenance and operations, finance, and inventory management effectively manage the balance between critical asset inventory cost and service levels. • Maintenance parts personnel have the proper experience, training, and education.
3	Stores and logistics (warehousing)	• Review the warehouse layout (aisle-shelf-bin numbering system) to ensure it is clearly marked and works well. Look at where the parts are physically located (busy items close to the issuing station, for example). • Make sure all material requirements are on every work order. • Do regular cycle counts based on an ABC analysis.
4	Metrics	• Ensure your inventory management performance indicators are developed from and support your maintenance and asset lifecycle cost strategy. • Measurements are developed to serve business drivers and management levels.
5	Source and refurbish	• Ensuring purchasing strategies align with corporate culture. • Used parts return process, used parts sort process, re-utilization strategies. • Ensure repair standards are "like new" so that repaired parts can return to inventory.
6	Inventory planning	• Inventory effectiveness measurements by asset type or part type. • (Multiple replenishment) inventory policies. • Having min/max reorder policy automatically generates orders to suppliers daily.
7	Systems support	• Ensure that there is an accurate listing of **all** stores in the EAM (each with a standardized parts description). • Include a "where used" and a BOM cross-reference (especially for critical spares needed for critical equipment).
8	Asset lifecycle alignment	• Complement the parts stocked to support the maintenance strategy with an understanding of where the parts will be stocked (i.e., which stockroom). • Understand where you are in the lifecycle of all assets that require parts support and adjust accordingly.
9	Optimizing inventory	• Design engineering, maintenance, operations, and inventory management teams work together to optimize asset and parts inventory effectively. • Make sure stores' records are visible company-wide (consuming in-house inventory is usually much smarter than buying new).
10	Re-Think Process	• Staff are empowered to drive quality program improvements as a regular part of their culture. • A corporate discipline and culture exists to identify all parts with a system-recognized part number (including direct purchase, disbursements from inventory, used parts, new returns, and overhaul kit BOMs).

6.8 CONCLUSIONS

How the maintenance parts management area is organized and divided responsibilities can vary from organization to organization. The maintenance parts management group reports to Procurement and Finance. In others, it reports through the Maintenance and Operations group.

Regardless of where it reports, the mission should be focused on serving the company's expected support role for the assets that drive the value for the organization's reason for existing in the first place. At times, a maintenance organization can presume that Operations are king and drive too much inventory, or Procurement and Finance are king and drive down the dreaded inventory levels and drive up inventory turns. Optimal inventory in a leading operation drives a balanced inventory that understands its role and mission. It also drives an inventory and service-level balance that complements the value of the company's assets are there to optimize.

An example of a world-class parts operation is described in detail in the Appendix of this chapter. Look there for an actual example of what is stressed in this chapter.

Implementing an optimization methodology is essential for achieving the benefits identified in the cost/benefit analysis. However, as with any improvement, unless the policies, practices, and culture support the organization's change to sustain optimization, the benefits will be short-lived.

You must carefully address the "What's in it for me" issue with employees. They must understand their importance in supporting new initiatives and sustaining change. You will need to conduct training and education sessions, but, most importantly, the message must be top-driven. Senior management must communicate its support throughout the organization. The message must be repeated over time to encourage continuous improvement. In addition, conduct performance audits to ensure that optimization is always achieved.

Once you have evaluated your organization's maturity level and settled on long-term goals and expectations, you can apply the appropriate optimization methodologies:

- First, analyze your current inventory and materials sources. Then, create a sustainable inventory strategy.
- Second, evaluate and adopt inventory control techniques to optimize stocking decisions and apply Performance Measurements.
- Third, optimize demand and supply management, emphasizing strategic sourcing and long-term planning reliability.
- Fourth, review procurement strategies for operations material and implement better supplier management and spend control.
- Last, realize that change is only sustainable if its employees are prepared for it.
- Communication is driven from the top down, and adequate training is essential to optimize maintenance parts management.

MAINTENANCE PARTS MANAGEMENT OPTIMIZATION EXAMPLE

EXAMPLE OF A LEADING MAINTENANCE PARTS OPERATION

IBM Canada's parts operation supports all IBM-manufactured and OEM-supported products on a warranty or maintenance agreement within Canada. Its client base is primarily corporate but can also be small businesses and home users. IBM strives to service its client's equipment through the warranty period and planned functional lifecycle. In other words, when a client determines to purchase equipment to perform certain functions for a specific period, IBM will contract to support this equipment through the planned installed lifecycle. An example would be a major bank that elects to set up thousands of teller terminals across Canada using monitors, printers, keyboards, base personal computers, and network equipment with identical/similar distributed software support. The bank may plan to purchase a specific configuration of equipment and then support this new footprint for 6–10 years, even though the original manufacturer may have created the equipment with a three-year lifecycle plan.

The challenge for a maintenance organization is to provide a timely forecast and repair service so that:

- the functionality of the client's equipment can be sustained
- the client will experience a high level of satisfaction with the service provided
- the service will result in the client wanting to renew service contracts when they come due.

To meet the service-level expectations of its clients, IBM has placed parts stations in twenty critical cities across Canada. Two of the twenty locations operate 24 hours a day, seven days a week (Toronto, Montreal), with the central distribution center located in the Greater Toronto area. However, just having the stockrooms is not enough. They need to effectively determine what part to stock and where before installing the supported assets while the asset is installed and during the sunset phase of the client footprint. They also need to coordinate the delivery of new parts and return of used parts within the technician's work order process to effectively manage the part and the technician's time.

Within Canada, IBM supports more than 2000 products and the potential of 500,000 different part numbers. The number of part numbers stocked in Canada is 60,000; however, via their system, they can pass on an order for any part to another parts location in North America, a supplier, or an internal plant and escalate the delivery from these referred locations. Their technicians utilized a radio frequency-supported personal terminal to inquire about parts availability and to order parts or receive an order and shipping status "live." This order process allows for an internal target of 15 minutes from when the part is ordered to when it is picked, packed, and handed to a courier for a local emergency delivery. The technicians have a map of their city that declares how long they should wait to receive a part in stock. Typically stocked parts are received within an hour of being ordered in any of the 20 cities in Canada that have an IBM parts stockroom. This high-quality order fulfillment process allows the technician to feel that their order will arrive in an acceptable time-frame. It supports the concept that they do NOT need to hide inventory or do other unnatural acts to support the assigned assets and clients. This represents over 300 deliveries of emergency parts each business day in the Toronto area. To enhance delivery, each parts station has couriers available on-site during their operating hours to complete the process and send a referred part on the next flight out or leverage a car or air charter if the need calls for it with 15 minutes notice.

Managing used parts returns and refurbishment or disposition management is critical to IBM's parts management philosophy. Initial parts are procured through suppliers or other IBM plants; however, the bulk of the used parts (more than 70% in financial value) are returned to be refurbished as new and returned to stock. This is a significant cost saving in parts unit cost and procurement cost. Refurbished parts have been statistically proven to be more reliable than new parts. This is particularly true with electronic components.

Along with the typical inventory metrics such as inventory turnover and inventory reserves management, the organization focuses on five key areas of excellence to achieve its key goals.

- Parts availability/parts acquisition time
- Parts costs
- Systems availability
- Distribution quality
- Parts quality

They accomplish this with a clear focus on delivery (as previously discussed), a focus on the elements of leading inventory management, and the cost of sourcing each part.

Within inventory management, they centrally manage and control all the planned adds and deletes of the actual spare parts inventory across Canada and the inventory stocking policy set in the system for automated replenishment. They create year-over-year plans for inventory and report on the system's monitored KPIs of inventory and the overall spare maintenance parts management operation.

For new asset/product additions, they work with maintenance to develop an ISPs strategy to support the asset and monitor the sparing strategy's performance over the asset's life. Existing inventory items set the safety stocking levels and economic reorder quantities based on asset criticality, supplier lead times, and expected service levels. They manage a spare parts master to ensure standardization of parts identification and harmonization of the parts stocked so that multiple pieces of the same parts are not in the spare parts network multiple times for the same asset. They leverage systems support to flag critical stock situations and "alert" an inventory planner that a critical situation needs to be reviewed for expediting.

Like most inventory systems, they utilize a dynamic min/max and EOQ solution, complemented by understanding the required safety stock to buffer supplier lead times. They also have a dynamic ISPs strategy to complement the process and leverage system-managed date parameters, such as when the date part was first added or last used as part of their replenishment algorithm.

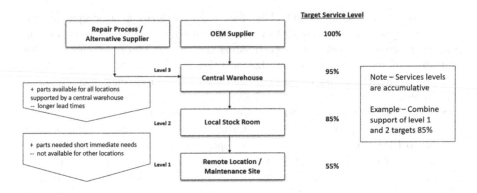

FIGURE 6.10   Example of a leading practice stock echelon hierarchy.

Their algorithm (or "inventory policy") determines target inventory levels for each part at each stocking location and the appropriate replenishment policy for each part (replenishment planning). Some parts are identified to be tactically supported from a second stocking tier (echelon) rather than the stock location closest to the asset. Expensive items or slow, non-critical parts movers are often effectively held at central locations to reduce safety stock carrying costs. The planner in the inventory management group forecasts or simulates inventory requirements at the location and piece level and can predict the service levels these pieces or locations generate. Through parts planning, they could effectively set up declining levels of support from their central warehouse to a local warehouse and through to a remote storeroom leveraging an echelon support structure. This could allow (or target) 55% of parts availability support at the remote storeroom and 85% of parts availability support at the local warehouse, and, if desired, 95% support at the central warehouse. They can run "what if" scenarios to determine the best action for a sparing strategy or asset mix change. They can also develop an "end-of-life" stocking strategy to respond to supplier "end-of-life" or "end of support" notifications.

The planner in the inventory management group forecasts inventory surplus management tactics, including surplus re-distribution within Canadian operations.

Systems usability and availability are essential to the success of this inventory management set of processes.

The Unit cost sourcing focus for IBM Canada has been primarily around **supply** and **repair** elements. This includes inbound inventory processes such as formal procurement and managing used parts returns through parts repair.

Within IBM Canada, the purchasing process supports strong relationships with suppliers and partners.

The procurement processes include:

- Ensuring purchasing strategies align with corporate culture;
- Having standardized naming/numbering conventions for items purchased and suppliers used;

- Analyzing commodity regular spending to confirm the best sourcing strategy and alignment to corporate culture;
- Having a defined and standard set of RFI/RFQ processes for commodity types or business environments;
- Supplier targets linked to total maintenance parts management objectives;
- Leveraging systems/technology to automate and, as a result, minimize the need for procurement workload in the procurement transaction process;
- Monitoring supplier performance and reviewing in regularly held joint supplier/company meetings; and
- Leveraging communications and technology in an integrated fashion to optimize inventory within the enterprise.

IBM's **repair** practices look to optimize the scope of parts they can repair based on the forecasted usage of parts and treat each part procured as a potential opportunity lost for savings in parts costs from repair. Parts are usually repaired to a "like new" state and returned to inventory at a value not fully burdened as a new part. (i.e., cost of repair plus 25% of a new part can be a total cost less than 50% of a new part). Credits generated from the repair action typically become a credit to expense against the original parts usage. For IBM, over 70% of all parts used (dollar value) are repaired/refurbished and returned to stock as new, generating credit to the original expense. This action lowers the average cost of each part and the overall value of the inventory and creates the opportunity to lower overall parts usage costs.

This process also tracks warranty returns to the original equipment manufacturer (OEM). In this case, the expense credit is the total value of the credit given by the OEM, and the piece is returned to stock (when a replacement part is provided) at the total value.

IBM plans capacity against the potential parts they can repair and their capacity to fulfill the repair. They have online access to the OEM for engineering documents to assist in their repair strategies. In addition, they can farm out the repair to another organization when it is more financially viable.

They have forecast visibility to the volumes of the expected parts to assist in planning and scheduling their resources (parts/people/set-up). Their repair lead times can be integrated into the replenishment systems of the inventory network they supply.

The inventory management system within IBM Canada also supported the following:

- Content management and substitution management;
- Used parts return process;
- Used parts sort process;
- Re-utilization strategies;
- Working with procurement to have vendor-managed inventory for the component parts used in repair;
- Limited supplier "service level agreements (SLAs)";
- Warranty identification process; and
- Supplier warranty redemption process management.

PERMISSIONS

* Pages 133–157, *Asset Management Excellence: Optimizing Equipment Life-Cycle Decisions*, 2nd Edition by Editor(s), John D. Campbell, Andrew K. S. Jardine, Joel McGlynn Copyright (2011) by Imprint. Reproduced by permission of Taylor & Francis Group.

1 Adapted from copyright from *Physical Asset Management Lecture Notes* (Part 1 and Part 2), UofT (2019), edited by (Barry). Reproduced by permission of Asset Acumen Consulting Inc.

BIBLIOGRAPHY

Don Barry, *Maintenance Parts Management Excellence: A Holistic Anatomy*, Boca Raton, CRC Press, 2023.

Section II

Managing Equipment Reliability

7 Assessing and Managing Risk*

Don M. Barry
Previous contributions from Siegfried F. Sanders,
J. Kaderavek, and G. Walker

The risks inherent in asset management are coming under ever greater scrutiny. Because of increasing competition and intolerance of environmental impacts (waste), managing assets effectively becomes an additional market differentiator as companies strive to create value and thrive in a global marketplace. There are four basic groups of business assets: financial, human, intellectual, and physical. For decades, successful businesses have managed the first three well, but recently, businesses have recognized that managing physical assets is the next improvement opportunity. As a result, the term *asset management* has emerged to describe managing a business's physical assets. In a broad sense, asset management manages physical assets from the cradle to the grave. Maintenance manages asset risk during the physical assets' productive life. This chapter will explore the issue of risk management in maintenance and describe several effective methods to help assess and manage risk.

What is maintenance risk management? The noun *risk* can be considered "the likelihood of injury, damage, or loss." As a verb, it means "to expose to the chance of injury, damage, or loss." Thus, risk management identifies the chance and reduces the exposure of "injury, damage, or loss." The reader will learn the nature of asset risk and risk management processes, including a proven method for identifying critical equipment. Managing asset risk includes the function of the asset itself, the safety of workers, and the prevention of adverse environmental effects. Risk management includes identifying risk and reducing unacceptable levels of risk.

Assessing risk is what organizations decide what to do in the face of today's complex landscape. Threats and vulnerabilities are everywhere. They could come from external influences or a careless user.

Leadership must understand the impact and urgency of the organization's asset risks and how much mitigation efforts will cost. Asset risk assessments help set these priorities. They help to evaluate the potential impact and probability of each risk. Leadership can quantify which mitigation efforts to prioritize within the asset's operating context and the organization's strategy, budget, and timelines.

Risk assessments can be applied across a business or an asset's operating context. The following table offers examples of risk assessment approaches in business that can be applied to systems or assets.

DOI: 10.1201/9781032679600-9

TABLE 7.1

Risk Assessment Techniques That could be Applied to Assets and Systems

Assessment Approach	Approach and Benefits	Challenges
Quantitative	• Provide analytical rigor to the process • Assets and risks receive dollar values • Often presented in financial terms • Cost-benefit analyses let decision-makers prioritize mitigation options	• Not easily quantifiable • It may require judgment calls – undermining the assessment's objectivity • It can also be quite complex • Communicating results may be difficult beyond the boardroom • Expertise may be limited
Qualitative	• A scientific approach to risk assessment • Polling people to understand perspectives across the organization • Understanding how the processes would work should a system become unavailable • Assessors categorize risks (High, Medium, or Low)	• Approaches are inherently subjective • Must develop easily explained scenarios, questions, and interview methodologies • Must avoid bias and then interpret the results • It may be challenging to prioritize without a cost-benefit agreed to approach
Semi-Quantitative	• Organizations will use a numerical scale to assign a numerical risk value (e.g., 1–10 or 1–100) • Grouping Risk scores by thirds or fifths to score as high, medium low etc. • It can be more objective and provide a basis for prioritizing risk items	• Blending quantitative and qualitative methodologies creates a more analytical assessment
Asset-based	• Inventory all assets and prioritize • Confirm how existing controls are working and being effective • Understanding haw each vital asset could functionally fail • Understanding the impact and frequency of an asset's functional failure • Leverage an asset risk-based approach that aligns with the organization's culture	• Asset-based approaches typically do not produce complete risk assessments. • Risk data is always captured for system analysis • Policies, processes, culture, and other "people" factors can expose the organization to failure risks
Vulnerability-based	• Expands risk assessment scope beyond an organization's assets • Examines the known weaknesses and deficiencies within organizational systems or environments • Identifies possible threats that could exploit vulnerabilities and potential consequences • Demonstrates effective risk management and vulnerability management	• Often leverages like asset failure analysis that may not apply to the specific assets' operating context, organization's culture, or external threats

(Continued)

TABLE 7.1 (CONTINUED)
Risk Assessment Techniques That could be Applied to Assets and Systems

Assessment Approach	Approach and Benefits	Challenges
Threat-based	• Evaluates the conditions that create risk • This would include asset audits, operating context, and the scenarios contributing to threats • Looks for threats beyond the physical asset/infrastructure • Assessments may re-prioritize mitigation options • A threat-based assessment may find that increasing training frequency reduces risk at a lower cost	• Threat-based methods can supply a complete assessment of an organization's overall risk posture • Training may be required • May prioritize systemic controls over employee training

Many methods systematically identify and develop asset hazard and risk scenarios. Two comprehensive methods are:

- failure modes, effects, and criticality analysis (FMECA), and
- hazard and operability studies (HAZOPS).

Both methods identify hazards during the asset's life cycle and guide for reducing operation, people, and environmental risks. HAZOPS has been in use in the chemical industry since the early 1980s. Also included is a list of relevant national and international methods, standards, and regulations for determining asset reliability and risk assessment.

7.1 INTRODUCTION

Asset management in maintenance is about making decisions. It involves determining the optimum maintenance policy, including preventive maintenance activities, spare parts to keep, worker skills to maintain, tools to provide, and decisions on repair or replacement. While managers try to base these decisions on the best data available and a rational understanding of the issues and trade-offs involved, there is always some uncertainty involved, and with uncertainty comes risk. Therefore, the nature of the maintenance manager's job is to manage risk with a community of stakeholders.

Risk refers to an event where the outcome is uncertain, and the consequences are generally undesirable. Strictly speaking, buying a lottery ticket is a risk since it is unknown in advance whether it will win even with published odds. Every action or inaction has some risk, as every event may end with an undesirable consequence. For example, jumping off the top of a tall building seems risky since it is almost certain that this will lead to instant death, but there have been instances of people falling from great heights and surviving. The chance of winning the lottery jackpot and surviving a fall from a tall building may be similar, but more people buy lottery tickets than jump from tall buildings.

Risk is the product of probability and consequence. Therefore, two situations, one with a high probability and low consequence (i.e., tripping on an uneven floor and being injured) and one with a low probability and high consequence (i.e., an aircraft crashing and killing everyone), can have similar risks values. For example, an air system leak in maintenance may have the same risk as a shaft failure on an air compressor. There is a high probability of joint failure in a compressed air system, but the resulting minor air leak has a small cost, safety, and environmental consequences. On the other hand, there is a low probability that the compressor shaft will fail, but the resulting loss of plant air would cause significant process disruption and be expensive to repair.

It is only human to focus more on high-consequence events than those with little impact, even when they are unlikely. Even experienced maintenance professionals have this bias. One of the most significant benefits of a total productive maintenance (TPM) program is that it addresses conditions such as leaks and routine adjustments, which are of minor consequence but happen frequently and still deserve attention.

This chapter will explore ways to analyze and deal with different risks consistently and rationally.

7.2   MANAGING MAINTENANCE RISK

In any operation, there is always some degree of risk. All activities expose people or organizations to potentially losing something of value. For example, maintenance's impact is typically on outcomes from equipment failure, human safety, or environmental damage. Risk involves three issues:

- The frequency of the loss.
- The consequences and extent of the loss.
- The perception of the loss to the ultimate interested party.

A significant equipment failure represents a significant need to manage maintenance risk. The production downtime could delay product delivery to the customer and cost the business loss of sales or even market share. There could be further losses if the equipment failure threatened the safety of employees or adversely affected the environment. A critical, high-profile failure could also create the impression that the business is out of control and tarnish its reputation in the marketplace. An example of this was the Perrier water incident in 1990. Perrier had a reputation for purity and for promoting health. A minor maintenance failure led to traces of Benzene contaminating the product and damaging Perrier's reputation. Recalling and destroying millions of bottles of product from countries across Europe and the United States was a vast expense. More significant was the damage to the company's reputation, causing the company to launch a high-profile and expensive public relations campaign to reassure customers that the product was safe. Other significant examples of failure to manage the risk that damaged either the firm or its supplier include the following:

- Union carbide, the toxic release of methyl isocyanate, Bhopal, India, 1989*, 15,000 deaths, $3 billion legal claims[1]
- BP oil refinery explosion, Texas City, Texas, 2005[2]
- American Airlines Flight 191, Chicago, Illinois, 1979[3]
- Prudhoe Bay oil field shutdown, Alaska, August 7, 2006,[4] 400,000 barrels per day lost production for several weeks
- BP Deepwater Horizon of 2010

Each of these examples can be researched in detail via the internet. The examples illustrate the consequences of making a wrong maintenance decision. The consequences can be many: lessened plant reliability and availability; reduced product availability; decreased product quality; and increased total operating costs, as well as potential environmental damage, loss of life, and legal claims.

How does the maintenance manager reduce the risk? One accepted premise is that increasing preventive maintenance results in less downtime and increased production, thus reducing risk and lowering overall maintenance costs. However, this is not entirely true; Figure 7.1 shows an optimum point where the combined preventive and downtime costs are minimum. This should determine the maintenance policy and the amount of preventive work that will lead to the lowest total cost.

While this is a valuable concept, one can never know the trade-off between additional preventive maintenance and its impact on downtime. The information on which risk decisions are based will never be exact or accurately foretell the future. Uncertainty is inevitable, so setting the maintenance policy is a question of managing risk and balancing risk versus cost.

Good data can improve the quality of decisions and confidence about the optimum maintenance point. Most risk management decision processes use historical data to predict future events. For equipment, this starts with manufacturers' recommended maintenance practices. Generally, that is the only guide available to predict the future, based on the premise that history predicts future events. Over time, you can use local history to modify risk management decisions, but conditions change over

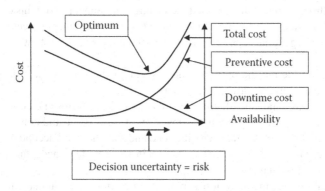

FIGURE 7.1 Preventive maintenance optimization curve.

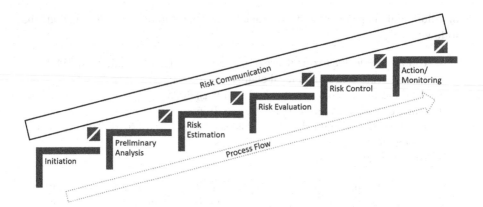

FIGURE 7.2   Risk management process.

time, so assumptions made in the past may not still apply. Thus, the prudent risk manager must continuously manage this uncertainty and its resulting risk.

Generally, there is little acknowledgment that conditions change. Systematically planning for risk, including changing conditions, can dramatically improve decision-making. To make the process and resulting decisions more credible, there must be adequate curation of the data used to make estimates, analysis methods, and correct unknown factors associated with the data. In addition, because of changing conditions, you should regularly revisit risk assessments to ensure that original conditions and assumptions apply.

Risk management aims to identify significant risks and take appropriate actions to minimize them reasonably. You must balance risk control strategies, effectiveness and cost, and stakeholders' needs, issues, and concerns to reach this point. Communication among stakeholders throughout the process is critical.

To assist in the risk management process, numerous tools and standards can help. However, there is also a trade-off between the effort put into the analysis and its possible benefit. Getting this wrong can lead to ridiculous situations, such as a chemical company that used an extensive hazard study for a microwave oven in the mess room. The investigators demanded additional interlocks, regular condition, and radiation monitoring – even though home kitchens used the identical appliance! Avoid this extreme solution by seeking help to determine an appropriate risk management process.[5]

There are six general steps in the risk management process, as shown in Figure 7.2:

1. Initiation: define the problem and associated risks, form the risk management team, and identify stakeholders.
2. Preliminary analysis: identify hazards and risk scenarios and collect data.
3. Risk estimation: estimate the probability of occurrence and consequences.
4. Risk evaluation: estimate benefits, cost, and stakeholders' acceptance of risk.
5. Risk control: identify options and obtain stakeholders' acceptance of controls and residual risk.
6. Action/monitoring: develop a plan to implement risk management decisions and monitor their effectiveness.

The risk team and stakeholders need to communicate frequently to make effective decisions about the risk management process. There must be an open dialogue to validate each step of the team's hypothesis and ensure all stakeholders are involved. Since evaluating risk can be time-consuming and expensive, investing only in the assets and processes identified as most critical is vital.

7.3 IDENTIFYING CRITICAL EQUIPMENT

There is generally far more to analyze than time or resources available to reduce risk. In addition, as described in the microwave oven example, not all equipment deserves the same degree of analysis. It is crucial to understand equipment criticality (i.e., how critical the asset is to the business) to determine where the risk management effort should go. Generally, determine the answer by the consequences if it fails. The elements that make up criticality include safety, health, environmental, and financial consequences. Quantify the financial and non-financial impacts to provide a common base for analysis. Commercial criticality analysis (CCA) is valuable for focusing on and prioritizing day-to-day maintenance and ongoing improvements that manage risk.

For example, an oil terminal used CCA to rank its 40 main systems for safety hazards and the cost per hour of downtime. Maintenance then used CCA ranking information to decide the priority of work. CCA also helped determine which units justified flying in spares when a breakdown occurred and for which cheaper, standard delivery was adequate. The CCA study also helped refocus the maintenance organization, providing 100% skills coverage 24 hours daily. Without any additional risk, significant savings resulted from switching to 24-hour coverage only for the critical units. In addition, when the reliability-centered maintenance (RCM) study started, CCA findings ensured that maintenance focused on areas of maximum impact. However, to achieve the most from RCM, assessing the long-term risks in asset maintenance is essential. Therefore, it is vital to consider risks to asset capability and possible shortcomings in information records and analysis methods.

Assessing criticality is specific to each facility. Even within the same industry, what is essential to one business may not be necessary to another. Issues such as equipment age, performance, design (a significant consideration), technologies using hazardous chemicals, geological issues (typically for mining and oil and gas industries), supplier relationships, product time to market (product cycle), finished goods inventory policy, information technology (IT) infrastructure, and varying national health and safety regulations all influence equipment criticality decisions.

Generally, all businesses need to consider the following asset criticality measures:

- Asset performance (reliability, availability, and maintainability).
- Cost (i.e., direct maintenance and engineering costs, indirect costs of lost production).
- Safety (i.e., lost time incidents or accidents, disabling injuries, fatalities).
- Environment (number of environmental incidences, cost of environmental cleanup, environmental compliance).

7.4   CASE STUDY: A MINERAL PROCESSING PLANT

We examine the first two criticality measures with a real case study involving a mineral processing plant. Management wanted a maintenance strategy with accurate equipment performance, cost measures, and a site-wide improvement program.

7.4.1   EQUIPMENT PERFORMANCE

Rule 1: Talk to the people who know the plant.

Initially, production statistics provided equipment downtime, which was clear and accurate to a process line level. However, breaking the picture down further required meetings with maintenance and production crews to determine what occurred. The main concern was equipment causing whole plant downtime for the first analysis. The downtime statistics are shown in Figure 7.3.

Using Pareto analysis, a simple and ranked presentation style, presenting the data quickly shows that the first three equipment types contributed 85% of total plant downtime. Judging from this, conveyors, pumps, and the clarifier were obvious targets for a program for equipment performance improvement.

The production and maintenance personnel checked the data (see Rule 1). One incident caused the clarifier downtime when a piece of mobile equipment struck it, causing substantial damage. This unusual event was unrelated to standard operating and maintenance and thus was eliminated from the operating risk analysis. After investigation, the plant implemented several remedial actions to prevent this incident from recurring. Conveyors, pumps, and screens then became the top three improvement targets. Even though eliminating the clarifier failure was justified, it is wise to remember that not all risk comes from normal operating and maintenance activities. External events such as a car striking a power pole, civil unrest, or terrorist activity can still lead to significant risk.

The next step was identifying, agreeing on, and implementing conventional reliability, availability, and maintainability (RAM) measures.

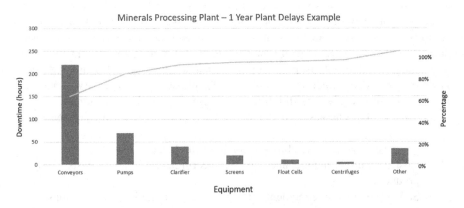

FIGURE 7.3   Pareto analysis of downtime caused by equipment failures.

7.4.2 Costs

Maintenance-collected breakdown and failure work order cost data from the existing work order system. Target equipment types and subtypes organized the costs. Even though this analysis concentrated on lost time and repair costs, maintenance could consider other factors such as capacity, quality, safety, or environmental effects. The results of the chemical plant analysis are shown in Figure 7.4. The facility concentrated on the three highest-cost pumps, screens, and conveyor types to increase reliability and reduce maintenance costs. This illustrates the 80/20 rule that 20% of your effort should reduce 80% of the cost. Although this was a systematic analysis of the failures, remember Rule 1: ask the operators and maintenance people. If you ask the maintenance people, it is very common that they can tell you the top three maintenance failures using anecdotal evidence. Don't discount this input to your analysis process. This information can confirm or expand your statistical analysis to determine the highest-cost failures.

Using this simple analysis, management identified critical equipment by the following:

- Poor overall performance (most downtime).
- Highest failure costs.

The facility reduced risk by launching a strategic reliability improvement program using the data. As a result, the organization has since reduced its risk, saved over $1 million annually, and significantly improved equipment performance by concentrating on the worst of the "bad actors."

7.5 SAFETY AND ENVIRONMENTAL RISK: DUTY OF CARE

Maintenance significantly impacts safety and the environment, the other two major areas of risk management. But isn't maintenance just supposed to keep the plant running? As you will see in this section, the general responsibilities of maintenance engineers can be extensive. Why should a business be concerned with safety? The legal reasons date to a 1932 ruling by Lord Atkin of the British Privy Council that the Stevenson soft drink company was liable for an injury sustained by Scottish widow Mary O'Donahue after drinking a contaminated soft drink. The ramifications of this ruling include the modern "duty of care" legislation that is the basis of most national occupational health and safety (OHS) legislation. The ruling ensures that all employers have a duty of care to their employees, that employees have a duty to each other, and that employees have a duty to their employer. An organization's standard of care and reasonableness should follow the national standard. An increase in the number and strength of these regulations is expected. This duty of care also extends to protecting the public and the environment. Thus, managing risks that may affect the safety or the environment is part of the responsibility of every employee, including maintenance. Maintenance through its action or inaction can significantly impact managing risk. The examples mentioned earlier illustrate this fact.

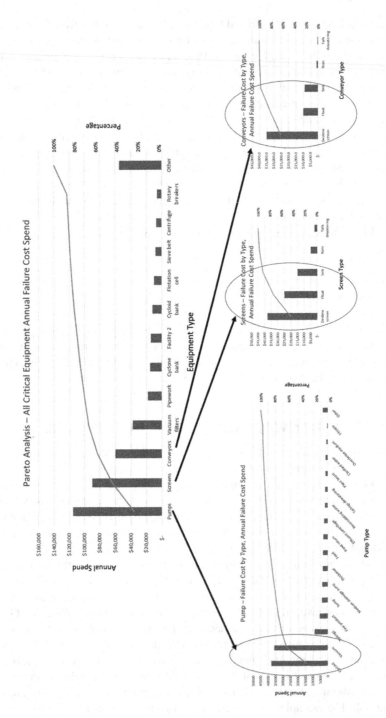

FIGURE 7.4 Pareto analysis of the cost of equipment failures.

7.6 MANAGING LONGER-TERM RISKS

Over the operating phase of the life cycle of an asset, maintenance needs to maximize reliability and minimize risk. However, certain conditions can increase risk in the long term. Use failure and reliability analytical methods to develop effective maintenance plans for asset reliability. Failure modes and effects analysis (FMEA), or the closely related FMECA and RCM, use multidisciplinary teams to develop maintenance plans for existing assets. Although these techniques are invaluable, there are potential long-term risks:

- *Slow degradation failures* are often difficult to predict and model, especially for equipment early in its life cycle. The probability of fatigue-related failures increases with every operating cycle of the asset. Predicting these failures using conventional reliability methods can be tricky, even impossible. One railway freight provider, aware that low-cycle fatigue increases failure rates and costs in its wagon fleet, conducted an accelerated reliability test on several wagons. The aim was to model the failures ahead of time to develop the appropriate condition and time-based tactics. Some wagons were fatigue tested to 10 times their current lives.
- *Incomplete execution of reliability methods*: if all plausible failure modes are not identified, then the analysis is not complete, nor are the maintenance plans. As a result, the asset will display unpredicted failure modes. Review maintenance plans periodically to evaluate the effectiveness of the reliability improvement.
- *Change in the operating environment*: the rate of effort (i.e., operating hours/year), physical environment (i.e., different geological composition for mining businesses), and operating procedures and techniques can all trigger new failure modes not predicted by the original analysis. Examples are airplane routes or truck services changing from highway transport to city delivery.
- *Change in maintenance environment*: maintenance tactics (i.e., servicing intervals), maintenance personnel quantity and quality (number, skill sets, and experience), and support and condition monitoring equipment can all lead to unanticipated failures.
- *Modifications and capital upgrades*: existing or new assets can also affect existing assets and result in unpredicted failures.
- *Change in plant operations*: this places different demands on installed assets; thus, the criticality of assets may either increase or decrease.

7.7 IDENTIFYING HAZARDS AND USING RISK SCENARIOS

Risk management starts by identifying hazards. Start by analyzing past events, incidents, or lost performance relating to assets, people, and the environment. This could be information from the assessed facility or events at other facilities. However, there are limitations to only using history, as already mentioned. Selecting an experienced team of engineers and process managers is also essential for a thorough understanding of the whole process or system and potential hazards or failures.

Develop risk scenarios by identifying hazards and evaluating direct and consequential loss. Therefore, a risk scenario is a sequence of events associated with probability and consequences. There are a variety of approaches and methods to identify and analyze risks. Some are observation and experienced judgment-based, while others are systematic analyses. Two analytical methods typically applied in asset management are FMEA, FMECA, and HAZOPS.

7.8   FMEA AND FMECA

The FMEA identify potential system failures and their effects. Criticality analysis (CA) ranks failure severity and probability of occurrence. When performing both steps, FMEA and CA, the result is a failure mode, effects, and criticality analysis.

There are two primary ways of doing FMEA. One is the hardware approach, which lists the effects on the system. The other is functional, based on the premise that every item in the system is designed to perform several functions that can be classified as outputs. List and analyze outputs for functional FMEA to determine their system effects. Variations in design complexity and available data usually dictate the analysis method. When detailed design information is available, the hardware is generally studied. Use the functional approach when in the conceptual design stages.

7.8.1   FMEA Objectives

FMEA and FMECA are integral to the design process and should be updated regularly to reflect design evolution or changes. FMEA provides inputs to product reviews at various levels of development. Also, FMEA information can minimize risk by defining special test considerations, quality inspection points, preventive maintenance actions, operating constraints, and other pertinent information and activities. This may include identifying redundancy, alarms, failure detection mechanisms, design simplification, and derating. FMEA can also be used to:

- Compare various design alternatives and configurations.
- Confirm the system's ability to meet its design reliability criteria or requirements.
- Provide input data to establish corrective action priorities and trade-off studies.

7.8.2   FMEA and CA Methodology

FMECA methodology consists of two phases: FMEA and CA. To perform FMEA:

- Define the system and its performance requirements.
- Define the assumptions and ground rules to be used in the analysis.
- List all individual components or functions at the required indenture level for the analysis.

- Develop a block diagram or other simple model of the system.
- Devise an analysis worksheet to provide failure and effects information for each component and other relevant information.

CA ranks each potential failure identified by FMEA according to its combined severity and probability of occurring. The CA may be performed qualitatively or quantitatively.

FMEA is very versatile and valuable for risk analysis. It will be easy to rank failures for their severity and probability of occurrence if CA is included. Then the best corrective actions can be determined and prioritized. The FMEA worksheet should tailor the analysis's detailed information to fit the situation. However, the FMEA method has a significant shortcoming, requiring much time and effort and making it expensive.

The analysis becomes even more complex if the effects of multiple (i.e., two or more simultaneous) failures are considered. It is easy to overlook both human and external system interactions. Too much time and effort are often spent analyzing failures that have a negligible effect on the system's performance or safety. To help with the analysis, several computer packages automate the FMECA process. Even if using a computer, FMECA should be performed by people intimately familiar with the analyzed design or process to curate the findings and ensure accuracy and effectiveness.

In conclusion, a rigorous FMECA analysis is a highly effective method to detail risk scenarios. As a result, the risk management team and stakeholders will better understand risk levels and controls.

7.9 HAZOPS

HAZOPS was developed in the process industry to identify failure, safety, and environmental hazards. It evolved in the chemical industry in the 1970s, particularly under Trevor Kletz, an employee of the U.K.-based chemical company ICI. HAZOPS is the fourth in six hazard studies covering a new plant development life cycle, from initial plant concept to commissioning and effective operation.

The early studies consider product-manufacturing hazards, such as a material's toxicity or flammability, and the pressures and temperatures required. The later studies check that the plant was built according to the requirements of the earlier studies and that operating conditions comply.

Hazard study 4, the hazard and operability study, examines the hazards inherent in the plant's design and any deviations in painstaking detail. This is particularly relevant to asset maintenance. A distressingly high proportion of serious and fatal chemical accidents occur when normal conditions in the plant are temporarily disturbed. Refer to the Bhopal and Texas City risk management failures mentioned earlier. Maintenance is essential as both an input and an output of the HAZOP – input because the maintenance of plant assets may require changes to standard plant operating conditions and as an output in numerous ways. The study could conclude, for instance, that maintenance is done only under certain conditions or that some plant items must meet specified standards of performance or reliability. The HAZOP helps determine the importance of various equipment and the need for either a CCA or an FMECA/RCM analysis.

7.10   METHOD

Small teams carry out a hazard and operability study. Typically, the team requires the following:

- An operations expert is familiar with how the plant works (which, in the chemical industry, is probably a chemical engineer).
- An expert on material hazards and handling methods (typically, in the chemical industry, a chemist).
- An expert in the way the plant equipment itself may behave (likely the plant engineer).
- A trained facilitator, usually a safety expert.

Call other experts to advise on such things as electrical safety and corrosion.

The essential starting point for the HAZOP is an accurate plant diagram (in the process industry, piping, instrumentation, or piping and instrumentation diagram P&I diagram) and processing instructions that spell out how the product is manufactured.

Start the study by reviewing the inherent manufacturing hazards and standards. This includes briefing the team on the flammability, toxicity, and corrosion characteristics of all the materials and process characteristics, such as high temperatures or pressures, potential runaway reactions, and associated hazards. In addition to managing the inherent risks of normal operations, the present rise in international terrorism means that asset managers must now consider how terrorist attacks may cause a failure of processes. This is especially important for processes containing dangerous materials that can be released and cause mass casualties.[6]

After doing a desktop study of plant processes to determine potential areas of high risk, the team must physically inspect the risky manufacturing processes, considering

TABLE 7.2
Hazard Checklist Parameters

Hazard	Parameter Change	Parameter
Corrosion	more of	pressure
Erosion	less of	temperature
Abrasion	none of	flow
Cracking	reverse of	Ph
Melting	other than	quantity
Brittleness		concentration
Distortion		mixture
Leakage		bombing
Perforation		impact
Rupture		

the plant areas identified as potentially high risk, down to individual items, and even to each section of pipework. For every process considered, follow a sequence of prompts. First, address regular plant operation to understand each hazard and how it is controlled. Then consider deviations from normal operations.

Maintain a hazard checklist of things like those in Table 7.1; consider what changes could occur in each relevant parameter. Use the topics in the table to ask questions such as, "Can a *leak* occur because there is a *reverse* in the normal *flow* of the process?" Perform the analysis, documenting hazardous results and the appropriate actions to reduce the risk. This could include plant redesign or changes to operating and maintenance procedures. For example, more checking or calibration may be required in a facility that relies on automatic safety control systems to ensure the instruments operate correctly or create a purchasing specification for replacement parts. Follow this approach until every part of the plant and the process has been considered and documented.

In summary, HAZOP provides plenty of benefits. The plant risk assessment will be rigorous, comprehensive, and done to high safety standards. Besides reducing risk, this process will also reveal other opportunities for improvement.

But there are some limitations to keep in mind. HAZOP can be time-consuming, typically taking several full days for the team to go through a single process in a complex plant. This level of thoroughness is costly, tedious, and resource-consuming. Consider the cost/benefit ratio before initiating a significant effort like HAZOP.

7.11   STANDARDS AND REGULATIONS

7.11.1   INTERNATIONAL STANDARDS RELATED TO RISK MANAGEMENT

Several international and national standards and regulations are available for a risk management program. The following section lists the number and name of the standards and provides a brief overview of their content. The standards cover basic reliability disciplines, risk management, environmental and safety regulations, cost of quality, and software reliability concepts.

These standards provide excellent further reading and will guide asset managers.

7.11.2   CAN/CSA-Q631-97(R2002) RAM DEFINITIONS

7.11.2.1   Scope

This standard lists terms and basic definitions to describe reliability, availability, maintainability, and maintenance support. The terminology is engineering but is also adapted to mathematical modeling techniques.

7.11.2.2   Application

RAM addresses general concepts related to reliability, availability, and maintainability – how items perform over time and under stated conditions and concerns all the lifecycle phases (concept and definition, design, and development, manufacturing, installation, operation and maintenance, and disposal).

Note the following regarding this terminology:

- Reliability, availability, and maintainability are qualitative "abilities" of an item.
- Maintenance support is defined as the qualitative ability of an organization.
- Such general abilities can be quantified by suitable "random variable" conditions, such as "time to failure" and "time to repair."
- You can apply mathematical operations to these random variables using relations and models. The results are called "measures."
- The significance of variables and measures depends on the data collected, the statistical treatment, and the technical assumptions made in each circumstance.

In RAM, one essential ability measured is uptime. How to define and measure uptime has been a hot industry debate topic. Management spends considerable time grappling with interdepartmental issues such as determining operations versus maintenance caused downtime, equipment handover time, and logistics and administrative delays. Often, it is hard to reach an explicit agreement on these issues. Figure 7.5, based on CAN/CSA-Q631-97 "Reliability, Availability, and Maintainability (RAM) Definitions," shows one method to define the time available for maintenance and operations. As clear as the diagram is, there is room for disagreement on the nomenclature definitions and assigning events to one of the downtime categories. These terms will have slightly different meanings in any individual operation, and management must

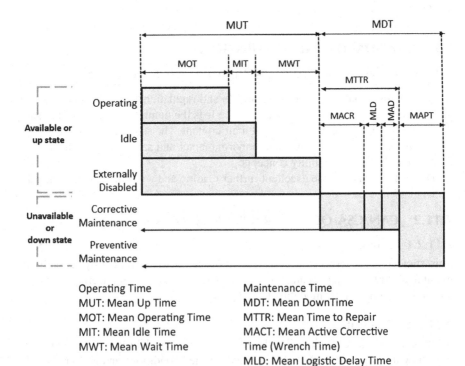

Operating Time   Maintenance Time
MUT: Mean Up Time   MDT: Mean DownTime
MOT: Mean Operating Time   MTTR: Mean Time to Repair
MIT: Mean Idle Time   MACT: Mean Active Corrective
MWT: Mean Wait Time   Time (Wrench Time)
                                       MLD: Mean Logistic Delay Time

FIGURE 7.5   Asset availability dynamics diagram.

agree to a common definition. This is more important than assigning responsibility; having good data to manage risk is crucial. Define and measure downtime correctly; otherwise, risk analysis will use uncertain data resulting in more risk.

RAM is also excellent in establishing the performance of standard asset management measures across an organization. This would include specific equipment, process line, systems, and the maintenance organization's ability to support that performance. The precise definition and detail make these measures easily communicated to stakeholders. This reduces the risk of "miscommunication" and "miscomprehension" plagues most managers when establishing an "apples for apples" performance comparison.

7.11.3 CAN/CSA-Q636-93(R2001) Guidelines and Requirements for Reliability Analysis Methods – Quality Management

7.11.3.1 Scope

This standard guides asset managers in selecting and applying reliability analysis methods. Its purpose is to do the following:

1. Describe some of the most common reliability analysis methods representing international standard methods.
2. Guide you in selecting analysis methods, depending on technology and the system or product used.
3. Establish how results will be documented.

7.11.3.2 Application

No reliability analysis method is comprehensive or flexible enough to suit all situations. Consider the following factors to select an appropriate model:

- Analysis objectives and scope.
- System complexity.
- Consequences of system failure.
- Level of detail in design, operation, and maintenance information.
- Required or targeted level of system reliability.
- Available reliability data.
- Specific constraints such as regulatory requirements.
- Staff, level of expertise, and resources available.

Appendices A–F of the standard contains detailed descriptions of the most common reliability analysis methods. The following is an overview of how to apply these methods:

A. **Fault tree analysis:** this method may be suitable when one or more of these conditions apply:
   - A detailed and thorough system analysis is needed with a relatively high level of resolution.
   - There are severe safety and economic consequences of a system or component failure.

- The reliability requirements are stringent (i.e., system unavailability ≤ units?).
- Many staff and resources, including computer facilities, are available.

B. **Reliability block diagram:** consider this method if one or more of these conditions apply:
  - A rudimentary system study or a higher hierarchical level is needed (although the method may be used at any level or resolution).
  - The system is relatively simple.
  - The analysis must be simple and straightforward, even if some details are lacking.
  - Reliability data can be obtained at a block level, but data for more detailed analysis are either unavailable or not warranted.
  - The reliability requirements are not very stringent.
  - There are limited staff and resources.

C. **Markov analysis: This method may be best if one or more of these conditions apply:**
  - Multistate or multiple failure modes of the components will be modeled.
  - The system is too complex to be analyzed by simple techniques such as a reliability block diagram (which may be too difficult to construct or solve).
  - The system has unique characteristics, such as the following:
    - A component can't fail if some other specified component has already failed.
    - You can't repair a component until a particular time.
    - Components don't undergo routine maintenance if others have already failed.

D. **FMEA: may be suitable when one or more of these factors apply:**
  - Ranking the failure modes' relative importance is required.
  - All possible failure modes and their effect on system performance must be detailed.
  - The system components aren't dependent on each other to any significant degree.
  - The prime concern is single-component failures.
  - Many staff and resources are available.

E. **Parts count: consider this method if one or more of these conditions apply:**
  - Only a very preliminary or rudimentary conservative analysis will be performed.
  - The system design has little or no redundancy.
  - There are limited staff and resources.
  - The system being analyzed is in a very early design stage.
  - Detailed information on components is unavailable, such as part ratings, part stresses, duty cycles, and operating conditions.

F. **Stress analysis: you may prefer stress analysis if one or more of the following conditions apply:**
  - A more accurate analysis than the parts count method is desired.

- Many staff and resources, including computer facilities, are available.
- The system being analyzed is in an advanced design stage.
- Detailed information on components, such as parts ratings, part stresses, duty cycles, and operating conditions, is available.

7.11.4 CAN/CSA-Q850-97 (R2009) Risk Management – Guidelines for Decision-Makers

7.11.4.1 Scope

The standard helps to effectively manage all types of risks, such as the risk of injury or damage to health, property, the environment, or something else of value. The standard describes a process for acquiring, analyzing, evaluating, and communicating information for decision-making.

Note: The Canadian Standards Association has a separate standard to address risk analysis (CSA Standard CAN/CSA-Q634) and environmental risk assessment (CSA Standard Z763).

7.11.4.2 Application

This standard provides a comprehensive decision process to identify, analyze, evaluate, and control all risks, including health and safety. Due to cost constraints, risk management priorities must be set, which this standard encourages.

7.11.5 AS/NZS 4360:2004 Risk Management

7.11.5.1 Scope

The standard helps establish and implement risk management, including context, identification, analysis, evaluation, treatment, communication, and ongoing monitoring of risks.

7.11.5.2 Application

Risk management is an integral part of good management practice. It is an iterative process consisting of steps that, in sequence, continually improve decision-making. Risk management is about identifying potential risks and implementing ways to avoid or mitigate losses.

This standard may be applied at all activity, function, project, or asset stages. Maximum benefits will be gained by starting the process at the beginning. It is usual to carry out different studies at various project stages.

The standard details how to establish and sustain a risk management process that is simple yet effective.

Here is an overview:

- *Establish the context*: Establish the strategic, organizational, and risk management context in which the rest of the process will occur. Define criteria to evaluate risk and the structure of the analysis.
- *Identify risks*: Identify what, why, and how problems can arise as the basis for further analysis.

- *Analyze risks*: Determine the existing controls and their effect on potential risks. Consider the range of possible consequences and the probability of occurrence. Risk can be estimated and measured against pre-established criteria by combining consequence and probability.
- *Evaluate risks*: Compare estimated risk levels with pre-established criteria. They can then be ranked to identify management priorities. For low-risk scenarios, no action may be required.
- *Treat risks*: Accept and monitor low-priority risks. For higher-priority risks, develop a specific management plan with sufficient funding to implement risk reduction measures.
- *Monitor and review*: Monitor and review how the risk management system performs, looking for ways to improve continuously.
- *Communicate and consult*: Communicate and consult with internal and external stakeholders about the overall process at each stage.

Appendix B of the standard details the steps to develop and implement a risk management program:

- Step 1: The support of senior management.
- Step 2: Develop the organizational policy.
- Step 3: Communicate the policy.
- Step 4: Manage risks at the organizational level.
- Step 5: Manage program, project, and team risks.
- Step 6: Monitor and review.

CAN/CSA-Q850-97 describes the risk communication process, including stakeholder analysis, documentation, problem definition, and general communications. At the same time, AS/NZS 4360-1999 has a well-developed and articulated risk management process model and provides a valuable summary for implementing a risk management program. If both are available, the risk manager can use both standards to decide an appropriate process for a particular maintenance environment.

7.11.6   ISO 14000 ENVIRONMENTAL MANAGEMENT SYSTEMS

ISO 14000 is a series of international, voluntary environmental management standards. Given that environmental damage is one of the critical factors in risk management, it is included here. Developed under ISO Technical Committee 207, the 14000 series of standards address the following aspects of environmental management:

- Environment Management Systems (EMS)
- Environmental auditing and related investigations (EA&RI)
- Environmental labels and declarations (EL)
- Environmental performance evaluation (EPE)
- Lifecycle assessment (LCA)
- Terms and definitions (T&D)

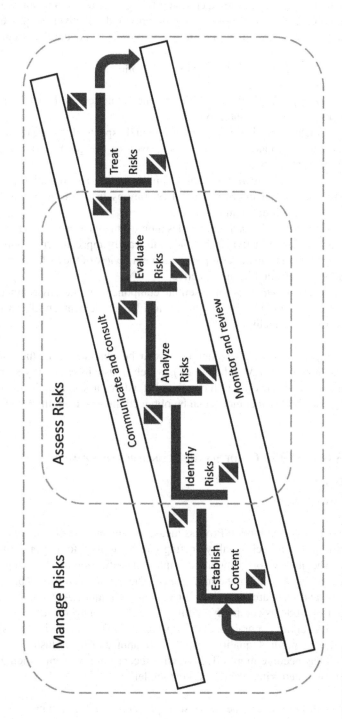

FIGURE 7.6   The steps in a risk assessment and management program.

The ISO series of standards provide a common framework for organizations worldwide to manage environmental issues. They broadly and effectively improve environmental management, strengthening international trade and overall environmental performance.

The key elements of an ISO 14001 EMS are as follows:

- *Environmental policy*: includes the environmental policy and how to pursue it via objectives, targets, and programs.
- *Planning*: includes analyzing the environmental aspects of the organization (i.e., processes, products, and services as well as the goods and services used by the organization).
- *Implementation and operation* include implementing and organizing processes to control and improve critical operations from an environmental perspective (i.e., products and services).
- *Checking and corrective action*: includes monitoring, measuring, and recording characteristics and activities that can significantly impact the environment.
- *Management review*: includes top management review of the EMS to ensure it continues to be suitable and effective.
- *Continual improvement*: is an essential component of the environmental management system. It completes the cycle: plan, implement, check, review, and improve continually.

ISO 14000 standards and related documents can be obtained from the National Standards Association (ISO Member Body), which is usually a country's primary ISO sales agent. In countries where the national standards association is not an ISO member body, ISO 14000 documents can be obtained directly from the ISO Central Secretariat.

7.11.7   BS 6143-1990 Guide to the Economics of Quality

7.11.7.1   Scope

BS 6143 has two parts:

- Part 1 – Process cost model: Process measurement and ownership are critical in using this model. Quality costing can be applied to any process or service. The quality cost categories simplify classification by clarifying the cost of conformance and nonconformance. The method involved is process modeling, and there are guidelines for various techniques. As well, the process control model is compatible with total quality management.
- Part 2 – Prevention, appraisal, and failure model: This is a revised version of traditional product quality costing in manufacturing industries. This approach has become more effective with recent improvements, though it may be combined with the process cost model.

This standard will help determine the cost of preventing defects, appraisals, internal and external failures, and quality-related cost systems for effective business management.

7.11.7.2 Application

For asset managers unfamiliar with this standard, it deals with a manufacturing cost structure that can readily be applied to direct maintenance charges. Costs are defined as follows:

- *Prevention cost*: any action to investigate, prevent, or reduce the risk of nonconformity or defect.
- *Appraisal cost*: evaluating quality requirement achievements, such as verification and control performed at any stage of the quality loop.
- *Internal failure cost*: the cost of nonconformities or defects at any stage of the quality loop, including, for example, scrap, rework, retest, reinspection, and redesign.
- *External failure cost*: nonconformities or defects after delivery to a customer/user. This can include claims against warranty, replacement, consequential losses, and evaluating penalties.
- *Identifying cost data*: quality-related costs should be identified and monitored. The way data must be classified is relevant and consistent with other accounting practices within the company. Otherwise, comparing costing periods or related activities won't be easy.
- *Quality-related costs*: are a subset of business expenses, and it is helpful to maintain a subsidiary ledger or memorandum account to track them. By using account codes within cost centers, the quality cost of individual activities can be better monitored. Allocating costs is essential to prevent failures and should not be done solely by an accountant. The analyst may need technical advice as well.

Quality costs alone do not give managers enough perspective to compare them with other operating costs or identify critical problem areas. To understand how significant a quality cost is, compare it with other regularly reported organizational costs.

7.11.8 EPA 40 CFR 68 Chemical Accident Prevention Provisions

7.11.8.1 Scope

The regulation requires U.S. facilities with more than a threshold quantity of certain chemicals to develop and publish a risk management plan to mitigate the effects of an accidental release, fire, or explosion on the surrounding public. Section G of this program even for those facilities not covered by this regulation.

7.11.8.2 Application

Facilities must analyze risk at least every five years using approved methods of analysis, including the following:

- What-if
- Checklist
- What-if/checklist
- HAZOP
- FMEA
- Fault tree analysis

The process hazard analysis must address the hazards of the process, past incidents, controls to prevent failures, consequences and health effects of failure, location of hazards, and human and safety factors.

A team with engineering and process operations expertise must perform the process hazard analysis. The team must include at least one employee with experience and knowledge specific to the evaluated process and one member knowledgeable in the specific process hazard analysis methodology.

Most importantly, though, a system must address and resolve the team's findings and recommendations promptly. Documentation of action taken is vital for compliance. This follows the simple rule:

- Say what you will do;
- Do what you say;
- Document that you did it.

Review the hazard analysis for facilities covered by this regulation at least every five years after completing the initial process hazard analysis.

7.11.9   OSHA 29CFR 1910.119 Process Safety Management of Highly Hazardous Chemicals

7.11.9.1   Scope

The regulation requires U.S. facilities with more than a threshold quantity of certain chemicals to develop and publish a process safety management (PSM) plan to protect the employees of a covered facility from the effects of failures of systems containing highly hazardous chemicals.

7.11.9.2   Application

The application of PSM is very similar to the U.S. Environmental Protection Agency's (EPA's) 40 CFR 68 risk management plan. Even a provision in 40 CFR 68 allows the Occupational Safety and Health Administration (OSHA) PSM plans to qualify for inclusion in 40 CFR 68 risk management plans. Both regulations have similar requirements for the following:

- Qualified personnel must use recognized hazard analysis methods to develop the PSM plan.
- Action plan to act on the findings and recommendations.
- Documentation of findings, plans, and actions are taken.
- Review of plans at least every five years.

7.11.10   ANSI/AIAA R-013-1992 Software Reliability

7.11.10.1   Scope

Software reliability engineering (SRE) is an emerging discipline that applies statistical techniques to data collected during system development and operation. The purpose is to specify, predict, estimate, and assess software-based systems' reliability. This is recommended for defining SRE, becoming more important to industrial plants as more process equipment and instrumentation include software for control and maintenance.

7.11.10.2 Application

The techniques and methods in this standard have been successfully applied to software projects by industry practitioners to do the following:

- Determine whether a specific software process will likely produce code that satisfies a software reliability requirement.
- Determine the need for software maintenance by predicting the failure rate during operation.
- Provide a metric to evaluate process improvement.
- Assist software safety certification.
- Determine whether to release a software system or stop testing it.
- Estimate when the subsequent software system failure will occur.
- Identify elements in the system that most need a redesign to improve reliability.
- Measure how reliably the software system operates to make changes where necessary.

7.11.10.3 Basic Concepts

There are at least two significant differences between hardware and software reliability. First, the software does not experience fatigue, wear out, or burnout. Second, because software instructions within computer memories are accessible, any line of code can contain a fault that could produce a failure. The failure rate of a software system over time is generally decreasing due to fault identification and removal. Software failures are unlikely to reoccur after being identified and removed (Figure 7.7).

7.11.10.4 Procedure

The following is an 13-step generic procedure for estimating software reliability. Tailor this to the specific project and the current lifecycle phase. Not all steps will be used in every application, but the structure provides a convenient and quickly remembered standard approach. The following steps are a checklist for reliability programs:

1. Identify the application.
2. Specify the requirement.
3. Allocate the requirement.
4. Define failure: the testers, developers, and users usually negotiate a project-specific failure definition. It is agreed upon before the test begins. Most importantly, the definition is consistent over the project's life.
5. Characterize the operational environment, including system configuration, evolution, and operating profile.

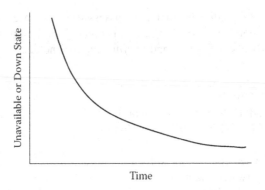

FIGURE 7.7   Software reliability measurement curve.

6. In modeling software reliability, remember that systems frequently evolve during testing. As a result, new code and components can be added.
7. Select tests: SRE often involves operations and collecting failure data. Operations should reflect how the system will be used. The standard includes an appendix of information to help determine failure rates.
8. Select models: included various reliability models. We recommend that you compare several models before making a selection.
9. Collect data: learn from previous lessons to make an effective reliability program. This does not mean you must keep all the information about the program as it evolves. Also, clearly define data collection objectives. When data is required, it will affect the people involved. As a result, cost and schedule can suffer, too.
10. Two additional points to consider when collecting data: (1) Motivate the data collectors; and (2) Review the collected data promptly. If this advice is not followed, quality will suffer.
11. Estimate parameters: there are three techniques in the standard to determine model parameters – method of moments, least squares, and maximum likelihood.
12. Validate the mode: to validate a model and address its assumptions adequately. This can be done effectively by choosing appropriate failure data items and relating specific failures to particular intervals or changes in the life cycle.
13. Perform analysis: after the data has been collected and the model parameters have been estimated, perform the appropriate analysis. The objective may be to estimate the software's current reliability, the number of faults remaining in the code, or when testing will be complete.

Be careful about combining a software reliability value with a system reliability calculation. The risk analysis may require a system reliability figure, while execution time is the basis for software reliability. In that case, it must be converted to calendar time to be combined with hardware reliabilities. After converting to standard units, one can calculate the system reliability value.

7.12 AP-913 STANDARDS FOR ASSET RISK AND RELIABILITY

It is a generic process to facilitate a reliability program culture and discipline for nuclear power plants across an asset lifecycle.

The application of AP-913 has been against the North American nuclear industry, where perfection is barely tolerated. Their reliability program must ensure that stewardship of these critical assets meets their defined design and performance criteria at their accepted levels of reliability throughout the lifecycle of the asset and facility. This lifecycle is defined as design, construction, commissioning, start-up, operation, and decommissioning.

Considerable discipline is needed for a successful reliability program, and one would expect a nuclear environment to be the place for this. However, a reliability program could and should be in place for any asset-intensive organization. Enterprises such as aerospace, chemical, mining, process, pharmaceutical, petrochemical, and other high-hazard industries should be invested in a working asset reliability program to mitigate safety, environmental, and business risks.

Formally AP913 includes elements to accomplish its objective of enhancing plant availability and safety (Figure 7.8):

- performance monitoring
- performance evaluation
- problem prioritization
- problem analysis and corrective action recommendation
- corrective action implementation and feedback

Like RCM, the AP913 process identifies structures, systems, and components (SSC). It helps to prioritize asset systems and assets. Each assessed asset looks at intended functions and functional failures and performs a proactive failure modes and effects analysis (FMEA). Next, it confirms the safety function(s), the consequence of failure, the likelihood that the SSCs will be required to perform the safety function, and its potential frequency.

Risk and reliability calculations will be made to measure and rationalize tolerable risk against the asset and the assessed scenario.

Risk mitigation activities will be declared for proactive inspection cycles, preventive maintenance (PMs), and predictive maintenance (PdM). Otherwise, a "run-to-failure" (RTF) or one-time change (Redesign) mitigation maintenance strategy will be implemented.

The AP-917 process formally starts with scoping and identification of critical components.

- Identify important functions
- Identify critical components
- Identify non-critical components
- Identify RTF components

AP913 Input/Output Process Relationship Matrix

Elements	Scoping and Identifying Critical Components	Performance Monitoring	Continuing Equipment Reliability Improvements	PM Implementation	Corrective Action	Lifecycle Management
Scoping and Identifying Critical Components	Process Start	Input	Input			
Performance Monitoring	Output to		Input	Output to	Input/Output	Input
Continuing Equipment Reliability Improvements	Output to	Output to		Input	Output to	Input
PM Implementation		Input	Output to			
Corrective Action		Input/Output	Input			Input
Lifecycle Management		Output to	Input		Output to	Process End

Reads as topline elements' relationship to other elements listed on the side.
For example, PM Implementation (Topline) receives input from Continuing Equipment Reliability Improvements and provides output to Performance Monitoring

FIGURE 7.8 High-level process matrix relationship flow of the AP913 activities.

Continuing equipment reliability improvement
- Development and use of PM templates
- Continuing adjustments to PM task and frequency based on a station and industrial equipment operating experience
- Documentation of the PM technical bases
- Consideration of alternative maintenance strategies to ensure reliable equipment
- Continuous improvement from plant staff recommendations
- Performance monitoring
- System performance
- Component performance
- Predictive trending results
- Operations rounds monitoring
- Monitor testing and inspection results

PM Implementation

- Preventive maintenance (PM)
- Document equipment "as found" equipment condition
- Equipment condition feedback
- Standard post-maintenance test

Corrective Action

- Corrective maintenance
- Failure cause and corrective action
- Prioritization of equipment problems

Lifecycle Management

- A long-term strategy for system and component health
- Prioritization of improvement activities
- Integration of long-term plans with the station business strategy

These processes are not materially different from a well-executed RCM program (see Chapter 8). They could be applied across any asset-intensive organization. However, the safety and environmental impacts of a poor reliability outcome in nuclear make the program uniquely critical should a catastrophic outcome arise. The process itself is critical. Having the entire organization committed to understanding and executing this set of processes is essential for their success and the community safety for which they work.

7.13   RISK-BASED INSPECTION STANDARDS API 580 AND 581

Risk-based inspection (RBI) is primarily applied against compression vessels such as pipes and boilers.

API 580 documents how RBI can be used as a quantitative measure. RBI methods and work processes can be reviewed to confirm the level at which they meet the quality prescribed in the standard. API 580 was initially published in 2002 and has been updated even over the past six years.

API 581 Risk-based inspection guides provide RBI programs' suggestions on fixed equipment (primarily pipes and compression components) in petrochemical, refining, chemical process plants, and production facilities. API 581 introduces principles and presents minimum general guidelines for RBI. In addition, AP 581 provides quantitative calculation methods to determine an inspection plan.

This standard includes a method for calculating a probability of failure (POF) and the consequence of failure (COF). In RBI, failure is defined as a loss of containment from the pressure boundary resulting in leakage to the atmosphere or rupturing of a pressurized component.

The risk increases as damage accumulates during in-service operation as the risk tolerance or target is approached. Therefore, an inspection is recommended of sufficient effectiveness to quantify the damaged state of the component accurately.

The inspection action does not reduce the risk; however, it reduces uncertainty and allows for a mitigating action to be introduced while collecting an accurate quantification of the damage present in the component.

A thoughtful execution of an RCM assessment could facilitate the standards required from the RBI set of processes.

7.14 CONCLUSION

This chapter has shown that managing asset risk must remain competitive in the changing global marketplace. Companies that ignore asset risk management place themselves where every day becomes a roll of the dice. Will a significant failure occur that can cost millions, lose customers, cause fatalities, or damage the environment? Management must decide that asset risk management is a priority. Maintenance professionals should embrace risk management because risk management improves maintenance effectiveness.

After concluding this chapter, the reader should understand risk and risk management better. This includes how to define risk and several proven methods to manage risk. These include methods such as FMECA and HAZOPS. There are accepted standards to help the risk manager with the process of identifying and managing risk.

Management must accept asset risk management as a vital maintenance process in managing a company. Without a good risk management process, the next edition of this book may include another company in the list of major failures. Do not let it be yours.

PERMISSIONS

BIBLIOGRAPHY

1 Q631-97 (R2002) Reliability, Availability, and Maintainability (RAM) Definitions.
2 CSA Q636-93 (R2001) Guidelines and Requirements for Reliability Analysis Methods.
3 CAN/CSA Q850-97 (R2009) Risk Management: Guideline for Decision Makers.

4 AS/NZS 4360:2004 Risk Management.
5 ISO 14000 Environmental Management Systems.
6 BS 6143-2:1990 guide to the economics of quality.
7 EPA-40 CFR Part 68 Chemical Accident Prevention Provisions.
8 OSHA 29CFR 1910.119 Process Safety Management of Highly Hazardous Chemicals.
9 ANSI, R-013-1992 Software Reliability.
10 Reliability Programs for Nuclear Power Plants, Canadian Nuclear Safety Commission.

8 Reliability by Design*
Reliability-centered Maintenance

Don M. Barry
Previous contributions from Don M. Barry and Jim Picknell

The technological world is evolving through Industry 4.0. Assets are even more critical for an asset-intensive organization's value position. Assets are now recognized as an essential contributor to operational excellence. With the supporting tools, the asset can now be calculated as a "risk constraint" in the growing mechanization and automation and the increased focus on cost, productivity, and risk: this demands precise asset management strategies and maintenance tactics.

Reliability-centered maintenance (RCM) is the pre-eminent disciplined technique for establishing the best maintenance task in a scheduled maintenance program. This chapter introduces RCM, describes it in detail, and explores its history. We discuss who should be using RCM and why. RCM is increasingly important as society becomes less tolerant of risk, more interested in holding real people responsible for business failures (including safety, environmental or operational), and productivity demands increase. RCM can be deployed to improve plant availability and reliability, product quality effectively, returns on (equipment) assets, and equipment life. When well-executed, RCM can be effective for proactively defining maintenance to ensure safe and environmentally friendly plants. In addition, a demonstrated effective maintenance program can help plants qualify for lower commercial insurance risk.

The RCM process is described in this chapter with the essential factors to consider as you work through an RCM analysis. The chapter includes a flow diagram for a suggested process that complies with the SAE standard for RCM programs and a simplified decision logic diagram for selecting appropriate and adequate maintenance tactics. We explain the deliverables and how to get them from the vast amount of data usually produced by the process. The scope of RCM projects is also described to get a feel for the effort involved. We thought it would be helpful to include an effective RCM implementation showing team composition, size, time, and effort required, and tools to make the task easier.

Of course, RCM has been used in several environments and in numerous ways. Some are slight variations of the thorough process, others are less rigorous, and some are downright dangerous. These methods are discussed along with their advantages and disadvantages. We also examine why RCM programs fail and how to recognize

DOI: 10.1201/9781032679600-10

and avoid those problems. As responsible maintenance and engineering professionals, we want to improve our organization's effectiveness. You will learn to gradually introduce RCM successfully, even in the most unreceptive environments.

This chapter is likely to generate some controversy and discussion. That is just what we intend. In law, the concept of justice and legal realities sometimes conflict. Similarly, you will find that striking a balance between what is right and what is achievable is often a difficult challenge.

8.1 INTRODUCTION

RCM is the pre-eminent method for establishing the best maintenance task in a scheduled maintenance program. It has been highly effective in civil and military aviation, military ship and naval weapon systems, utilities, mining, petrochemical, and the chemical industry. It's mandated in civil aircraft and often, as well, by government agencies procuring military systems. Increasingly, companies select RCM when reliability is vital for safety, environmental reasons, or keeping the plant running at maximum capacity.

The published SAE Standard JA1011, "Evaluation Criteria for Reliability-Centered Maintenance (RCM) Processes," outlines a process's criteria for RCM. This standard determines whether a process is RCM through seven specific questions, although it doesn't specify the process itself.

This chapter describes a process designed to satisfy the SAE criteria. We've included several variations that don't necessarily answer all seven questions but are still called RCM. Consult the SAE standard for a comprehensive understanding of the complete RCM criteria.

In our increasingly litigious society, we are more and more likely to be sued for "accidents" that at one time would have been accepted as being out of our control. Today, the courts take a harsh stand with those who haven't done all they could to eliminate risks. In recent decades, many examples of disastrous "accidents" could have been avoided, such as the carnage at Bhopal, the Challenger explosion, and the tendency of the original Ford Pinto gas tanks to explode when rear-ended.

The incident at Bhopal triggered sweeping changes in the chemical industry. New laws were established, such as the "Emergency Planning and Community Right to Know Act" passed by the US Congress in 1986 and the Chemical Manufacturers Association's "Responsible Care" program. Following the Pinto case and others, consumer goods manufacturers are held to more stringent safety standards.

Despite its wide acceptance, RCM has been criticized as too onerous and expensive to solve the relatively simple determining of what maintenance to do. These criticisms often come from industries where equipment reliability, environmental compliance, and safety are not significant concerns. Sometimes, they also result from failing to manage the RCM project properly instead of flaws within the RCM process itself. Alternative methods to RCM are covered later in this chapter, although we don't recommend using them. A full description of their risks is included.

We all want the maximum from our transportation and production systems, infrastructure, and plants. They're costly to design and build, and downtime is costly. Downtime is needed to sustain operations and logical breaks in production runs or transportation schedules. RCM helps eliminate unnecessary downtime, saving valuable time and money. RCM contributes to operational excellence.

RCM generates the mitigation tasks for a scheduled maintenance program that logically anticipates specific failure modes. It can also effectively help to:

- detect failures early enough for them to be corrected quickly and with little disruption
- eliminate the cause of some failures before they happen
- eliminate the cause of some failures through design changes
- identify those failures that can safely be allowed to happen

This chapter provides an overview of the different types of RCM:

- Aircraft vs. military vs. industrial
- Functional vs. hardware
- Classical (thorough) vs. streamlined and "lite" versions

We describe the basic RCM processes step by step. This includes a brief overview of critical equipment and failure modes effects causal analysis (FMECA), covered thoroughly in Chapter 7 (Assessing and Managing Risk).

8.2   WHAT IS RCM?

RCM is a logical, technical process determining which maintenance tasks will ensure a reliable, "as designed" system, under specified operating conditions, in a specified operating environment. Each of the various reference documents describing RCM applies its definition or description. We refer readers to SAE JA1011 for a definitive set of RCM criteria.

RCM takes you from start to finish, with well-defined steps arranged sequentially. It is also iterative – it can be carried out several ways until initial completion. RCM determines how to improve the maintenance plan based on experience and optimizing techniques. As a technical process, RCM delves into how things work and what can go wrong. Using RCM decision logic, you select maintenance interventions or tasks to reduce the number of failures, detect and forecast when one will be severe enough to warrant action, eliminate or accept it, and run until failure.

RCM aims to make each system as reliable as it was designed. Each component within a system has its unique combination of failure modes and failure rates. Each combination of components is unique; failure in one component can cause others to fail. Each "system" operates in its environment: location, altitude, depth, atmosphere, pressure, temperature, humidity, salinity, exposure to process fluids or products, speed, and acceleration. Depending on these conditions, certain failures can dominate. For example, a level switch in a lube oil tank will suffer less from corrosion than if it were in a saltwater tank. An aircraft operating in a temperate maritime climate will corrode more than in an arid desert. The environment and operating conditions can significantly influence what failures will dominate the system.

The operating environment's impact on a system's performance and failure modes makes RCM valuable. Technical manuals often recommend a maintenance program for equipment and systems, and sometimes, they include the effects of the operating environment. For example, car manuals specify different lubricants and anti-freeze densities that vary with ambient operating temperature. However, they don't usually address the wear and tear of such things as driving style (aggressive vs. timid) or how the vehicle is used (taxi or fleet vs. weekly drives to visit grandchildren). In industry, manuals are not often tailored to any particular operating environment. For example, an instrument air compressor installed at a sub-arctic location may have the same manual and dew point specifications as one installed in a humid tropical climate. RCM specifically addresses the environment experienced by the fleet, facility, or plant.

8.3   WHY USE RCM?

Because it works, is cost-effective, aligns your team to the influences of achieving operational excellence, and prepares your organization for the data-dependent Industry 4.0 competitive environment and future.

The latest boasted benefits of an RCM-implemented program and culture can suggest:

- Increased asset reliability, availability, and productivity
- Optimized maintenance and avoidance of unnecessary costs
- A boost in general safety awareness and environmental integrity
- A team generated an enhanced understanding of equipment (asset) behavior by operating context
- A program that exceeds the traditional SAE standards broadens existing capabilities and aligns with ISO management systems (ISO55000 and ISO 31000)
- Insights that can be integrated with other business risk management solutions
- A sharing alignment of information across the enterprise
- A set of data insights that can be used now and prepare the enterprise for an automated future

RCM has been around for more than 50 years, since the late 1960s, beginning with studies of airliner failures carried out by Nowlan and Heap. United Airlines wanted to reduce the amount of maintenance for what was then the new generation of larger wide-bodied aircraft. Previously, aircraft maintenance was based on experience and influenced by known and obvious safety concerns. As aircraft grew larger, with more parts and therefore more things to go wrong, maintenance requirements similarly grew, eating into the flying time needed to generate revenue. In the extreme, achieving safety could have become too expensive to make flying economical. But the airline industry has developed almost entirely proactive maintenance thanks to RCM and United's willingness to try a new approach. This resulted in increased flying hours, with a drastically improved safety record.

In fact, aircraft safety has been consistently improving since RCM was introduced. In addition, RCM has reduced the number of maintenance resource hours needed for new aircraft per flight. Why? RCM identifies functional failures that can be caught through monitoring before they occur. It then reveals which failures require some usage or time-based intervention, develops failure-finding tests, and indicates whether a system redesign is needed. Finally, it flags failures that can be left to occur because they cause only minor problems. Where aircraft are concerned, this is a small number indeed. As a result, frequent fliers seldom experience delays for mechanical or maintenance-related problems, and airlines can usually meet their flight schedules.

RCM has also been used successfully outside the aircraft industry. Those managing military capital equipment projects, impressed by the airlines' highly reliable equipment performance, often mandate the use of RCM. In one shipbuilding project, the total maintenance workload for the crew was cut by almost 50% compared to other similarly sized ships. At the same time, the ship's service availability improved 60% to 70%. The amount of downtime needed for maintenance was significantly reduced. In that project, the cost of performing RCM was high in the millions of dollars, but the payback justified it in hundreds of millions.

The mining industry operates in remote locations far from parts, materials, and replacement labor sources. Consequently, miners, like the navy and airline industry, want high reliability – minimum downtime and maximum productivity from the equipment. RCM has been a considerable benefit. It's made fleets of haul trucks and other equipment more available while reducing maintenance costs for parts and labor and planned maintenance downtime.

RCM has been successful in process industries in chemical plants, oil refineries, gas plants, remote compressor and pumping stations, mineral refining and smelting, steel, aluminum, pulp and paper mills, tissue converting operations, and food and beverage processing and breweries. RCM can be applied anywhere where high reliability and availability are essential.

A decade-old study on asset management strategies revealed that "improved" reliability was the top motivator for improving maintenance processes over focusing on costs or asset uptime. However, when asked in the same survey where they were in their proactive planning of maintenance tasks, less than 30% indicated that they had engaged in RCM.

Figure 8.1 shows that almost 90% of polled companies reported that Preventive Maintenance is their most commonly used proactive maintenance strategy. While Predictive Maintenance and RCM have also proven to produce desirable results, many companies have yet to leverage this as part of their living program in maintenance.

This suggests that more than 70% of companies still have yet to justify how they can get at the real benefits of RCM for them. As a result, many companies do too little of the proper maintenance or too much of the wrong maintenance (i.e., Preventive Maintenance) and leave money and risk on the table when it comes to leveraging their assets.

RCM has expanded over the past decade, but too many organizations still have not committed to the discipline or recognized how essential it is to succeed as they move into a competitive Industry 4.0/5.0 marketplace.

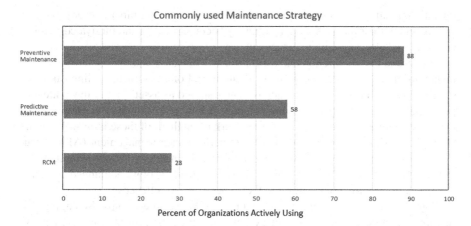

FIGURE 8.1   Adopted proactive maintenance strategies.

A more recent study of over 100 large asset-intensive companies suggests that their executives want insights that bring an understanding of how to:

- Achieve "return on assets" across their asset lifecycle
- Predict and mitigate conditions and behaviors that lead to asset failure
- Reduce the consequences of asset failure, safety, environmental, and operational
- Improve decision-making through integrated and consistent enterprise asset-related information and insights
- Leverage existing asset-related data
- Develop strategies to obtain required additional data

RCM enables maintenance and operations to respond quickly and positively to business dynamics. It can support new technologies such as digital electronics, pneumatics, hydraulics, and automated Internet of Things (IoT) enabled production systems. RCM recognizes that successful maintenance demands a complete understanding between the operators, maintainers, and design engineers. RCM also recognizes the evolution of Industry 4.0/5.0.

Properly executed RCM will eliminate maintenance tasks without value in the equipment's expected functionality. They will generate a comprehensive understanding of the required maintenance tasks and the frequency and resources (skills, tools, and spare parts) required to perform them. It will direct the team to identify real opportunities and tasks to improve:

- safety and environmental integrity;
- plant operating performance;
- maintenance cost-effectiveness; and
- the length of an asset's life cycle.

Teams that focus on RCM typically understand how each area of their plant contributes to the value of their business. They team up with their stakeholder group (i.e., operations) and collect insights and data to drive a highly competitive organization to compete in their markets. As a result, they become focused and motivated to influence a more "effective" maintenance culture.

The results in a production and asset priorities database and maintenance requirements insights and thresholds provide an audit trail of assessment findings and recommendations. It helps a plant adapt to change should its asset's operating context or personnel change. It defines the maintenance that should be done or not done, and it provides a framework for helping a plant work through all of these issues. When the RCM data insights are captured in an asset performance management (APM) solution, it enables the production to become more agile in an automated way. Maintenance can add the percentage of work predicted to the percentage of work planned as a critical production metric contributing to their operations excellence.

Commercial insurance plays a critical role in the world economy. Managing risk requires an in-depth knowledge of systems and processes utilized by a plant. As insurance companies look to manage a plant's risk.

In short, leading maintenance organizations do an RCM analysis, implement RCM as a prime influence on which maintenance task will be placed into their enterprise asset management systems (EAM) or asset performance management systems (APM) and workflow, and leverage RCM as part of their living program.

Examples of some of the data elements collected for an RCM analysis are shown in Table 8.1. This data can be applied to the equipment mitigating tactic decision grid and the ongoing alignment to support operational excellence.

Many perceive RCM as onerous and resource-intensive. However, the payback can be a few months or weeks, depending on the application and the team's effectiveness in working the RCM process.

8.4   WHO SHOULD USE RCM?

RCM should become a discipline and culture where productivity supporting operational excellence is crucial at every plant, fleet, or building.

That includes companies that can sell everything they produce, where uptime, high equipment reliability, and predictability are paramount to complete. It also includes anyone producing to meet tight schedules, such as just-in-time parts delivery to automotive manufacturers, where equipment availability is critical. Availability means physical assets (equipment, plant, fleet, etc.) are there when needed. The higher equipment availability leads to more productive assets and operational excellence. Availability is measured by dividing the time assets available by the total time needed to run.

If a failure causes damages, there's a growing trend in our increasingly litigious society for those affected to sue. Practically everyone, then, can benefit from RCM. There is no better way of ensuring that the correct maintenance is done to avoid or mitigate failures.

$$Ao = Uptime/Total\ Time$$

Uptime is Total Time minus Downtime. Downtime for unplanned outages is the total time to repair the failures or mean time to repair (MTTR). Reliability takes into

TABLE 8.1
Example of the Data Types That Could Be Extracted from an RCM Analysis[1]

RCM Step	Example of Data Collected
Asset operating context	Asset data: • location • risk references • operating context
Desired functions	Functions • primary standards • secondary standards • protective, detective functional notation
Failure mode effects analysis	Failure modes, failure effects • frequency • severity, impact • HSE Impacts • operator awareness • early warnings • likeliness • predictability • operational costs • maintenance costs, parts costs • diagnose timelines • repair timelines • secondary damage
Consequence	Impact is: • safety, environmental • hidden • operational or non-operational
Prescribed mitigation	Proactive prescribed solution is: • technically feasible and worth doing Outcomes: • predictive • preventive • run to failure • inspection • redesign • combination of above
Other reliability process data created	Prescribed solution • aligned to asset location and risk prioritization Solution application: • frequency • assignment • risk weighting • work order base • prediction thresholds • support requirements • training, tools • safety needs • parts/procurement

account the number of unplanned downtime incidents you suffer. Reliability is, strictly speaking, a probability. A commonly used interpretation is that high mean time between failures (MTBF) values indicate higher reliable systems. Generally, plants, fleets, and buildings benefit from greater MTBF, which means fewer disruptions.

RCM is essential in achieving maximum reliability – the longest MTBF. For most systems, MTBF is long, and repair time (MTTR) is short. Reducing MTTR requires a high level of maintainability. Adding the two (MTTR+MTBF) gives the total time a system could be available if it would never break down. Availability is the portion of this total time the system is in working order and available to do its job.

Availability can be rewritten as:

$$\text{Ao} = \text{MTBF}/(\text{MTBF} + \text{MTTR})$$

Mathematically, we can see that maximizing MTBF and minimizing MTTR will increase Ao.

Generally, an operation is better off with fewer downtime incidents. However, additional measures are needed if downtime threatens the manufacturing process, delivery schedule, and overall productivity. Reliability is paramount. RCM will help you maximize MTBF while keeping MTTR low.

8.5   THE RCM PROCESS

RCM has seven steps to meet the criteria of the published SAE standard:

1. Prioritize/Identify the equipment/system to be analyzed.
2. Determine its functions and asset operating context.
3. Determine what constitutes failure of those functions.
4. Identify what causes those functional failures.
5. Identify their impacts or effects.
6. Use RCM logic to select appropriate maintenance tactics.
7. Document the final maintenance program and refine it as operating experience is gained.

If you want to perform the new RCM3 version, an additional step is injected into the seven, which extracts more data points in the failure-effects step and breaks out more options to mitigate risk. This will be expanded on later in this chapter.

Steps 3 through 5 constitute the Failure Modes and Effects Analysis (FMEA) portion of RCM. FMEA is discussed in greater detail in Chapter 7. Some practitioners limit their FMEA analysis to failures, ignoring those that can be effectively prevented. But FMEA, as used in the RCM context, must consider all reasonably likely failures – the ones that have occurred can currently be prevented, and those that haven't yet happened.

In the first step of RCM, you decide what to analyze. A plant usually contains many different processes, systems, and equipment. Each does something different; some are more critical to the operation than others. Some equipment may be essential

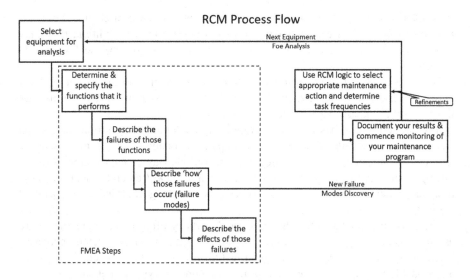

FIGURE 8.2   RCM process overview.

for environmental or safety reasons but have little or no direct impact on production. For example, if a wastewater effluent treatment system that prevents untreated water from being discharged from the plant goes down, it doesn't stop production. The consequences can still be significant if environmental regulations are flouted. The plant could be closed, and the owners fined or jailed.

Start by establishing criteria to determine what is essential to the operation. Then, use them to decide which equipment or systems are most important, demanding the most significant attention. There are many possible criteria, including:

- Personnel safety
- Environmental compliance
- Production capacity
- Production quality
- Production cost (including maintenance costs)
- Public image

When a failure occurs, the effect on each of these criteria can vary from "no impact" or "minor impact" to "increased risk" or "major impact." Each criterion and how it's affected can be weighted. For example, safety is usually rated higher than production capacity. Likewise, a significant impact is weighted higher than no impact. For each system or equipment considered for RCM analysis, imagine a "worst-case" failure, then determine its impact on each criterion. Multiply the weights of each criterion and add them together to arrive at a "criticality score." Items with the highest score are prioritized to be analyzed first.

There are both active and passive functions in each system. Active functions are usually the obvious ones for which we name our equipment. For example, a motor control center controls the operation of various motors. Some systems also have less obvious secondary or even protective functions. For instance, a chemical process loop and a furnace have a secondary containment function. In addition, they may include protective functions such as thermal insulation or chemical corrosion resistance.

Keep in mind that some systems, such as safety systems, do not become active until another event. Unfortunately, normally passive state failures are often difficult to spot until too late.

Each function also has a set of operating limits, defining its "normal" operation and failures. It has failed when the system operates outside these "normal" parameters. Our system failures can be categorized in various ways, such as: high, low, on, off, open, closed, breached, drifting, unsteady, and stuck. Remember that the function fails when it falls short of or exceeds its operating environment's specified parameters.

Determining functions for the individual parts is often easier than the entire assembly. Functions are more easily identified as the hardware detail increases. There are two ways to analyze the situation. One is to look at equipment functions at a reasonably high assembly level. You must imagine everything that can go wrong. This works well for pinpointing major failure modes. However, you could overlook some less obvious possible failures with severe consequences.

An alternative is to look at "part" functions. This is done by dividing the equipment into assemblies and parts, similar to taking the equipment apart. Each part has its own functions and failure modes. Breaking the equipment into smaller chunks (parts) makes it easier to identify all failure modes without missing any. This is more thorough, but it does require a bit more work.

To save time and effort or prioritize their approach, some practitioners perform a "Pareto" analysis of the failure modes to filter out the least common ones. In RCM, though, all reasonably likely failures should be analyzed. You must be confident that a failure is unlikely to occur before it can be ignored.

Plant   Assembly   Component

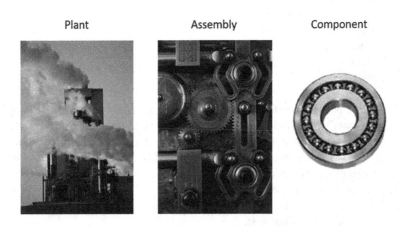

Functions are more easily identified as the hardware detail increases

FIGURE 8.3   Asset detail versus function detail.

A failure mode is physical. It shows "how" the system fails to perform its function. We must also identify "why" the failure occurred. The root cause of failures is often a combination of conditions, events, and circumstances.

For example, a cylinder may be stuck in one position because its hydraulic fluid lacks lubrication. As a result, the cylinder has failed to stroke or provide linear motion. "How" it fails is the loss of lubricant properties that keep the sliding surfaces apart. There are many possibilities, though, for "why." For instance, the cause could be a problem with the fluid, either using the wrong one, leaking, dirt, or surface corrosion due to moisture. Each can be addressed by checking, changing, or conditioning the fluid.

Not all failures are equal. They can affect the rest of the system, plant, and operating environment. The cylinder's failure above could be severe if excessive effluent flows into a river by actuating a sluice valve or weir in a treatment plant. Or, the effect could be as minor as failing to release a "dead-man" brake on a forklift truck stacking pallets in a warehouse. In one case, it's an environmental disaster; in the other, only a maintenance nuisance. But if an actuating cylinder on the brake fails in the same forklift while it's in operation, there could be a serious injury.

By knowing the consequences of each failure, we can determine what to do, whether it can be prevented, predicted, avoided altogether through periodic intervention, eliminated through redesign, or requires no action. Then, we can use the RCM process (and logic) to choose the appropriate response.

RCM helps classify failures as hidden or obvious and whether they have safety, environmental, production, or maintenance impacts. These classifications lead the RCM practitioner to default actions if appropriate predictive or preventive measures cannot be justified. For example, a fire sprinkler cannot be detected or predicted while it is in normal operation (dormant), but by designing redundancy of sprinklers, we can mitigate the consequences of failure (Figure 8.5). More severe consequences typically require more extensive mitigating actions.

Normally, a fire sprinkler is "dormant". We only know of failure when it's too late!

Protection devices typically act as "hidden" failures

FIGURE 8.4   Protection device behavior.

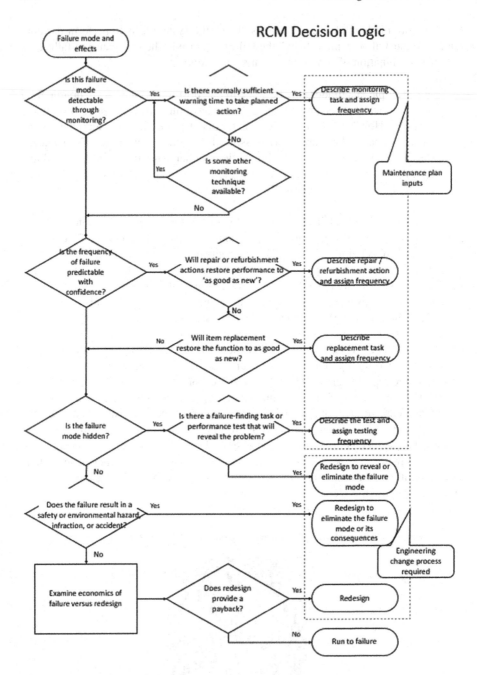

FIGURE 8.5   RCM decision logic.

Most systems failures involving complex mechanical, electrical, and hydraulic components will fail randomly. You can't confidently predict them. Still, many are detectable before functional failure takes place. For example, if a booster pump fails to refill a reservoir providing operating head to a municipal water system, the system doesn't have to break down. If you watch for the problem, it can be detected before municipal water pressure is lost. That is the essence of condition monitoring. We look for the failure that has already happened but hasn't progressed to the point of degrading the system. Finding failures in this early stage helps protect overall functional performance.

Since most failures are random, RCM logic first asks if it's possible to detect the problem in time to keep the system running. If the answer is "yes," condition monitoring is needed. You must monitor often enough to detect deterioration, with enough time to act before the function is lost. For example, check the booster pump's performance once daily if you know it takes a day to repair it and two days for the reservoir to drain. That provides a buffer of at least 24 hours after detecting and solving the problem before the reservoir system is adversely affected.

If you can't detect the problem in time to prevent failure, RCM logic asks if it's possible to reduce the impact by repairing it. Some failures are entirely predictable, even if they can't be detected early enough. For example, we can safely predict brake wear, belt wear, tire wear, erosion, etc. These failures may be difficult to detect through condition monitoring in time to avoid functional failure, or they may be so predictable that monitoring for the obvious isn't warranted. Why shut down equipment monthly to monitor for belt wear if you know it will not appear for two years? You could monitor every year, but in some cases, it's more logical to replace the belts automatically every two years. There is the risk, of course, that a belt could either fail earlier or still operate well when replaced. If you have good failure history, you may be able to perform Weibull analysis (see Chapter 10) to determine whether the failures are random or predictable.

If the above approach isn't practical, you may have to replace all the equipment. Usually, this makes sense only in critical situations because it typically requires an expensive sparing policy. The cost of lost production would have to be more than entirely replacing equipment and storing the spares.

Because safety and protective systems are typically inactive, you may be unable to monitor for deterioration. If the failure is also random, replacing the component on a timed basis may not make sense because the new part might fail as soon as it has been installed. We can't tell again because the equipment won't reveal the failure until switched to active mode. In these cases, some sort of testing may be possible. We sorted these hidden ones from the rest in our earlier classification of failures. If condition monitoring and usage or time-based intervention aren't practical, you can use RCM logic to explore functional failure-finding tests. These tests can activate the device and reveal whether or not it's working. If such a test isn't possible, redesign the component or system to eliminate the hidden failure. Otherwise, there could be severe safety or protective consequences, which are deemed unacceptable.

A version of the decision logic is depicted in the following diagram:

You can either redesign or accept the consequences for non-hidden problems that cannot be prevented through early prediction, usage, or time-based replacements. Redesign is the best decision for safety or environmental cases with unacceptable consequences. For production-related cases, whether to redesign or run to failure may depend on the cost of the consequences. For example, if production is likely to shut down entirely for a long time, redesigning would be wise. If the production loss is negligible, run to failure is appropriate. If there aren't any production consequences, but there are maintenance costs, the same applies. In these cases, the decision is based on economics – the cost of redesign versus failure (i.e., lost production, repair costs, overtime, etc.).

Task frequency is often challenging to determine. It's invaluable when using RCM to know the failure history, but it isn't always available. For instance, there isn't any operating history for a system still being designed. You're against the same problem with older systems where records haven't been kept. An option is to use generic failure rates from commercial or private databases. Failures, though, don't happen exactly when predicted. Some will be random, and some will become more frequent late in the "life" of the equipment, and so on. Allow some leeway. Recognize also that the database information may be faulty or incomplete. Be cautious and thoroughly research and deal with each failure mode in its own right.

Once RCM is completed, you must group similar tasks and frequencies to apply the maintenance plan in an actual working environment. You can use a slotting technique to simplify the job. This consists of a pre-determined set of frequencies: daily, weekly, monthly, every shift, quarterly, semi-annually, and annually, or by units produced, distances traveled, or the number of operating cycles. Choose frequency slots closest to those that maintenance and operating history show are best suited to each failure mode. Later, these slots can group tasks with similar characteristics in a workable maintenance plan.

After running the failure modes through the above logic, consolidate the tasks into a maintenance plan. This is the final "product" of RCM. Then, those maintaining and operating the system must continually strive to improve the product. The original task frequencies may be overly conservative or too long. If too many preventable failures occur, proactive maintenance isn't frequent enough. If there are only small failures or the preventive costs are higher than before, maintenance frequencies may be too high. This is where optimizing task frequency is vital.

The output of RCM is a maintenance requirements document, which describes the condition monitoring, time or usage-based intervention, failure-finding tasks, and the redesign and run-to-failure decisions. It is not a "plan" in the true sense. It doesn't contain specific maintenance planning information like task duration, tools, test equipment, parts and materials requirements, trades requirements, and a detailed sequence of procedures. These are the logical next steps after RCM has been performed.

The maintenance requirements document is not just a series of lists transferred from the RCM worksheets or database into a report. RCM takes you through a rigorous process that identifies and addresses individual failure modes:

- for each plant, there are numerous systems
- for each system, there can be numerous pieces of equipment
- for each piece of equipment, there are several functions
- for each function, there may be several failures
- for each failure, there may be many failure modes with varying effects and consequences
- for each failure mode, there is a task, and this task may have assigned parts to be considered

In a complex system, there can be thousands of tasks. To get a feel for the size of the output, consider a typical process plant with spares for only about 50% of its components. Each may have several failure modes. The plant probably carries 15,000 to 20,000 unique part numbers (stock-keeping units) in its inventory. That means there could be around 40,000 equipment components with one or more failure modes. The maintenance requirements document would be substantial if we analyzed the entire plant.

Fortunately, a limited number of condition monitoring techniques are available – just over 50. These cover most random failures, especially in complex mechanical, electrical, and hydraulic systems. Tasks can be grouped by technique (i.e., vibration analysis), location (i.e., the machine room), and sub-location on a route. Hundreds of individual failure modes can be organized, reducing the number of output tasks for detailed maintenance planning. It's essential to watch the specified frequencies of the grouped tasks. Often, we don't precisely determine frequencies, at least initially. It may make sense to include tasks with frequencies like "every four days" or "weekly" but not "monthly" or "semi-annually." This is where the slotting technique described earlier comes in handy.

Time or usage-based tasks are also easy to group. For example, you can assemble all the replacement or refurbishment tasks for a single piece of equipment by frequency into a single overhaul task. Similarly, multiple overhauls in a single area of a plant may be grouped into one shutdown plan.

Another way to organize the outputs is by who does them. Tasks assigned to operators are often performed using the senses of touch, sight, smell, or sound. These are often grouped logically into daily, shift, or inspection rounds checklists. Ultimately, a complete listing should tell what maintenance to do and when. The planner determines what is needed to execute the work.

8.6   WHAT DOES IT TAKE TO DO RCM?

RCM very thoroughly examines plants and equipment. It involves detailed knowledge of how equipment and systems operate, what's in the equipment, how it can fail to work, and the impact on the process, plant, and environment.

You must practice RCM frequently to become proficient and maximize its benefits. To implement RCM:

- select a team of practitioners
- train them in RCM
- teach other "stakeholders" in the plant operation and maintenance what RCM is and what it can achieve
- select a pilot project to demonstrate success and improve upon the team's proficiency
- roll out the process to other areas of the plant

Typically, it is best to demonstrate RCM's success through a pilot. Before the RCM team begins the analysis, determine the plant baseline reliability and availability measures and proactive maintenance program coverage and compliance. These measures will be used later to compare what has been changed and how successfully.

8.6.1 THE TEAM, SKILLS, KNOWLEDGE, AND OTHER RESOURCES NEEDED

A multi-disciplinary team is essential, bringing in specialists when needed. The team needs to know the day-to-day operations of the plant and equipment, along with detailed knowledge of the equipment itself. This dictates at least one operator and one maintainer. They must be hands-on and practical, willing and able to learn the RCM process, and be motivated to make it a success. The team must also be versed in plant operations, usually supplied by a senior operations person, such as a supervisor who has risen through the ranks. The team needs to know planning, scheduling, and overall maintenance operations and capabilities to ensure the tasks are worth doing in the plant. You may have to contract work to qualified service providers, particularly infrequent yet critical equipment monitoring. Someone can provide this expertise with a supervisory maintenance background.

Finally, detailed equipment design knowledge is essential. The maintainers will know how and why the equipment is put together, but they may have difficulty quantifying the reasons or fully understanding the engineering principles. A design engineer or senior technician/technologist from maintenance or production, usually with a strong background in either mechanical or electrical, is needed on the team.

RCM is very much a learning process for its practitioners. Facilitating five team members is optimum to fulfill the requirements. Too many people slow the progress. Too few means too much time will be spent understanding the systems and equipment.

The team will need help to get started. Someone in-house may already have done RCM, but you will likely have to look outside the company. Training is usually followed up with a pilot project, producing an actual asset risk/reliability assessment product.

RCM is thorough, and that means it can be time-consuming. Training the team takes approximately three days but can last as long as a month, depending on the approach. Training of other stakeholders can take from a couple of hours to three days, depending on their interest and need to know. Senior executives and plant managers should be involved to know what to expect and what support is needed. Operations and maintenance management must understand the time demands on their staff and what to expect in return. Finally, operators and maintainers must also be informed and involved. Since their co-workers are on the RCM team, they are probably shouldering additional work.

The pilot project time can vary widely, depending on the complexity of the equipment or system selected for analysis. A good rule of thumb is to allow the team a month for pilot analysis to ensure they learn RCM thoroughly and are comfortable using it. On average, each failure mode takes about half an hour to analyze. Determining the functional failure to task frequency can take six to 10 minutes per failure mode. Using our previous process plant example, a comprehensive analysis of all systems comprising at least 40,000 items (many with more than one failure mode) would entail over 20,000 person-hours (nearly ten person-years for an entire plant). When divided by five team members, the analysis could take two years. It's a big job.

8.7 IS RCM AFFORDABLE?

8.7.1 WHAT TO EXPECT TO PAY FOR TRAINING, SOFTWARE, CONSULTING SUPPORT, AND STAFF TIME

In the above example, RCM requires a lot of effort. That effort comes at a price. Ten person-years at an average of $70,000 per person adds up to $700,000 for staff time alone. The training for the team and others will require a couple of weeks from a third party at consultant rates. The consultant should also be retained for another month for the pilot project. Even though consulting rates are steep, running into thousands of dollars per day, it's worth it. The price is small considering that lives can be saved and environmental catastrophes or major production outages avoided. Experienced RCM experts have seen RCM prevent significant safety and production calamities.

Software is available to help manipulate the vast amount of analysis data you generate and record. Several databases step you through the RCM process and store results. Some software costs only a few thousand dollars for a single-user license. Some are just RCM training, while others are part of EAM/APM. Prices for these high-end systems that include RCM are typically hundreds of thousands of dollars. If you're working with RCM consultants, you'll find that many have their preferred software tools. A word of caution about RCM systems – there are many versions available. Some were created before the SAE standard was published, and some have been marketed since that event. They should be examined to see if they comply with and fit your asset management systems strategies.

To decide on task frequencies, you must know the plant's failure history, which is generally available through the maintenance management system. If not, you can obtain failure rates from the original equipment manufacturer (OEM) databases. Another option for documented failure rates is to inquire about the recent experience in your existing plant. Chances are they have a good idea of how often a failure has happened. Likely your parts keeper will know how often a part was ordered for a specific piece of equipment. You may need help to build queries and run reports. You can use external reliability databases, but they're often difficult to find and charge a user or license fee.

8.8 RCM VARIETIES

RCM comes in different varieties, depending on the application. These include:

- Aerospace (originally part of an Maintenance Steering Group (MSG) taskforce)
- Military (various for naval and combat aircraft – described in numerous US Military Standards)
- Commercial as described by John Moubray
- Streamlined versions, some of which don't meet the SAE JA1011 criteria and are no longer called RCM

There are two basic types (besides the terminology). First, structural components are thoroughly analyzed for stress, often using finite element modeling techniques.

As a result, you identify weaknesses in the airframe structure that must be regularly inspected and undergo non-destructive testing. In the second case, you follow the RCM logic questions about time-based methods, such as condition-based monitoring, once you decide on a maintenance plan. Severe safety concerns justify this conservative approach if an aircraft fails.

The military standards describe the same processes, using military equipment terminology examples from military applications.

The commercial versions are what we have described in this book.

8.9   CLASSICAL RCM, RCM2, RCM3, STREAMLINED AND ALTERNATIVE TECHNIQUES

8.9.1   CLASSICAL RCM

Classical RCM is a term that is sometimes used to describe the original process laid out by Smith and Moubray and referenced in the SAE standard. Classical RCM has proven highly successful in numerous industries, particularly at:

- reducing overall maintenance effort and costs
- improving system and equipment performance to achieve design reliability
- eliminating planned to-be-installed redundancy and reducing capital investment

Many cost and effort reductions have occurred in industries that were:

- over-maintaining (i.e., civil and military aircraft, naval ships, nuclear power plants)
- not maintained, with low reliability (i.e., thermal power plants, mining – haul trucks, water utilities)
- overly conservative in design practices (i.e., former public utilities that must now survive in deregulated environments, oil and gas/petro-chemical)

Despite these successes, many companies fail in their attempt to implement RCM. There are many reasons for failure, including:

- Lack of management support and leadership.
- Lack of vision about what RCM can accomplish (i.e., the RCM team and the rest of the plant don't know what it's for and what it will do for them).
- Lack of vision in how it will be needed to support an Industry 4.0/5.0 competitive world.
- There is no apparent reason for doing RCM (i.e., it becomes another "program of the month").
- Not enough resources to run the program, especially in "lean manufacturing."
- A clash between RCM's proactive approach and a traditional, highly reactive plant culture (i.e., RCM team members are pulled from their work to react to day-to-day crises).

- Giving up before RCM is completed.
- Lack of focus on the priority of critical equipment to be assessed.
- Continued errors in the RCM assessment process. Results that don't stand up to practical "sanity checks" by rigorous maintenance reviewers. This is often due to a lack of complete understanding of FMEA, Criticality, RCM logic, condition-based monitoring techniques, the distinction between condition-based maintenance (CBM) and time-based maintenance (TBM), and reluctance to accept run-to-failure conclusions.
- The lack of available information about the analyzed equipment/systems isn't necessarily significant but often makes people cold.
- Criticism that RCM-generated tasks seems the same as those already in prolonged use in the preventive maintenance (PM) program. It can be seen as an arduous exercise that merely proves what is already being done.
- Lack of measurable success early in the RCM program. This is usually because the team hasn't established a starting set of measures, an overall goal, and ongoing monitoring.
- Results don't happen quickly enough, even if measures are used. For example, doing the correct PM type often isn't apparent immediately. Typically, results are seen in 12 to 18 months.
- There is no compelling reason to maintain the momentum or even start the program (i.e., no legislated requirement, the plant is running well, and the company is making money despite itself).
- The program runs out of funding.
- The organization cannot implement RCM results (i.e., the belief that no system can trigger PM work orders on a pre-determined basis).

The above list is a blueprint for how to ensure failure. One criticism of RCM is that it's "the $1 million solution to the $100,000 problem." This complaint is unfounded. How you manage RCM is usually at fault, not the process itself. The reasons for the failure described above can be traced back to management and commitment flaws.

There are many solutions to the problems we've outlined. One that works well is using an outside consultant. A knowledgeable facilitator can help get you through the process and maintain momentum. Often, however, companies stop pursuing change as soon as the consultant leaves. Therefore, the best chance of success is using help from outside in the early stages of your RCM program. Successful consultants recognize the reasons for failure and avoid them.

8.9.2   RCM2

RCM2 is a process to determine what must be done to an asset system to preserve its functions (while minimizing or avoiding failure consequences). RCM2 complies fully with the minimum requirements of the SAE JA 1011 and SAE JA 1012 RCM standards.

RCM2 was the initial reliability method published by John Mobray and fully aligns with the SAE: JA1011 standard. Aladon now owns this process approach. This is no accident, as Mobray helped SAE establish its standard. The RCM2 process starts with understanding the asset's operating context. For example, SAE JA10111 mentions the operating context is important but not required. Next, this method defines vital performance standards of the assessed equipment (i.e., primary and secondary standards) and determines if the equipment is a protective device. Next, functional failures are acknowledged as "failed states," often categorized as total or partial failures. RCM2 then defines the failure mode and failure effect to a workable data detail. The failure effect is defined in a paragraph with a carefully drafted explanation of what will happen if the failure mode occurs and nothing is done to prevent it). A record of the physical effects of each failure will be documented by facilitating the answer to questions like:

- What evidence (if any) is that the failure has occurred?
- In what ways (if any) does the failure threaten the safety or the environment?
- How (if any) does the failure affect production or operations?
- What physical damage (if any) is caused by the failure?
- What must be done to repair it?

Failure consequences (safety/environmental, operational, non-operational, and a single category of hidden failure consequences) are identified. Multiple failure risk is considered in this process and calculated.

The RCM2 process treats all hidden functions the same. The decision logic considers predictive and PM tasks as proactive failure management strategies, failure finding, redesigns, and no scheduled maintenance as default actions. A combination of tasks is a default action, and consequence mitigation is achieved primarily through optimizing protective devices (protected functions). For a proactive task to be qualified as the prescribed action: the PM must be technically feasible (according to the failure characteristics) and worth doing (reducing the consequences to an acceptable level).

8.9.3   RCM3

RCM3 complies fully with the minimum requirements of the SAE JA 1011 and SAE JA 1012 Standards and goes beyond these requirements. It also aligns with ISO 55000 and ISO 31000 Management Systems.

RCM3 is a process used to determine what must be done to an asset system to preserve its functions while minimizing the risks associated with failures to a tolerable level. It further considers a probabilistic risk assessment at a component level when compulsory redesigns or one-time changes are required. Every reasonably likely failure mode is assessed and quantified in its inherent risk. Less likely failure modes are considered based on inherent risk. Aladon promoted this process approach and recommended it when clients look for an upgraded approach to risk and reliability.

RCM3 is the latest risk and reliability discipline from the original Mobray organization (now Aladon). RCM3 complies with and extends beyond the standards of SAE JA1011. RCM3 extracts more data to help the organization adjust its mitigations when and as the collected data presents itself. RCM3 supports the notion of operating context throughout its analysis as an essential insight. First, the asset's operating conditions are understood and used to build a defensible risk management program. Next, this method defines vital performance standards of the assessed equipment (i.e., primary and secondary standards) and determines if the equipment is a protective or detective device. It expands on understanding secondary functional definitions (i.e., cleanliness, regulations, regulatory requirements, recycle/repurpose/reuse). Next, "failed states" are defined and documented between a general failed state, failing state, failed state, or end state (as part of the failing state).

RCM3 defines a failure mode as a cause-and-effect mechanism that creates the failed state. It groups root causes consistently and can allow the failure mechanism to tie in with its degradation level (as in risk-based inspection – RBI). RCM3 expands the role of the failure effect by organizing the effect into three levels (local effect, subsequent higher-level effect, and end effect). Potential worse-case effect scenarios are extracted (where protection is also in a failed state – allowing for actual *zero-base* analysis). A record of the physical effects of each failure will be documented by facilitating the answer to questions like:

- When is the failure most likely to occur?
- How often would the failure occur if no attempt was made to prevent it?
- What evidence (if any) is that the failure has occurred?
- In what ways (if any) does the failure threaten the safety or the environment?
- How (if any) does the failure affect production or operations?
- What physical damage (if any) is caused by the failure?
- What must be done to repair it?
- Does it cause any secondary damage?
- What is the revenue loss (if any)?

Failure consequences are identified as evident physical and economic risks, and hidden risks are separated into two categories (hidden physical and hidden economic risks).

Note: Physical risks impact health, safety, or the environment, while economic risks impact operational capability and financial well-being.

While many RCM approaches have a risk matrix, RCM3 can specifically view the risk matrix as part of the assessment process to understand whether additional mitigation actions could take the calculated risk from a high or medium risk level to a lower risk level while working on the questions listed above.

RCM3 addresses risk using the ISO 31000 Standards of Risk Management discipline.

Focus is placed on the reliability of the protected function first. Failure-finding frequencies are optimized by increasing the protection function's reliability (when applicable) as the primary concern. As a result, dependency on protective devices is reduced. The *worth-doing* criteria for risk criteria differ significantly from the RCM2 decision logic. *Economic risks* are considered (first) and not cost only. The mitigation strategy must reduce intolerable operational risk (now quantified) to be considered. The RCM3 process leads to defensible risk mitigation.

FIGURE 8.6   Example risk matrix.

8.9.4   STREAMLINED RCM

There are now methods that shortcut the RCM process, which, where appropriate, can be effective. But ensure they're proper and responsible, especially with health or safety risks. All risks should be quantified and managed. Refer to Chapter 7 for a thorough discussion of risk management.

No matter how effective, shortcuts cannot be considered RCM unless they comply with the Society of Automotive Engineers (SAE) JA1011.

These shortcut RCM methods have become known as "Streamlined" or "Lite" RCM, among others.

In one variation, RCM logic tests an existing PM program's validity or failure modes. This approach doesn't recognize what potential failure modes were already missing from the program, which is why doing RCM in the first place. This is not RCM.

For example, suppose the current PM program extensively uses vibration and thermographic analysis but nothing else. In that case, it probably works well-identifying problems causing vibrations or heat but not failures such as cracks, fluid reduction, wear, lubricant property degradation, wear metal deposition, surface finish, and dimensional deterioration. This program does not cover all possibilities. Applying RCM logic will result, at best, in minor changes to what exists. The benefits may reduce PM effort and cost, but anything that isn't already covered or any failure mode that has not already been experienced or identified will be missed. This streamlined approach adds minimal value. In practice: it's irresponsible.

8.9.5   CRITICALITY

Asset Criticality focuses on determining which equipment to focus on first (perhaps as a pilot) or to help determine focus priorities, which work well when applied to RCM right in the beginning, as mentioned in the RCM process section of this chapter (section 8.5). However, prioritized criticality is another RCM variation that can be used to weed out failure modes from ever being analyzed. This must be done carefully. One approach is described thoroughly in MIL-STD 1629A, but several techniques exist. Failures are not analyzed if their effects are considered non-critical, occur in non-critical parts or equipment, or don't exceed the set criticality hurdle rate. Reducing the need for analysis can produce substantial savings.

When criticality is applied to weed out failure modes, there should be little risk of causing a critical problem. Because classical RCM analysis has already been performed, using criticality is relatively risk-free. The drawback is the effort and cost expended, deciding to do nothing. In the end, there are virtually no savings.

You can cut costs in both areas by reducing RCM analysis before most of it is done. Using a criticality hurdle rate indicates many possible equipment failures without documentation, which can mean significant maintenance savings. Even when you look at worst-case scenarios, the downside is that you risk a critical failure will slip through unnoticed. How well this technique works depends on how much the RCM team knows. The greater their plant knowledge and experience, the less risk. The experienced and motivated team members are often the plant's best maintainers and operators.

This approach may be the right choice if you confidently know and accept the consequences of failure in production, maintenance, cost, environmental, and human terms. For example, many failures in light manufacturing have relatively little fallout other than lost production time. But in other industries, you could be sued if a failure could have been prevented or was mitigated through RCM. The nuclear power, chemical processing, pharmaceutical, and aircraft industries especially are vulnerable. SAE JA1011 stipulates that the method used to identify failure modes must show what is likely to occur. Of course, the level at which failure modes are identified must be acceptable to the owner or user. There is room for judgment, and if done correctly, this method can meet the SAE criteria that define RCM.

Criticality helps you prioritize so that the most critical items are addressed first. It cuts the RCM workload that typically comes with large volumes, systems and equipment, and limited resources to analyze them.

Many companies suffer from the failures we've described and others. Without the force of law, though, RCM standards such as SAE JA-1011 are often treated as mere guidelines that don't have to be followed. Sometimes, the people making the decisions aren't familiar with RCM and its benefits. So, what do you do if you know that your plant could suffer a failure that can be prevented? You must responsibly and reasonably do whatever it takes to avert a potentially serious situation. Recognize the doors open to you and use them to get started.

Similarly, if you foresee that an RCM implementation will likely fail, you must eliminate it. Even if this doesn't always seem practical or easy, consider the consequences of not taking action:

Hypothetically, if a company ignores known failures and does nothing, it could be sued, get a lot of negative publicity, and suffer heavy financial loss. The gas tank problem in the early model Ford Pinto, dramatized in the 1991 movie *Class Action*, starring Gene Hackman, is an example where the court-assigned significant damages. In November 1996, a New Jersey court certified a similar nationwide class-action lawsuit against General Motors due to rear brake corrosion.

In another hypothetical situation, if you acknowledge a potential problem but discount it as negligible and then experience failure, you could be blamed for ignoring what was recognized as a risk.

Risk can be reduced but never eliminated entirely, but it can be lessened. Even if you can't fully implement RCM, take some positive action and reduce risk as much as possible. Simply reviewing an existing PM program using RCM logic will accomplish very little. Analyzing critical equipment, you'll gain more where it counts most. If you follow that up by moving down the criticality scale, you'll gain even more and eventually complete RCM.

If performing RCM is too much for your company, consider an alternative approach that you can achieve. You'll at least reduce the risk somewhat. If you do nothing, you could be branded as irresponsible later, which would be deadly for your business and professional reputation.

Your ultimate challenge is convincing decision-makers that RCM is their best course. You need to build credibility by demonstrating that RCM works. Often, however, maintenance practitioners have relatively little influence and control. A maintenance superintendent may be encouraged to be proactive if they don't ask the operators or production staff for help or need upper management approval for additional funding. Sound familiar?

RCM needs operator and production help, though, to succeed. The best attempts to implement an RCM program can flounder without this support.

What can a maintainer realistically do to demonstrate success and increase influence? Realize, first of all, that many companies do value at least some degree of proactive maintenance. Even the most reactive may do some PM. They know that an ounce of prevention is worth a pound of cure, even if they're not using it to their best advantage. So the climate may already exist for you to present your case.

The cost of using RCM logic can be a stumbling block. The bulk of the work – identifying failure modes – is where most of the analysis money is spent. Reduce the cost, and you'll generate more interest.

8.9.6　Capability Driven RCM

Even under severe spending constraints, you can still improve by being proactive. By using what's known as Capability Driven RCM, or CD-RCM, you:

- Reverse the logic of the RCM process, starting with the solutions (finite numbers) and looking for appropriate places to apply them. Since RCM progresses from equipment to failure modes, the opposite process can pinpoint failures through decision-making to a result, even if they aren't identified.
- Extend existing condition monitoring techniques to other pieces of equipment. For instance, vibration analysis on some equipment can also work elsewhere.
- Look specifically for wear-out failures and do time-based replacements.
- Check standby equipment to ensure that it works when needed.

These are examples of proactive maintenance that can make a huge difference. It's crucial, though, to do root cause failure analysis when a failure does occur so that you can take preventive action for the future. Among the benefits of using CD-RCM, it can be a means of building up to full RCM.

There are some risks involved in CD-RCM, though. Some items may be over-maintained, especially in cases where run-to-failure has previously been acceptable. Over-maintenance could cause failures if it disrupts production and equipment in critical operations. The risk, though, is relatively small. The bigger problem may be the cost of using maintenance resources that aren't needed.

There is also a risk of missing some failure modes, maintenance actions, and redesign opportunities that could have been predicted or prevented if the upfront analysis had been done. While the consequences could be significant, the risk is usually minimal if the techniques are broad enough. For instance, Nowlan and Heaps found that condition monitoring is effective for airlines because 89% of aircraft failures in their study were not time-related.

CD-RCM could be used similarly with other complex electro-mechanical systems using complicated controls and many moving parts. For instance, looking for wear-out failure modes in industrial plants is like the traditional approach to maintenance. Most failures are influenced by operating time or some other measure. Fewer items will be examined by searching for failure conditions only where parts and the process materials are in contact with each other. The potential failures are generally evident and easy to spot. This approach makes it possible to over-maintain, especially if the equipment you decide to perform TBM on has many other random failure modes.

In RCM, failure finding is used for hidden failures that either are not detected using CBM or avoided using TBM. One favored method is to run items that are usually "normally off" to test their operation. This is done under controlled conditions to detect a failure without significant problems. For example, there may be a failure risk on the start-up of a system or equipment, but there is also control over the consequences because the check is done when the item isn't needed. Correcting the failure reduces its consequences during "normal" operation when it would be needed if the primary equipment or device failed. Doing failure-finding tasks without knowing what you're looking for may seem foolish, but it's not. Failures often become evident when the item is operated outside its normal mode (often "off"). Although all "hidden failures" may not be found, many will.

Again, this is not as thorough as a complete RCM analysis, but it's a start in the right direction. By showing successful results, the maintainer may be able to extend his proactive approach to include RCM analysis. CD-RCM is not intended to avoid or shortcut RCM. It is a preliminary step that provides positive results consistent with RCM and its objectives.

To be successful, CD-RCM must:

- Ensure that the PM work order system works (i.e., PM work orders can be triggered automatically, and the work orders get issued and carried out as scheduled). If this is not in place, help is needed beyond the scope of this chapter.
- Identify the equipment/asset inventory (this is part of the first step in RCM).
- Identify the available conditioning monitoring techniques that may be used (which are probably limited by plant capabilities).
- Determine the kinds of failures that each of these techniques can reveal.
- Identify the equipment where these failures dominate.

- Decide how often to monitor and make the process part of the PM work order system.
- Identify which equipment has dominant wear-out failure modes.
- Schedule regular replacement of wearing components and others that are disturbed.
- Identify all standby equipment and safety systems (alarms, shutdown systems, redundant standby equipment, backup systems, etc.), normally inactive but needed in particular circumstances.
- Determine appropriate tests to reveal failures that can only be detected when the equipment runs. The tests implement in the PM work order system.
- Examine failures experienced after the maintenance program is implemented to determine their root cause so appropriate action may be taken to eliminate them or their consequences.

The result of using CD-RCM can be:

- CBM techniques like vibration analysis, lubricant/oil analysis, thermographic analysis, visual inspections, and non-destructive testing are extensive.
- Limited use of time-based replacements and overhauls.
- In plants where redundancy is common, extensive "swinging" of operating equipment from A to B and back, possibly combined with equalization of running hours.
- Extensive testing of safety systems.
- Systematically capturing and analyzing information about failures to determine the causes and eliminate them in the future.

All of these actions will move the organization to be more proactive. As CD-RCM targets proven methods that make sense, it builds credibility and enhances the likelihood of implementing full-blown RCM.

8.10 HOW TO DECIDE?

8.10.1 SUMMARY OF CONSIDERATIONS AND TRADE-OFFS

RCM is a lot of work. It is also expensive and worth the investment! The results, although impressive, can take time to accomplish. One challenge of promoting a

		Risk rating				
		Insignificant	Minor	Medium	High	Major
Probability	Certain	11	16			
	Likely	7	12			
	Possible	4	8	13		
	Unlikely	2	5	9	14	
	Rare	1	3	6	10	15

FIGURE 8.7   It can feel like a lot of work.

full-blown RCM program is justifying the cost without being able to show what will be the tangible savings.

The cost of not using RCM, however, may be much higher. Some alternatives to RCM are less rigorous and downright dangerous. RCM is a thorough and complete approach to proactive maintenance that achieves high system reliability. It addresses safety, and environmental concerns, identify hidden failures, and appropriate failure-finding tasks or checks. It identifies where the redesign is appropriate and whether run-to-failure is acceptable or desirable.

Simply reviewing an existing PM program with an RCM approach is not an option for a responsible manager. Too much can be missed that may be critical, including safety or environmental concerns. It may be the start of a reliability program for someone newly assigned to the task, but it is not RCM.

Streamlined or "Lite" RCM may be appropriate for industrial environments where prioritizing criticality is an issue. RCM results can be achieved on a smaller but well-targeted subset of the critical equipment and systems failure modes. While this is a form of RCM, care must be taken to ensure it meets the SAE criteria for RCM.

If RCM investment is not immediately feasible, the final alternative is to build up to it using CD-RCM. This adds a bit of logic to the old approach of applying new technology everywhere. In CD-RCM, you take stock of what you can do now and use it as widely as possible. Then, once success is demonstrated, you can expand upon the program. Eventually, RCM can be used to make the program complete.

8.10.2 RCM Decision Checklist

Throughout this chapter, you have had to consider specific questions and evaluate alternatives to determine if RCM is needed. We summarized them here for quick reference.

1. Can the plant or operation sell everything it can produce? If the answer is "Yes," high reliability is essential, and RCM should be considered. Skip to question 5. If the answer is "No," focus on cost-cutting measures.
2. Does the plant experience unacceptable safety or environmental perfor-mance? If "Yes," RCM is probably needed. Skip to question 5.
3. Is there already an extensive proactive or PM program in place? If the answer is "Yes," consider RCM if the program costs are unacceptably high. If "No," consider RCM if maintenance costs are high compared with others in the same business.
4. Are maintenance costs high relative to others in your business? If "Yes," RCM is right. Proceed to question 5. If not, RCM won't help. Stop here. Consider root cause failure analysis as your living program. Confirm that all assets have been considered in prioritization. At this point, one or several of the following apply:
   - a need for high reliability
   - safety or environmental problems
   - an expensive and low-performing PM program
   - no significant PM program and high overall maintenance costs

5. RCM is right. Next, you need to ensure that the organization is ready for it. Is there a "controlled" maintenance environment where most work is predictable and planned? Like PM and PdM, does planned work generally get done when scheduled? If "Yes," the organization passes this basic readiness test, the maintenance environment is under control. Proceed to question 6. RCM won't work well if the decided tasks are not applied in a controlled environment. If this is the case, RCM alone isn't enough. Get the maintenance activities under control first. Stop here.

6. RCM is needed, and the organization is ready for it. But senior management support is still likely required for the investment of time and cost in RCM training, piloting, and rollout. If that is not forthcoming, consider the alternatives to full RCM. Proceed to question 7.

7. Can senior management support be obtained for investing time and cost in RCM "Lite" training and piloting? This investment will require about one month of team time (5 people) plus a consultant. If "Yes," consider RCM "Lite" to demonstrate success before attempting to roll RCM out across the entire organization.

8. If "No," you must prove credibility to senior management with a less thorough approach that requires little upfront investment and uses existing capabilities. The remaining alternative here is CD-RCM and a gradual build-up of success and credibility to expand on it.

8.11   RCM AND EMBRACING THE TECHNOLOGY DISRUPTION

The key to embracing the expected continuous technology evolution and disruption is accepting the influences of change and resetting elements your organization may have allowed to be abandoned.

Huddle and counsel with your stakeholders and determine asset management priorities in adopting your forecasted operating models and adapting with change enablement.

- Get your organization's priorities right!
- Get your leaders aligned.
  - Include your staff to understand the priorities
- Understand where you are and what the journey should include
  - Near-term and longer-term
- Get IT, OT, and the Maintenance team aligned to support each other
  - It takes a community
- Get started!

An influential and empowering culture is a surer route to organizational success.

The risk and reliability culture health check is an excellent organizational middle checkpoint. If you are fully engaged in a reliability culture and embracing the insights and data points that can be collected, you are likely en route to becoming a leading practice. If you have not started a reliability discipline or culture or did start but abandoned the initiative, this is a great place to challenge the reasons.

Many organizations may declare they have a reliability culture. In one example, it was suggested that they had foundational reliability habits such as:

- Create an asset baseline health metric
- Identify "bad actor" assets and address
- Create "defect elimination" teams
- Perform "lessons learned" on recent urgent/critical outages
- Challenge current tactics and maintenance plans
- Review backlogs
- Encourage empowered total productive maintenance (TPM) activities
- Challenge missed scheduled activities

They made some of the listed activities above in one location but were inconsistent. This is not a leading reliability program. A leading reliability set of disciplines would include the following process and disciplines that were proactive RCM activities:

- Identify critical assets for a risk and reliability assessment
- Create reliability teams with maintenance, operations, and design engineering
- Calibrate the standards (through training) for the reliability teams of RCM methods introduction
- Create RCM facilitators in your company
- Pilot a few reliability analyses (with FMEA outputs)
- Create the knowledge and related data to identify the best tactic for each asset's operating context
- Create the thresholds for flagging actions
- Document the reliability data and tactic "plans" in your EAM solution
- Get management commitment to the new plans

This would prove that an organization sought to bridge the foundational approach to asset reliability and RCM. Unfortunately, this is also where many leading organizations stall as they identify some assets and implement them to achieve good results but feel the effort was too resource-intense to justify the asset-specific initiative. The key here is to recognize the organization's journey as they move into Industry 5.0 and the convergence of IT and operations technology (OT) data. The data elements of an RCM analysis are broad, dependable, and, most notably, preparing the organization for its future. Examples of the type of data that should be collected are described earlier in Table 8.1.

The RCM data outcomes are five options:

- **Predictive Maintenance (PdM)** – condition monitoring
- **Preventive Maintenance (PM)** – age or usage-based restoration or replacements
- **Failure-Finding Tasks (FF)** – periodic checks to see if typically "dormant" devices are still functional
- **No Scheduled Maintenance (NSM)** – run the asset to failure if consequences (risks, costs, customer disruption) are more acceptable than being proactive (i.e., it costs less and has an acceptable reliability impact)
- **One-time changes** – design, procedural, or training outputs that generally avoid the failures altogether or manage the consequences better than maintenance

An IT/IoT fully enabled risk and reliability culture would take you to the ultimate level of the asset risk and reliability culture. Not starting or stopping before you collect and enable the data in your EAM/APM solution does not send your organization on a leading practice pathway.

8.11.1 RECOGNIZING WHERE YOU ARE IN THE RISK AND RELIABILITY CONTINUUM

Many organizations do not see the need for a risk and reliability program. As suggested earlier, RCM is viewed to be a lot of work. With the evolving technology and the tools continuing to enable IT and OT convergence and integration throughout the organization's value chain: ignoring asset reliability while technology enablers are being used in your competitive market will be a going-out-of-business strategy. You cannot ignore your market's progress while you accept doing little to nothing. The evolution of accepting a risk and reliability culture could be described as something like this:

- Minimal Risk and Reliability Culture
- Foundational Reliability Culture
- Reliability-centered Maintenance Culture
- IT and IoT-enabled Reliability Culture

If you are not doing anything, you will fail. If you believe you have a foundational reliability culture, you are not doing enough and will fail in a later cycle. If you have a fully embraced RCM culture, you need to ensure you cover all the critical assets and plan to fully leverage all the data collected (e.g., Table 8.1) With the world moving to Industry 5.0, you must drive to an IT and IoT-enabled reliability culture to expect to compete in your organization's future.

Table 8.2 could help an organization recognize its current working level.

A fully implemented RCM data technology leveraged by a fully supported reliability culture will efficiently provide your data scientists with a curated understanding of the following:

- Asset operating contexts
- FMEA data
- Mitigating tactics and thresholds
- How to automate decisions from Reliability data insights

FIGURE 8.8 Technology, data, and a reliability culture continuum.[1]

TABLE 8.2

Observed Elements of Each Risk and Reliability Progression[1]

Minimal Reliability Culture	Foundational Reliability Culture	RCM Culture	RCM Data Technology Leveraged the Reliability Culture
• Manual • Maintenance execution efficiency focus • No formal program	• Asset baseline health metric • "Bad actor" assets identified • Defect elimination teams • Outage reviews • Backlog reviews • TPM culture activities • Schedule compliance	• Identify/prioritize critical assets and create/train Reliability Teams • Include Maintenance, Operations, and Design Engineering • Calibrate RCM methods introduction • Create RCM Facilitators and Pilot Reliability analysis • Document the asset's operating context • Create insights to prescribe the best mitigation tactic • Create the thresholds for flagging actions • Document data and tactic "plans" in your EAM solution • Get Management commitment	• An "IT Integrated" Reliability culture • Integrated and Automated • Asset Health, Transaction and Reliability threshold data • Market and external data • Information Technology (IT) and Artificial Intelligence (AI) to automate a first-pass RCM analysis

8.11.2 DATA IS THE FOUNDATION FOR ACHIEVING A LEADING PRACTICE

Data is the foundation for achieving a leading practice. This is particularly true when collecting the benefits from a risk and reliability initiative. The benefits declared are shown in Table 8.3:

TABLE 8.3

Benefits of a Successful RCM Legacy[1]

Improved Productivity	Increased Maintenance Effectiveness	Reduces Costs
• Optimizing ROA • Reduces operating costs • Improves sustainable reliability • Improves customer satisfaction • Improves the "asset value" proposition	• Enhances compliance activities • Reduces business interruptions • Increases asset utilization • Increases uptime • Increases staff commitment and understanding	• Manages lifecycle costs • Optimizes maintenance and inspection intervals • Prevents unwanted events • Creates fact/data-driven decisions • Empowers reliability teams driving "asset value"

Imagine the benefits when you have a fully engaged and implemented risk and reliability culture with the data leveraged in your integrated EAM/APM solution and the operations production solution!

Your organization would be able to harvest the benefits such as listed above plus recognize the following:

- Asset health, transaction, and reliability threshold data
- Market and external data
- IT and AI to automate a first-pass RCM analysis, and
- An "IT integrated" reliability culture

Plus, you would be able to engage your data scientist and the analyst community with the reliability curated data better understand:

- Asset operating contexts
- FMEA data
- Mitigating tactics and thresholds
- How to automate decisions from reliability data insights

While the priority is to keep assets running that contribute to how your organization creates value. This reliability culture, leveraged with the help of IT, will drive asset and production returns. In addition, it will reinforce the priority to help make the maintenance technician environment efficient, effective, and safe – it is not removing crafts altogether.

8.12 MAKING RCM A LIVING PROGRAM

As indicated earlier in this chapter, doing an early prioritization of your assets and piloting RCM will get things started. Working through your prioritized assets with trained subject matter experts will allow you to experience the benefits of determining the most effective maintenance.

Introducing new assets or technologies within your operation would be very well served by performing an RCM analysis as part of an asset's lifecycle design phase. Leveraging this process at this early design cycle stage would:

- verify design integrity;
- identify any required modifications;
- identify risks;
- develop failure management policies and maintenance tasks;
- identify required spares to support the maintenance tasks; and
- develop operating strategies.

One of the biggest reasons RCM is perceived to have failed is that little attention was given to implementing the tasks identified after the hard part of deciding on the most appropriate task and frequency, given a specific failure mode and asset operating context. Many organizations keep their RCM data separate from their EAM, APM system, or workflow processes. If an RCM panel determines that TBM is the best action for a specific failure mode and asset operating context, postponing or canceling that PM should not be easy. As with many projects, they should be started with the end in mind. An RCM analysis will help you pick the most effective maintenance. The next logical step is documenting these tasks in your process or system. In other words, RCM helps you pick the most effective thing to do in a given failure mode, and a well-set-up EAM/APM helps you execute the effective task efficiently. Leveraging the RCM data in an automated environment can contribute to operational excellence.

Software companies now exist to help us take the RCM decisions made and integrate them directly with an EAM/APM. Many EAM/APM solutions today can directly accept the RCM data (functional failure, failure mode, and associated task and resource requirements) into their aligned data fields to implement the selected tasks efficiently. For example, processes can be set up so that should a task be canceled by an operator or maintenance supervisor, the maintenance planner would have immediate access to review the original RCM assigned task and frequency and determine the risk of accepting a maintenance postponement or cancelation.

An organization that has completed a full RCM analysis for its critical assets and set up the assigned tasks to be automated in its EAM/APM would be considered a leading maintenance organization. Suppose their entire maintenance program is well defined through RCM, assigned tasks are set up in their EAM/APM, and their business performance is better than desired. Then, they likely do not require further RCM work except when considering new assets or a new operating context.

Low-priority assets that do not directly contribute to a company's business goals may not be deemed necessary for a full RCM assessment. In these cases, perhaps a root cause failure analysis is all that is required to keep their reliability program living. However, suppose an early assessment indicates that safety, environmental, or operational impact is at risk. In that case, a leading company should assemble a trained RCM group of experts to assess these assets and automate the tasks and resources into the workflow managed by their EAM/APM.

PERMISSIONS

* Page 189–218, *Asset Management Excellence: Optimizing Equipment Life-Cycle Decisions*, 2nd Edition by Editor(s), John D. Campbell, Andrew K. S. Jardine, Joel McGlynn Copyright (2011) by Imprint. Reproduced by permission of Taylor & Francis Group.

1 Adapted from copyright *IT and IoT in Asset Management Training* (2019) edited by (Barry). Reproduced by permission of Asset Acumen Consulting Inc.

BIBLIOGRAPHY

Aberdeen Group, December 2006, "Collaborative Asset Maintenance Strategies – Redefining the Roles of Product Manufacturers and Operators in the Service Chain."

MIL-STD 1629A, Notice 2. Procedures for Performing a Failure Mode, Effects and Criticality Analysis, Washington, DC, Department of Defense, 1984.

MIL-STD 2173 (AS), Reliability-Centered Maintenance Requirements for Naval Aircraft, Weapons Systems, and Support Equipment, US Naval Air Systems Command; NAVAIR 00-25403, Guidelines for the Naval Aviation Reliability Centered Maintenance Process, US Naval Air Systems Command; S9081AB-GIE-O1*O/MAINT*-Reliability-Centered Maintenance Handbook, US Naval Sea Systems Command.

John Moubray, *Reliability-Centered Maintenance II*, 2nd ed., Oxford, Butterworth-Heinemann, 1997. www.aladon.com, January 2022.

MSG-3. Maintenance Program Development Document, Revision 2. Washington, DC, Air Transport Association, 1993.

F. S. Nowlan, H. Heap, Reliability-Centered Maintenance. Report *ADI*A066-579, National Technical Information Service, December 19, 1978.

SAE JA1011, Evaluation Criteria for Reliability-Centered Maintenance (RCM) Processes, Society of Automotive Engineers, August 1999.

Anthony M. Smith, *Reliability Centered Maintenance*, New York, McGraw-Hill, 1993.

9 Reliability by Operator*
Total Productive Maintenance (TPM)

Don M. Barry
Previous contributions from Doug Stretton and
Patrice Catoir

Total Productive Maintenance (TPM) is an essential cross-functional working culture focusing on asset productivity while contributing to operational excellence. TPM is a compelling working philosophy and part of an organization's culture for managing maintenance, operations, and engineering in a plant environment. It harnesses the power of the entire workforce to increase the productivity of the company's physical assets, optimizing man/machine interaction. It is an internal continuous improvement process that meets increasingly challenging market demands and provides mass customization for individual enterprises. A company can achieve this with a highly flexible value production process and workforce.

This chapter will expose the fundamental functions of TPM, what they mean, and how they are used and integrated into a comprehensive program. When wholly implemented, TPM becomes more than a program to run the plant or operation – it becomes part of the operating culture.

We also explore the implementation issues you can expect and compare a TPM approach with typical legacy environments. This will dispel some of the many myths about TPM.

Finally, we link TPM to other optimizing methodologies discussed elsewhere in this book. With their combined effect, you can implement a proper continuous improvement environment.

In some maintenance management cultures worldwide, TPM and a focus on overall equipment effectiveness (OOE) have facilitated a simplified "Quality" mindset approach to maintenance. This approach promotes maintenance and operations working together to drive an optimized Return on Asset (ROA) and optimized set of resources.

9.1 INTRODUCTION: WHAT IS TPM?

TPM is a company-wide program for improving equipment effectiveness. However, improving equipment effectiveness is something maintenance alone cannot do.

Perhaps the simplest way to describe TPM is to suggest that it is an extension of Total Quality Management (TQM) but focuses on asset maintenance that requires

DOI: 10.1201/9781032679600-11

cross-functional departments to work together for a common outcome. The earliest version of TQM and TPM seems to have become popularized in Japan during the post-war period (the 1950s forward).

TPM promotes the operator's activities to conduct maintenance activities they can safely perform, including cleaning, lubricating, retightening, and inspection, thereby raising production efficiency. Such activities will contribute to the early detection of natural equipment deterioration through operator inspection actions, fulfilling the early warning awareness requirements of a potential asset failed state and providing an opportunity for the operator to provide an equipment restoration action if practical. In the case of TPM, the expected outcome could be optimizing the exploitation of existing assets and resources for the organization's benefit.

Seiichi Nakajima, an engineer from Japan, is credited with coining the term TPM. He was credited with pulling together the elements of TPM into a highly effective process. Many of the elements of TPM, as we know it today, were merged into TPM from US origins.

TPM is the most recognized continuous improvement philosophy, but it's also misunderstood. TPM can change your organization and radically boost overall production performance.

Some claim that TPM reflects a specific culture and isn't applicable everywhere. That's been proven wrong countless times. Others maintain that TPM is just common sense, but many people with common sense haven't been successful using TPM. TPM is much more than this.

It has been suggested that TPM is thriving if it is evident in a single location of a multi-campus, multi-country enterprise. While it is partially exploited, TPM is not thriving if only evident in a few locations. A thriving TPM program would be a leadership-sponsored, fully executed culture across the enterprise.

The objectives of TPM are to optimize the relationship between human/machine systems and the quality of the working environment. What confuses skeptics is TPM's approach to eliminating the root causes of waste in these areas.

TPM recognizes that engineering, operations, and maintenance roles are interdependent. It uses its combined skills to restore deteriorating equipment, maintain essential equipment and operating standards, improve design weaknesses, and prevent human errors. The old paradigm of "I break, you fix" is replaced with "Together we succeed." This is a radical change for many manufacturing and process organizations.

TPM is more about changing your workplace culture than adopting new maintenance techniques. For this reason, it can be agonizingly difficult to implement TPM, even though its concepts seem so simple. As a result, the published TPM methodologies are associated with implementation techniques. While technical change is rapid, social change takes time and perseverance. Most cultures need an external stimulus. Modern manufacturing philosophies around process improvement, specifically Just-In-Time (JIT) and Total Quality Management (TQM), are market-driven – they force an organization to make a cultural change. TPM has grown with the need for flexible manufacturing to produce a range of products to meet highly variable customer demands. Once TPM is in place, it continues to develop and grow, promoting continuous improvement. A TPM organization drives change internally.

9.2   BENEFITS OF A TPM CULTURE

TPM has been credited in many companies for minimizing their "loss impacts" from the targeted optimal asset utilization and value production capacity. Loss categories can be assigned with six primary types recognized as:

- Equipment breakdowns;
- Operational setup and equipment adjustments;
- Asset availability;
- Operational stoppages;
- Operational speed degradation;
- Defects and rework; and
- Start-up losses.

The recognized production and equipment benefits can include the following:

In companies that have developed a thorough understanding of TPM, it can stand for "Total Productive Manufacturing." This would recognize that TPM encompasses more than maintenance concerns, with the common goal of eliminating all waste in manufacturing processes. TPM creates an orderly environment where routines and standards are methodically applied. Combining teamwork, individual participation, and problem-solving tools maximize your equipment use.

What do you need to develop a TPM culture? Besides TPM tools, it requires production work methods, production involvement in minor maintenance activities, and

TABLE 9.1
Benefits of a Thriving TPM Culture

Tangible Benefits	In-tangible Benefits
• Reduced process defects, interruptions, and waste	• Higher levels of employee empowerment, engagement, and satisfaction
• Energy conservation	• Reduced safety issues
• Increased asset productivity	• Improved employee "ownership" of their
• Reduced maintenance and production costs	equipment production results
• Improved early warning of maintenance issues	• Employee participation in business results and feeling of contribution
• Engaged operators in setups and asset adjustments	• Diverse job skill development and application
	• Improved sustainment of a clean, structured,
• Improved understanding of equipment issues and mitigation	and attractive workplace
	• Team collaboration and ownership in
• Improved standards for job safety and asset reliability	business results
	• Improved employee understanding through
• Higher customer satisfaction (due to met delivery commitments)	training and skills development
	• The operator can manage many maintenance
• Increase equipment capacity	measures directly

A member survey by the Japan Automotive Manufacturers Association 1980 found an average Overall Equipment Effectiveness (OEE) of 43%. Five years later, they found that their members had an average OEE of 85% if they had adopted TPM methodologies.

teaming production and maintenance workers. Many believe the operator becomes crucial to equipment reliability rather than a significant impediment.

This concept must be taught, accepted, and applied at all levels of your organization, starting from the bottom up and nurtured by top management. The result is an organization committed to continuously improving its working environment and human/equipment interface.

One way to check if a TPM culture is in place is to observe the level at which it is recognized that maintenance and operations staff habitually support each other.

Evidence of a TPM culture being in place can include the following:

- Leaders regularly discuss performance and costs with their work team;
- Weekly and monthly operational (production) reviews are facilitated with both maintenance and operational personnel actively contributing;
- Participative organization where decisions are made at the lowest effective level;
- Empowered and self-directed Operator, Maintenance, and Engineering Work Teams perform much of the work;
- Operators perform minor maintenance activities;
  - Multi-skilled tradespeople are a feature of the organization: i.e., Electricians doing minor mechanical work, mechanics doing minor electrical work;
- Partnerships established with selected suppliers and contractors; and
- Continuous improvement teams are in place and active.

9.3   WHAT ARE THE FUNDAMENTALS OF TPM?

TPM has eight fundamental functions:

- Autonomous maintenance
- Equipment improvement
- Focused process improvement
- Education and training
- Quality maintenance
- Asset lifecycle management
- Health, safety, and environment
- TPM program management

9.3.1   AUTONOMOUS MAINTENANCE

Many people confuse autonomous maintenance (AM) with TPM, but AM is only one part of TPM, though a fundamental one. The confusion arises because, during a TPM implementation, AM directly affects most people.

AM is owned by the team that uses the equipment daily. It gets production workers involved in equipment care, working with maintenance to stabilize conditions and stop accelerated deterioration.

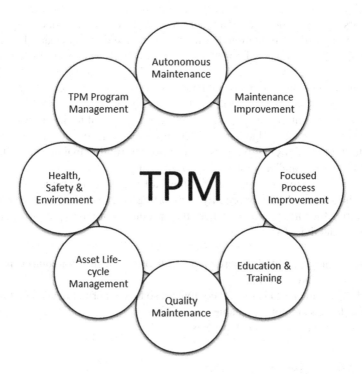

FIGURE 9.1 TPM functions summary.[1]

AM fundamentally:

- Establishes and maintains primary equipment conditions by eliminating the equipment deterioration causes and contamination sources and restoring equipment to a "like-new" operating functionality. This would include introducing operator/maintenance inspection, cleaning, lubrication, and tightening when needed or observed through an operator inspection cycle.
- Training operator/maintenance teams in the detailed operating principles of the equipment and then improving the standard essential conditions.
- Promoting operator ownership of the equipment as self-directed teams, continuously improving equipment condition and performance to reduce losses further.

Operators learn about equipment function and failures, including prevention through early detection and treating abnormal conditions. This can create conflict because of past work rules. For AM to succeed, operators must see improvements, strong leadership, and control elements delivering satisfactory service levels. Often, AM's impact on maintenance is overlooked. It helps your staff support the operators, make improvements, and solve problems. More time is spent on maintenance diagnostics, prevention, and complex issues. Operators perform routine equipment inspections and CLAIR (Cleaning, Lubrication, Adjustment, Inspection, and minor Repair) maintenance tasks critical to equipment performance.

The first AM task involves equipment maintainers and operators completing an initial cleaning and rehabilitation. During this time, the operators learn the details of their equipment and identify improvement opportunities. They learn that *cleaning is inspection*. Regular cleaning exposes hidden defects that affect equipment performance. Inspection routines and equipment standards are established. The net effect is that the operator becomes an expert on their equipment.

This development of the *operator as an expert* is critical to the success of TPM. An expert operator can judge abnormal from normal machine conditions and communicate the problem effectively while performing routine maintenance. It is precisely this "expert care" that maximizes equipment.

The term "Autonomous" implies that the staff may be empowered to make decisions at the source of the opportunity. Supporting an empowered culture promotes opportunities for further benefits where they promote optimizing the working environment productivity by:

- Developing an attitude where everyone feels responsible for equipment performance and condition
- Changing the attitude toward equipment so that abnormalities, breakdowns, and defects are not acceptable
- Emphasizing quality in all aspects of work

9.3.2 Maintenance Improvement

Typically managed by the maintenance team – maintenance improvement focuses on:

- Prioritizing equipment involves evaluating the current maintenance operating context, performance, and costs to set the focus for the area activity.
  - Support is provided to the AM area to establish a sustainable standard – essentially, the team focuses on preventing and eliminating the influences of equipment breakdowns.
- Enterprise Asset Management (EAM) systems provide detailed data on the maintenance processes and the use of spares. The team notionally identifies the best tactic for maintaining the equipment, starting with a preventive maintenance (PM or time-based maintenance) approach before considering a predictive maintenance (PdM or condition-based maintenance) approach where they are appropriate and cost-effective.
- Continuous improvement is a cultural staple where a tenacious team mindset drives to eliminate equipment outages and contribute to asset reliability.

An essential function of TPM is to focus the organization on a common goal. Since people behave how they are measured, developing a comprehensive performance measure for all employees is critical. Many programs have selected overall equipment effectiveness (OEE) as the key metric for TPM performance.

OEE combines equipment, process, material, and people concerns and helps identify where the most waste occurs. It focuses on maintenance, engineering, and production on the vital issue of plant output.

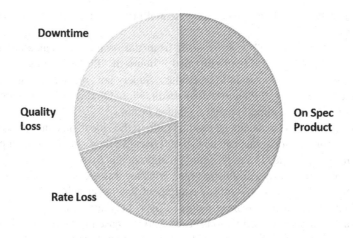

FIGURE 9.2 OEE- time allocation of activities.

Using OEE, you will better identify whether your operation produces a quality product. An operation consistently produces an on-spec product, or it is not. OEE forces the organization to address all the reasons for lost production, turning losses into opportunities for improvement. OEE is the measurement of all equipment activities in a given period. At any one time, equipment will always perform one of the following: On-Spec Product, Downtime, Quality Loss, or Rate Loss.

The size of the pie in Figure 9.2 is the amount of product produced at the ideal rate for a period of calendar time. This is the ideal state where the organization's efforts produce an on-spec product.

Often, an organization removes calendar time from the OEE calculation. For example, suppose your plant operates five days a week. You may want to eliminate the downtime caused by not operating on Saturday and Sunday, but remember that this OEE calculation is a subset of the OEE for the plant. In Figure 9.2, a decision has been made to reduce the size of the pie.

The following formula calculates OEE:

$$\text{OEE} = \text{Availability\%} \times \text{Production Rate\%} \times \text{Quality Rate\%}$$

Where
Availability = Production Time/Total Time
Production Rate = Actual Production Rate/Ideal Production Rate
Quality Rate = Actual On-Spec Production/Actual Total Production

Availability is simply the ratio of Production Time to Calendar Time. Measuring the downtime and performing some simple mathematics to arrive at the production time is more convenient. Downtime is any time the equipment is not producing. Equipment, systems, or plants may be shut down but available for production.

The downtime could be unrelated to the equipment, caused – for example – by a lack of raw materials to process. Count all downtime, including any you have scheduled.

Excluding some downtime violates the essential principle that no one should "play" with the numbers.

"Planned Maintenance" was excluded from the Availability calculation at one plant. For example, if an 8-hour shift had 2 hours of "Planned Maintenance," the Total time for that shift was said to be 6 hours. Supervisors reacted by calling almost any maintenance activity "Planned." As a result, while OEE went up, total output did not. Downtime is downtime, typically planned or not.

The production rate is the actual product ratio to the "ideal" instantaneous production rate. Setting the ideal rate can be difficult because there are different approaches to determining its value. The most common approach is to use either a rate shown to be achievable or the design production rate, whichever is greater. The design rate is a good target if your plant, system, or equipment has never achieved it – typical in newer plants. If the plant has been modified over the years and production capacity has increased, or you've used various "tricks" to increase production levels successfully, the demonstrated maximum production rate is applicable. In many cases, this corresponds to an upper control limit.

By looking at the production speed indicators, you can measure the production rate for continuous operation "on the fly." Batch processes, however, can be challenging to measure. In a batch process, cycle time, or average rate based on production time output, is used to measure production rate. Quality Rate is the "on-spec" product ratio to the actual production rate. "On spec" means producing what is needed in a condition that complies with product specifications. A product that does not meet specs may be saleable to some lower-spec if customers order significant enough quantities. An often-forgotten part of the Quality Rate: is rework, which shouldn't affect the production rate. Rework is fed back through the process (for example, steel scrap from the hot strip mill is fed back to the basic oxygen furnace at a steel mill), displacing virgin material that could be processed in its place.

Remember that knowing the OEE doesn't provide information to improve it. You need to determine what causes each loss and how significant it is. If you know where the most waste occurs, you can focus resources to eliminate it through problem-solving, root cause failure analysis (RCFA, see Chapter 2), reliability-centered maintenance (RCM, see Chapter 8), or some fundamental equipment care techniques.

Supporting an empowered culture promotes opportunities for further benefits where they promote optimizing the person/equipment or system in the workplace by:

- Restoring equipment to optimal operating conditions
- Maintain equipment in optimal operating condition
- Defining the responsibilities of operators, maintenance, and other technical staff to achieve optimal operating conditions
- Optimizing the quality of the working environment
- Developing an attitude where everyone feels responsible for equipment performance and condition
- Changing the attitude toward equipment so that abnormalities, breakdowns, and defects are not acceptable, and
- Emphasizing quality in all aspects of work.

9.3.3 FOCUSED PROCESS IMPROVEMENT

Fundamental process execution and improvement disciplines may be applied here. However, a TPM culture and an empowered staff: can support a "learning culture" and a faster path to action and a solution. This can follow the culture of a Plan, Do, Check, Act (PDCA) structured cycle. This PDCA cycle can be reinforced in the TPM team culture. When implemented, identified improvements can fit an organization of any size or complexity.

An empowered staff can apply a "5S" (6S with safety added) cultural mindset. 5S originates from the Japanese 5S program for empowered self-analysis of a working environment (SEIRI, SEITON, SEISO, SEIKETSU, SHITSUKE). In English, this translates to: (SORT OUT, SET IN ORDER, SHINE, STANDARDIZE, SUSTAIN).

A process focus can also support an automated process culture that may be executed through IT solutions that can come from an enterprise asset management (EAM), asset performance management (APM), internet of things (IoT), artificial intelligence (AI), and other emerging tools.

At its most superficial level: documenting existing processes and who performs each task in a process can help a cross-functioning team assess whether redundant or minimal value task steps unduly constrain process value. With the poor value, constraints are understood: an optimized process can be revised and agreed to in a shorter timeline. Many EAM and APM solutions have been designed to support optimal process execution. However, this may still require the process owners to agree on what will be automated and what may require in-process and interim approvals before a process can be fully automated.

A simple example of a focus process improvement could be how a maintenance part is assigned to a work order originating from an operator's service request. It could be decided that -in some situations- the operator could correct the process/ equipment defect directly if they could receive the parts in a time-effective manner. If such a process conforms to the organization's business controls, such a process could be agreed to and set up in an automated format to drive asset value (OEE).

9.3.4 EDUCATION AND TRAINING

Individual productivity is a function of skills, knowledge, and attitude. The cultural change in the operator's work environment makes education and training essential when implementing TPM.

This focus area ensures that staff is trained in the skills identified as essential for personal development and TPM's successful deployment in line with the organization's goals and objectives.

Activities here can include:

- Assessing the required knowledge and skills to carry out each job. This can include understanding the complexity of knowledge needed and the number of skilled people required to support the business needs. An "As Is" analysis assesses the current capabilities against the desired requirements, and a training plan is developed to close the gaps, where applicable. Like PDCA, the training plan is executed and evaluated to ensure the activity generates targeted improved capabilities.

- Designing, implementing, and improving a "Skill Development Plan/ System" to enable the ongoing development of all employees.
- Expanding as needed to cover broader roles and increasingly complex training needs.

In many organizations, operators must follow the supervisor's directions without question. The operator is trained never to deviate from a specific procedure. Often, the training method is the "buddy" system. However, the result can be operators who complete the minimum work required to perform the task without understanding their role in the overall operation. The operator may also learn bad habits from their "buddy."

In TPM, the operator is asked to participate in decision-making and constantly question the status quo. This is a fundamental requirement of continuous improvement. The initial impact, however, can be negative. The operator's first reaction is often that management is "dumping" work that has traditionally been the responsibility of maintenance or plant management. The operator may also worry about being unable to do what's required. So it's no wonder that the mention of TPM immediately mobilizes the union in many plants.

You must educate employees about the benefits of TPM and your business needs while training them to use TPM tools. Education develops an individual into a whole person, while training provides specific skills. But you must implement and update education and training at the same time. Therefore, each level must increase as the operator learns new concepts and skills.

Before starting TPM, the operators need to learn its philosophy and practices. They must also know about their company. If you involve operators in decision-making, they must understand the context.

The minimum training requirements are:

- An Introduction to TPM Culture
- General inspection techniques
- Diagnostic techniques
- Analytical problem-solving techniques
- Selected technical training

You need to appoint a special TPM team to teach operators and other personnel specific problem-solving methods such as Pareto Analysis, RCFA, and statistical process control to take a more proactive role. Training and education must be ongoing to ensure knowledge transfer and keep skills totally up to date.

9.3.5 Quality Maintenance

You have implemented OEE in your vital production areas and have a wealth of data about your losses. But how do you use that data to improve the OEE?

Although the OEE number is the focus, identifying what causes wasted availability and process rate is essential to improving it and applying quality Pareto Analysis. By working on the most significant losses, you make the most effective use of your resources.

Quality Maintenance aims to ensure zero defects in a production process. It understands the interactions between human resources, materials, machines, and methods that could enable defects. Once identified, this team approach works to predict/prevent defects from emerging before they happen. This should be the preferred scenario versus experiencing the equipment outage from an installed sensor that detects the defect after it has been produced. This approach should also optimize run-to-failure downtime outcomes when not optimal.

With AM, equipment improvement, and process improvement in place and supported by education and training, Quality Maintenance can become more of an overseeing living program. Operators and maintenance must understand the functional requirements of the equipment in their operating context to contribute to sustaining the desired equipment conditions.

Quality Maintenance is implemented by:

- First – aiming to eliminate quality issues by analyzing the work execution defects so that conditions can be defined that prevent defects from occurring; and
- Second – assess, analyze, and implement improvements. This second step supports and sustains the desired asset quality by standardizing the parameters and methods to achieve a zero-defect system.

There are many forms of waste that TPM can eliminate:

- Lost production time due to breakdowns
- Idling and minor stoppage losses from intermittent material flows
- Setup and adjustment losses (time lost between batches or product runs)
- Capacity losses from operating the process at less than maximum sustainable rates
- Start-up losses from running up slowly or disruptions
- Operating losses through errors
- Yield losses through less-than-adequate manufacturing processes
- Defects in the products (quality problems)
- Recycling losses to correct quality problems

You can apply OEE at the plant, production line, system, work cell, or equipment level. It can be measured yearly, monthly, weekly, daily, by shift, by the hour, by minute, or instantaneously. The measurement frequency must ensure that both random and systematic events are identified. You must report the data frequently enough to detect trends early on. A 90% OEE target is world-class. First, improve availability (essentially a maintenance and reliability effort) to successfully get there, then target production and quality rates.

However, as shown in Figure 9.3, using Pareto Analysis can be a helpful tool in developing priorities when improving OEE. Losses caused by operating at a

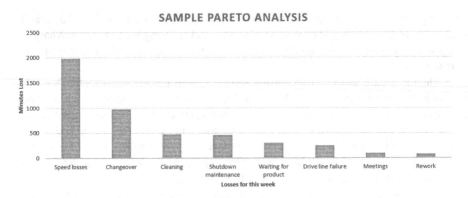

FIGURE 9.3   Sample pareto analysis.

less-than-ideal rate or producing off-spec products can be converted to time. In other words, a machine operating at 90% of the rated speed for 10 hours has lost the equivalent of 1 hour of production time. Pareto Analysis prioritizes the losses so that the organization focuses on the most prominent piece.

Most organizations have a narrow set of measures that zero in on defects or failures. Many organizations monitor mechanical downtime but not availability. For example, in Figure 9.3, the organization would try to correct the "drive line failure." However, completely removing the cause of "drive line failure" would be equivalent to a 10% reduction in the product lost to operating at a reduced speed.

In the figure, Pareto Analysis is critical to prioritizing OEE data. Note that the five highest causes in this example are losses often considered normal. In a non-TPM plant, "drive line failure" and "rework" would receive the most management attention.

The solution to many of these losses extends beyond maintenance, including production, engineering, and materials logistics. All elements of the plant's entire value chain or supply chain can impact how much quality product is produced. Correcting problems that lead to low availability, production rates, or quality can involve maintenance, engineering, and production process or procedural changes. Teamwork is essential to pulling these disciplines together.

The key to TPM is the use of teams. Usually, you organize the teams around production areas, lines, or work cells. They often comprise production and maintenance workers in a ratio of about 2 to 1. The teams work primarily in their assigned areas to increase equipment familiarity, a sense of "ownership," and cooperation among production and maintenance. Selecting the pilot area is vital. It must need change. You want impressive results to use later to "sell" the concept to other plant areas. Production and maintenance goals are the same through teaming because they are specific to the larger organization instead of a department or function.

9.3.6   Asset Lifecycle Management

The key to a successfully optimized asset value organization's execution is understanding the valued application of the asset throughout its lifecycle. This suggests that the stakeholders contributing to the organization's success are not just the maintenance and operations teams while the equipment is in planned production.

This section focused on new products and processes with steep ramp-up and minimized development lead time. The focus will typically capitalize on early equipment and product management and use the lessons learned from previous (like) equipment insights and experiences to eliminate the potential for losses through the planning, development, and design stages.

Early equipment management (i.e., during equipment commissioning) aims to introduce a loss and defect-free process so that equipment outages are minimized (zero breakdowns) and maintenance costs are considered and optimized from the commissioning stage into the operational stage, etc.

Product management aims to shorten product development lead times and commissioning cycles with teams working on simultaneous activities to achieve vertical start-up with zero quality loss (zero defects).

Finance, Design Engineering, and Maintenance Parts personnel – to name a few – also have critical roles, particularly during the asset design, commissioning, and the start of production phases and at an asset's end-of-life forecasted period. Too often, essential communication, data transfer, and transfer of asset ownership move from design to production, with many vital hand-offs ineffectively executed. This results in unrealized asset benefits once in production.

High-performing organizations extend the TPM culture to all the stakeholders that may influence the success of the asset's value creation, whether the influence is before the asset is in production or expected to be discontinued. These organizations will understand that there are many internal support costs, missed opportunities, and production loss implications when an aligned team culture is not well supported and executed across an asset's lifecycle.

The focus here can often be deployed after the first four areas. Asset lifecycle management builds on the learnings captured from other TPM area teams. It will publish and incorporate improvements into the following product and equipment design generation.

9.3.7 HEALTH, SAFETY, AND ENVIRONMENT

Of course, almost every organization will have an existing culture that declares its position on health, safety, and environment (HSE). Operating within these corporate HSE guidelines is a gate that must be met to operate. If HSE is not met, equipment OEE will not be permitted to start. The key to TPM is recognizing that the team has an HSE role. They must ensure that the HSE standards are adhered to and that these standards are understood and part of the fabric of every issue and solution they address.

HSE area activities aim to eliminate the root causes of incidents, proactively reduce future potential risks by targeting near misses and potential hazards, and prevent reoccurrences. The HSE team targets three key areas: people's behaviors, machine conditions, and the management system. In addition, all HSE area activities should align with relevant external quality standards and certifications.

Programs could include addressing a known or perceived risk through a process change or employee education. These risks would be included in any asset work prioritization discussion or process (i.e., if an equipment defect creates an increased risk of an HSE issue, the work priority should be exponentially weighted higher).

ESG is a growing concern for every organization in the world. More on the Environmental needs, and considerations are further described in Chapter 19.

9.3.8 TPM PROGRAM MANAGEMENT

Program management may be executed by focusing on any of the previously listed elements. It will also likely be a "living program" that will look for ways to tenaciously provoke opportunities for improvement, supporting a culture where maintenance and operations work together to improve OEE.

This area concentrates on all areas that provide administrative and support functions in the organization. The area applies the fundamental TPM principles to eliminate these departments' waste and losses. The area ensures that all processes support operational excellence within the manufacturing processes and are completed at optimal cost.

To establish sustainable, performing processes, the TPM Program Management team implements office versions of the following:

- Focused Process Improvement
- Autonomous Maintenance
- The Training and Education areas

They can deploy an agile (flexible) staffing workforce to allow departments to manage peak workloads without overstaffing. They can also prioritize improvement programs, by loss analysis, against the goals and objectives of the critical equipment in its operating context.

Progress reviews and reporting are typical in an organization with a thriving TPM culture. These reviews would likely be positioned weekly at the operator and maintenance craft level and monthly to include the operations and maintenance supervisors. A successful program will have a clear executive sponsor who will want to review these meeting summaries and support the constituents with change enablement support and training when needed.

For larger organizations with multiple (perhaps international) sites, the site's focus may differ depending on the maturity and readiness of a specific site's culture. Section 9.4 of this chapter suggests program considerations that may be injected into a division or site to start and maintain TPM as a living program.

9.4 HOW DO YOU IMPLEMENT TPM?

TPM is implemented in four major phases:

- Establish acceptable equipment operating conditions to stabilize reliability
- Lengthen asset life
- Optimize asset conditions
- Optimize life cycle cost

A phased approach is recommended because:

- As you cannot solve all cultural issues in one shot
- It is more economical to lengthen equipment life by eliminating accelerated deterioration (stop the bleeding first!)
- Until accelerated deterioration is arrested, legitimate design weaknesses usually are invisible
- Even if weaknesses are corrected, the overall effect may be masked by accelerated deterioration
- Application of systematic maintenance is most effective in a stable, predictable environment

The fundamentals continue and expand as you implement the four phases.

The first phase stabilizes reliability by restoring equipment to its original condition. This is done by cleaning the equipment and correcting any defects. Note major problems and establish a plan to resolve them. Ensure operators are trained to turn simple equipment cleaning into a thorough inspection to spot machine defects.

The second phase maintains the equipment's foundational operating condition. Standards to do so are developed. Begin data collection and set equipment condition goals. The operators perform minor maintenance activities to eliminate abnormal wear.

In phase three, improve the equipment's operation from its stabilized level. Cross-functional teams should target chronic losses to increase overall machine performance. Review and update standards. Find and analyze opportunities to increase equipment performance and operating standards beyond original capabilities.

Phase four is about optimizing the asset's cost over its entire life. You achieve this by extending equipment life, increasing performance, and reducing maintenance costs. Keep the machine in its optimal condition. Regularly review processes that set and maintain operating conditions. The new equipment will become part of the TPM process. Operator reliability is "built-in."

A cross-functional team approach should be used during all phases of TPM implementation. Building effective teams is a prerequisite to entrenching TPM ideas and behaviors. The principles and techniques are straightforward. The initial focus will be project management and carefully applying change management. Change and change management is central to TPM. If your organization isn't used to it or has a history of unsuccessful changes, this will be a significant hurdle to success. We recommend a pilot project to demonstrate success in one area before tackling TPM throughout the plant.

The choice of location for a TPM pilot is critical. The pilot area must improve and be visible to many people. Once you establish momentum, apply TPM to other areas. Divide it into manageable portions and implement them one at a time.

Successful TPM requires a transfer of responsibility between management and employees. It depends on an enthusiastic and dedicated management team that pays appropriate attention to change management issues.

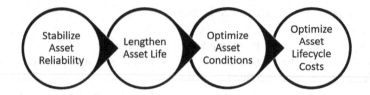

FIGURE 9.4   TPM implementation phases.[1]

The legacy culture presents the most significant change management problems when implementing TPM. See Table 9.2 for samples of typical change management issues.

TPM is implemented gradually over several (3–5) years. Once established, it becomes part of the plant's way of doing things – its culture. Since TPM is about changing the behavior of both workers and managers, it requires patience and positive reinforcement to achieve permanent change. Even poorly maintained plants can be "turned around" through TPM with the proper focus and commitment.

9.5   THE CONTINUOUS IMPROVEMENT OF WORKPLACE

Successfully implementing TPM creates an efficient, flexible, and continually improving organization. The process may be long and arduous, but it is as hard to remove once TPM has been accepted as the culture it replaced. This is significant because managers may change, but TPM will continue.

The TPM workplace is efficient because it follows tested procedures that are continually reviewed and upgraded. Change is handled fluidly because effective education and training prepare the workforce to participate in decision-making.

TPM embraces other optimizing maintenance management methodologies. RCM and RCFA are often very effective in a TPM environment. RCM impacts the preventive and predictive maintenance program, and RCFA improves specific problem areas. TPM impacts the working environment in virtually all respects – the way production and maintenance employees work, are organized, use other techniques like RCM and RCFA, solve problems, and implement solutions.

Your competitors may be able to purchase the same equipment but not the TPM experience. The time required to implement TPM makes it a significant competitive advantage that can't be easily copied.

9.6   TWO SIMPLE HIGH-LEVEL CASE STUDY EXAMPLES OF TPM

A TPM culture is often found when the business empowers maintenance and operations personnel to work together for the common goal of effective and efficient asset production. In both cases shown here, maintenance and operations personnel actively participated in scheduled review meetings and a continuous improvement mindset that supported the production influences of both operations and maintenance teams.

TABLE 9.2
Implementing TPM: The Legacy Culture

Legacy Approach	TPM Approach	Change Management Issues
Clear lines of responsibility exist between production and maintenance employees. When a machine breaks down, operators call maintenance.	Employees work together to solve problems. It is recognized that production and maintenance are inseparable, and problems need to be solved jointly.	Employees may believe that the goal of TPM is to eliminate maintenance jobs. Most people equate productivity gains to job losses. It is challenging to see TPM's objectives.
Supervisors direct employee actions. Employees do what they are told when they are told.	Self-directed teams develop and execute plans to achieve progressive goals.	Front-line supervisors have difficulty changing to their new "coach" role. Many don't trust their employees.
Management announces a new program to improve operations. Employees are trained. This is commonly referred to as the "Flavor of the Month."	Management announces that TPM will be implemented. TPM training is conducted, and a TPM team is formed. A pilot site is chosen, and work begins to improve the condition and performance of the pilot site.	At first, few employees recognized that TPM would be implemented. When the TPM team shows progress and starts to improve, some employees will see the benefits and want to participate. Others will fear the change and reject the process. Over time the nay Sayers convert or fade away. This is a challenging issue when a plant has failed at a TPM implementation.
Increased production level is achieved in two ways: employees are pushed to work faster, or equipment must be added.	Reducing losses due to availability, quality, and production rates increases production levels. The more significant the loss, the greater the potential benefit.	Operators believe that OEE is implemented to rate them and make them work harder. It is the losses that make them work harder. Eventually, they see that OEE helps them quantify problems they have always wanted to be corrected.
The relationship between the Union and Management is adversarial. Each tries to beat the other in negotiations. Grievances and disciplinary actions are used as negotiating tools.	Union and management work together to achieve the goals of TPM. Each represents their interests, but negotiations are successful if each side benefits.	TPM is a process that does what most Unions have always wanted. It gives employees a voice in their workplace and considers them a valuable resource. However, if the union is not involved initially, its distrust of management's intentions will be a significant hurdle for the TPM team.
Performance Measures exist for each department. Maintenance is evaluated on downtime production output.	OEE is the essential measurement. The organization is evaluated on OEE. Pareto Analysis prioritizes losses that affect OEE.	Invariably, when OEE tracking starts, significant losses are considered "normal" – changeovers. It is difficult for people to accept that something they have accepted for years can be changed.

A Taiwan-based semiconductor manufacturing company has a highly efficient operations and maintenance organization that considers maintenance activity part of the contribution (or potential constraint) to a successful production commitment cycle.

Their facilities and infrastructure production support are divided into four working groups that directly include the maintenance teams:

- Gas, chemicals, and slurry;
- Water treatment;
- Instrumentation and electrical;
- Mechanical.

They meet weekly as a cross-functional team – Maintenance and Operations – to review and forecast production and equipment issues and risks. At this type of meeting: the schedule is well established and efficiently executed for the stakeholders of a particular process or equipment issue. The team would enter the meeting room, be prepared to discuss the specific insights from a past issue, including lessons learned, and return to their regular business when the specific equipment element is completed. Including the maintenance and operations teams in this type of weekly (and perhaps in the summary and monthly operational review): promotes a culture of maintenance and operations working together to achieve the company's process and equipment goals and optimize the planned ROA. It also empowers both groups to work together behind the scenes of these weekly reviews to achieve this mutual goal.

A North American retailer's finance organization depended on its centralized computer systems to efficiently convert "sales activity to cash flow" and its dynamic financial position. They established a highly focused "Quality" program around the productivity and availability of its primary computing equipment. They promoted that the computer maintenance and operations teams should call out any potential issues before they become a physical constraint to their computing production. A shift-by-shift log and communications vehicle were created that operations would initiate, and maintenance would acknowledge and provide feedback. More than 97% of the communications were acknowledged, and 3% was often considered an early warning of an equipment issue that could be mitigated through proactive maintenance or operator training. They also had weekly and monthly operational reviews that included maintenance personnel. The operator performed many off-hour activities that could be safely done, as they best understood the 24-hour production dynamics and the criticality of their overnight production schedules. This allowed the maintenance personnel to be more available to guide operators and management during prime hours. This set of actions supported a *TPM culture and mindset*, as both maintenance and operations recognized that their value was best created when they supported each other to achieve (ROA) value for their organization.

PERMISSIONS

* Pages 217–228, *Asset Management Excellence: Optimizing Equipment Life-Cycle Decisions*, 2nd Edition by Editor(s), John D. Campbell, Andrew K. S. Jardine, Joel McGlynn Copyright (2011) by Imprint. Reproduced by permission of Taylor & Francis Group.

1   Adapted from copyright (Total Productive Maintenance Lecture Notes – TPM), (2021) edited by (Barry). Reproduced by permission of Asset Acumen Consulting Inc.

BIBLIOGRAPHY

Campbell, John Dixon, *UpTime: Strategies for Excellence in Maintenance Management*, New York, Productivity Press, 1995.
Barry, Don, *Total Productive Maintenance (TPM) Training Materials*, Physical Asset Management Course, Asset Acumen Consulting Inc., 2021.
Ishikawa, Karoo, *Guide to Quality Control*, Tokyo, Asian Productivity Organization, 1986.
Japan Institute of Plant Maintenance, *Autonomous Maintenance for Operators*, New York, Productivity Press, 1997.
Nakajima, Seiichi, *Introduction to TPM*, New York, Productivity Press, 1988.

Section III

Optimizing Maintenance and Replacement Decisions

10 Reliability Management and Maintenance Optimization*
Basic Statistics and Economics

Andrew K. S. Jardine
and Original by Murray Wiseman

As global industrial competitiveness increases, showing value, particularly in equipment reliability, is an urgent business requirement. Sophisticated, user-friendly software integrates the supply chain, forcing maintenance to be even more mission-critical. We must respond effectively to incessantly fluctuating market demands. All of this is both empowering and extremely challenging. Mathematical and statistical models are invaluable aids. They can help you increase your plant's reliability and efficiency at the lowest possible cost.

This chapter is about the statistical concepts and tools you need to build an effective reliability management and maintenance optimization program. We'll take you from the basic concepts to developing and applying models for analyzing common maintenance situations. Ultimately, given a defined situation, you should know how to determine the best course of action or general policy.

We begin with the relative frequency histogram to discuss the four main reliability-related functions: (1) the probability density function (PDF); (2) the cumulative distribution function (CDF); (3) the reliability function; and (4) the hazard function. These functions are used in the modeling exercises in this and subsequent chapters. We describe several common failure distributions and what we can learn from them to manage maintenance resources. The most useful of these is the Weibull distribution, and you'll learn how to fit that model to a system or component's failure history data.

This chapter uses the words maintenance, repair, renewal, and replacement interchangeably. The methods we discuss assume that maintenance will return equipment to "good-as-new" condition.

10.1 INTRODUCTION: THE PROBLEM OF UNCERTAINTY

Faced with uncertainty, our instinctive human reaction is often fear and indecision. We would prefer to know when and how things happen. In other words, we would

DOI: 10.1201/9781032679600-13

like all problems and their solutions to be *deterministic*. When timing and outcome depend on chance, problems are probabilistic or *stochastic*. Many problems, of course, fall into the latter category. Our goal is to quantify the uncertainties to increase the success of significant maintenance decisions. The methods described in this chapter will help you deal with uncertainty, but our aim is greater than that. We hope to persuade you to treat it as an ally rather than an unknown foe.

"Failure is the mother of success." "A fall in the pit is a gain in the wit." If your maintenance department uses reliability management, you'll appreciate this folk wisdom. In an enlightened environment, the knowledge gained from failures is converted into productive action. Achieving this requires a sound quantitative approach to maintenance uncertainty. We'll show you an easily understood relative frequency histogram of past failures.

In addition to the relative frequency histogram, we look at the PDF, the CDF, the reliability function, and the hazard function.

10.2   THE RELATIVE FREQUENCY HISTOGRAM

Assume that 48 items purchased at the beginning of the year all fail by November. List the failures in order of their failure ages. Group them, as in Table 10.1, into convenient time segments, in this case by month, and plot the number of failures in each one. The high bars in the center of Figure 10.1 represent the highest (or most probable) failure times: March, April, and May. By adding the number of failures occurring before April (i.e., 14) and dividing by the total number of items (i.e., 48),

TABLE 10.1
Failures in Month

Month	Jan	Feb	Mar	Apr	May	Jun	Jul	Aug	Sep	Oct
Failures	2	5	7	8	7	6	5	4	3	1

FIGURE 10.1   The relative frequency histogram.

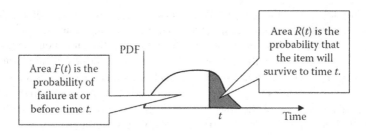

FIGURE 10.2   The probability density function.

the *cumulative probability* of the item failing in the first quarter of the year is 14/48. The probability that all items fail before November is 48/48, or 1.

Transforming the numbers of failures by month into probabilities, the relative frequency histogram is converted into a mathematical and more useful form called the *probability density function* (mentioned earlier). The data are replotted so that the *area* under the curve, between time 0 and any time *t*, represents the cumulative probability of failure. This is shown in Figure 10.2. (How the PDF plot is calculated from the data and then drawn is discussed more thoroughly in Section 10.6.).

The total area under the probability density function $f(t)$ curve is 1 because, sooner or later, the item will fail. The probability of the component failing at or before time *t* equals the area under the curve between time 0 and time *t*. That area is $F(t)$, the CDF. The remaining (shaded) area is the probability that the component will survive to time *t* and is known as the reliability function, $R(t)$. $R(t)$ and $F(t)$ can, themselves, be plotted against time.

The *mean time to failure* (MTTF) is

$$\int_{0}^{\infty} tf(t)\,dt$$

(shown in Appendix A.1). From the reliability, $R(t)$, and the probability density function, $f(t)$, we derive the fourth useful function, the hazard function, $h(t) = f(t)/R(t)$, which is represented graphically for four common distributions in Figure 10.3.

You have discovered the four key functions in reliability engineering in just a few short paragraphs. Knowing any one, you can derive the other three. Armed with these fundamental statistical concepts, you can battle random failures throughout your

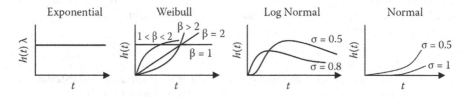

FIGURE 10.3   Hazard function curves for the common failure distributions.

plant. Although we can't predict when failures occur, we can determine the best times for preventive maintenance and the best long-run maintenance policies.

Once you're reasonably confident about the reliability function, you can use it and its related functions with optimization *models*. Models describe typical maintenance situations by representing them as mathematical equations. That makes it convenient to adjust certain decisions to get the optimum outcome. Optimization reduces long-term maintenance costs to the lowest point possible. Other objectives include the highest reliability, maintainability, and availability of operating assets.

10.3 TYPICAL DISTRIBUTIONS

In the previous section, we defined the four key functions once data have been transformed into a probability distribution. The prerequisite step of *fitting* or *modeling* the data is covered next.

How do you find the appropriate reliability function for a real component or system? There are two different approaches to this problem. In one, you estimate the reliability function by curve-fitting the failure data from extensive life testing. In the other, you estimate the parameters (unknown constants) by statistical sampling and numerous statistical confidence tests.[1] We'll take the latter approach.

Fortunately, we know from past failures that probability density functions (and their reliability, cumulative distribution, and hazard functions) of real maintenance data usually fit satisfactorily one of several mathematical equations already familiar to reliability engineers. These include exponential, Weibull, log-normal, and normal distributions. Each failure distribution is a "family" of equations (or graph curves) whose members vary in shape by their differing parameter values. So, their cumulative distribution functions can be fully described by knowing the value of their parameters. For example, the Weibull (two-parameter) CDF is

$$F(t) = 1 - e^{-\left(\frac{t}{\eta}\right)^{\beta}}$$

You can estimate the parameters β and η using the methods described in the following sections. Usually, through one or more of these four probability distributions, you can conveniently process failure and replacement data. The modeling process involves manipulating the statistical functions you learned about in Section 10.2 (the PDF, CDF, reliability, and hazard functions). The objective is to understand the problem, to forecast failures, and to analyze risk to make better maintenance decisions. Those decisions will impact the times you choose to replace, repair, or overhaul machinery and optimize many other maintenance management tasks.

The problem entails the following:

- Collecting good data
- Choosing the appropriate function to represent your situation and estimating the function parameters (e.g., the Weibull parameters β and η)
- Evaluating how much confidence you have in the resulting model

Modern reliability software makes this process easy and fun. What's more, it helps us communicate with management and share the common goal of business – implementing procedures and policies that minimize cost and risk while maintaining, even increasing, product quality and throughput. The most common failure rate or hazard functions are depicted in Figure 10.3. They correspond to the exponential, Weibull, log-normal, and normal distributions.

Real-world data most frequently fit the Weibull distribution. Today, Weibull analysis is the leading method in the world for fitting component life data,[2] but it wasn't always so. While delivering his hallmark paper in 1951, Waloddi Weibull modestly said that his analysis "…may sometimes render good service." That was an incredible understatement, but the initial reaction varied from disbelief to outright rejection. Then the U.S. Air Force recognized Weibull's research and provided funding for 24 years. Finally, Weibull received the recognition he deserved.

10.4 THE ROLE OF STATISTICS

Most enter an unacceptable state at some stage in life. One of the challenges of optimizing maintenance decisions is to predict when. Luckily, it's possible to analyze previous performance and to identify when the transition from "good" to "failed" is likely to occur. For example, while your household lamp may be working today, what is the likelihood that it will work tomorrow? Given historical data on the lifetime of similar lamps in a similar operating environment, you can calculate the probability of the lamp still working tomorrow or, equivalently, failing before then. Here's how this is done.

Assume that a component's failure can be described by the normal distribution illustrated as a PDF in Figure 10.4. You'll notice several interesting and useful facts about this component in the graph. First, the figure shows that 65% of the items will fail at some time within 507 to 693 hours, and 99% will fail between 343 hours and 857 hours. Also, since the PDF is constructed so that the total area under the curve

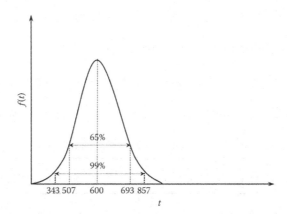

FIGURE 10.4   The normal (Gaussian) distribution.

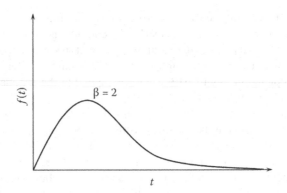

FIGURE 10.5   Skewed distribution.

adds to 1.0 (or 100%), there is a 50% probability that the component will fail before its mean life of 600 hours.

While some component failure times fit the normal distribution, it is too restrictive for most maintenance situations. For example, the normal distribution is bell-shaped and symmetrical, but many times the data are quite skewed. A few items might fail shortly after installation and use, but most survive for a fairly long time. This is depicted in Figure 10.5 – a Weibull distribution whose shape parameter β equals 2.0.

In this case, you can see that the distribution is skewed with a tail to the right. Weibull distribution is popular because it can represent component failures according to the bell-shaped normal distribution, the skewed distribution of Figure 10.5, and many other possibilities. Professor Weibull's equation includes two constants: β (beta), known as the shape parameter, and η (eta), known as the characteristic life. It is this flexible design that has made Weibull distribution such a success.

The hazard function, depicted in Figure 10.6, shows clearly the risk of a component failing as it ages. If the failure times have a beta value greater than 1.0, the risk increases with age (i.e., it is wearing out). If beta is less than 1.0, the risk declines (e.g., through work hardening or burn-in). If beta takes a value equal to 1.0, the

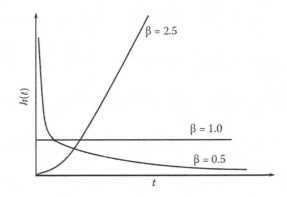

FIGURE 10.6   Weibull hazard function.

failure isn't affected by age (i.e., failures are purely random, caused by external or unusual stress). That is usually the case when a stone hits a car's windshield, severely cracking it, which is as likely to occur on a new car as an old one. In fact, many failures in industrial, manufacturing, process, and transportation industries are random or stress failures.

You want to optimize the maintenance decision to better know when to replace a component that can fail. If the hazard function is increasing (i.e., beta is greater than 1.0), you must identify where on the increasing hazard curve the optimal replacement time occurs. You do this by blending together the hazard curve and the costs of preventive maintenance and failure replacement, taking into account component outages for both. Establishing this optimal time is covered in Chapter 11, "Maintenance Optimization Models."

If the hazard function is constant (beta is equal to 1.0) or declining (beta is less than 1.0), your best bet is to let the component run to failure. In other words, preventively replacing such components will not make the system more reliable. The only way to do that is through redesign or installing redundancy. Of course, there will be trade-offs. To make the best maintenance decision, study the component's failure pattern. Is it increasing? If so, establish the best time to replace the component. Is it constant or declining? Then the best action, assuming there aren't other factors, is to replace the component only when it fails. There isn't any advantage, either for reliability or cost, in preventive maintenance.

Earlier, we mentioned the importance of reliability software in maintenance management. You can easily establish a component's beta value using such standard software. OREST[3] software was used in Figure 10.7, where the sample size is 10 with six

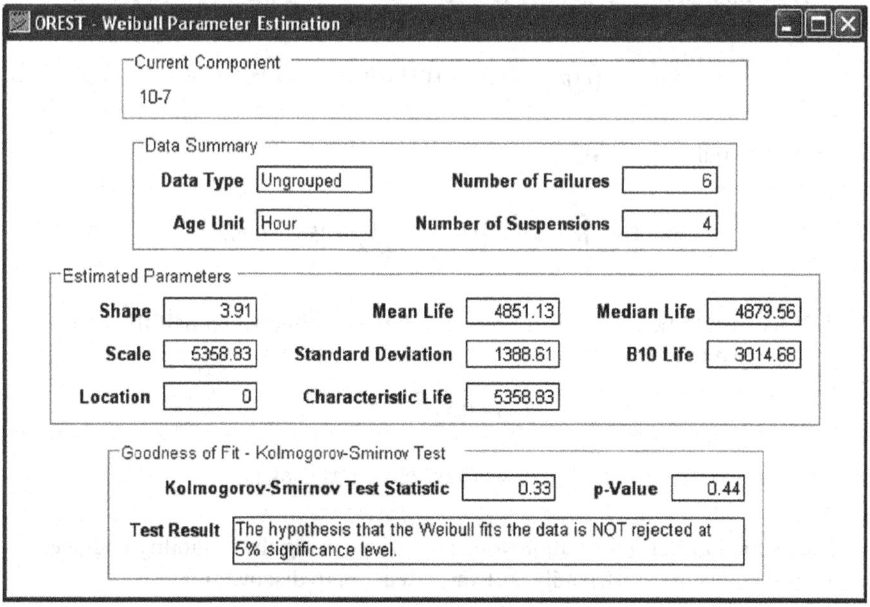

FIGURE 10.7 OREST–Weibull analysis.

failure observations and four suspensions, the beta (β) value is 3.91, and the MTTF of the item is 4851.13 time units. Additional aspects of the table are covered later in the chapter. Weibull ++[4] and SuperSMITH Weibull[5] perform similar functions.

We must stress that, so far, we have been focusing on items termed line replaceable units (LRUs). The maintenance action replaces or renews the item and returns the component to a statistically good-as-new condition. For complex systems with multiple components and failure modes, the form of the hazard function is likely to be the bathtub curve in Figure 10.8. In these cases, the three underlying causes of system failure are wear-out, quality, and random. Adding them creates the overall bathtub curve. There are three distinct regions: running-in period, regular operation, and wear-out.

Example

Here is an example illustrating how you can extract useful information from failure data. Assume (using the methods to be discussed later in this chapter) that an electrical component has the exponential cumulative distribution function, $F(t) = 1 - e^{-\lambda t}$, where $\lambda = 0.0000004$ failures per hour.

a. What is the probability that one of these parts fails before 15,000 hours of use?

$$F(t) = 1 - e^{-\lambda t} = 1 - e^{-0.0000004 \times 15000} = 0.006 = 0.6\%$$

b. How long until you get 1% failures? Rearranging the equation for $F(t)$ to solve for t:

$$t = -\ln(1 - F(t))/\lambda = -\ln(1 - 0.01)/0.0000004 = 25,126 \text{ hr.}$$

c. What would be the MTTF?

$$\text{MTTF} = \int_0^\infty tf(t)\,dt = \int_0^\infty t\lambda e^{-\lambda t}\,dt = \frac{1}{\lambda} = 250,000 \text{ hr.}$$

d. What would be the median time to failure (the time when half the number will have failed)?

$$F(T_{50}) = 0.5 = 1 - e^{-\lambda T_{50}}$$

$$T_{50} = \ln 2/\lambda = 0.693/0.0000004 = 1,732,868 \text{ hr.}$$

This is the kind of information you can retrieve using the reliability engineering principles in user-friendly software. Read on to discover how.

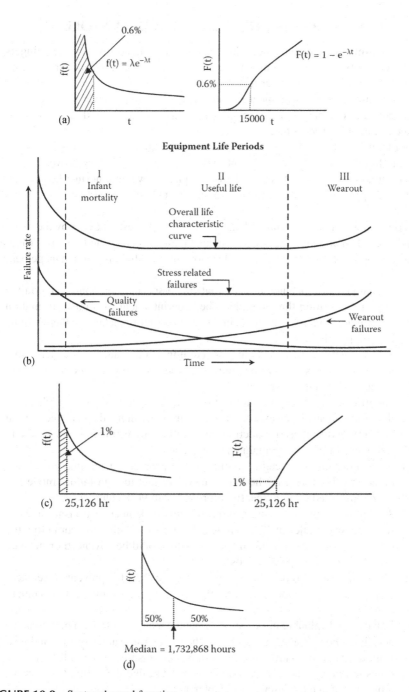

FIGURE 10.8 System hazard function.

10.5   REAL-LIFE CONSIDERATIONS – THE DATA PROBLEM

Ironically, reliability management data can slip unnoticed through your fingers as you relentlessly try to control maintenance costs. Data management can prevent this from happening and provide critical information from experience to improve your current maintenance management process. It is up to upper-level managers to provide trained maintenance professionals with ample computer and technical tools to collect, filter, and process data.

Without a doubt, the first step in any forward-looking activity is to get good information. In fact, this is more important than anything else. History proves that progress is built on experience, but countless examples show that ignoring the past results in missed opportunities. Unfortunately, many maintenance departments are guilty of this too.

Companies can be benchmarked against world-class best practices by the extent to which their data effectively guide their maintenance decisions and policies. Here are some examples of how you can make decisions using reliability data management:

1. A maintenance planner notices that an in-service component has failed three times within three months. The superintendent uses this information to estimate the failure numbers in the next quarter to make sure there will be enough people available to fix them.
2. When ordering spare parts and scheduling maintenance labor, determine how many gearboxes will be returned to the depot for an overhaul for each failure mode in the next year.
3. An effluent treatment system must be shut down and overhauled whenever the contaminant level exceeds a toxic limit for more than 60 seconds in a month. Avoid these production interruptions by estimating the level and frequency of preventive maintenance needed.
4. After a design modification to eliminate a problem, determine how many units must be tested, for how long to verify that the old failure mode has been eliminated or significantly improved with 90% confidence.
5. As stipulated by the manufacturer, a haul truck fleet of transmissions is routinely overhauled at 12,000 hours. A number of failures occur before the overhaul. Find out how much the overhaul should be advanced or delayed to reduce average operating costs.
6. The cost of lost production is 10 times more than for preventive replacement of a worn component. From this, determine the optimal replacement frequency.
7. You can find valuable information in the database to help with maintenance decisions. For instance, if you know the fluctuating values of iron and lead from quarterly oil analysis of 35 haul truck transmissions and their failure times over the past three years, you can determine the optimal preventive replacement age (examined in Chapter 12).

Obviously, it is worth your while to obtain and record life data at the system and, where warranted, component levels. When tradespeople replace a component, for example, a hydraulic pump, they should indicate which specific pump failed. It may be one of several identical pumps on a complex machine that is critical to operations. They should also specify how it failed, such as "leaking" or "insufficient pressure or volume." Because we know how many hours equipment operates, we can track the lifetime of individual critical components. That information will then become a part of the company's valuable intellectual asset – the reliability database.

10.6   WEIBULL ANALYSIS

Weibull analysis supported by powerful software is formidable in the hands of a trained analyst. Many examples and comments are given in the practical guidebook, *The New Weibull Handbook*.[2]

One of the distinct advantages of Weibull analysis is that it can provide accurate failure analysis and forecasts with extremely small samples.[2] Let's look closely at the prime statistical failure investigation tool, the Weibull plot. Failure data are plotted on Weibull probability paper, but, fortunately, modern software[3, 4, 6] provides an electronic version. The Weibull plot uses x and y scales, transformed logarithmically so that the Weibull CDF function

$$F(t) = 1 - e^{-\left(\frac{t}{\eta}\right)^{\beta}}$$

takes the form $y = \beta x$ + constant. By using Weibull probability paper, you get a straight line when you plot failure data that fit the Weibull distribution. What's more, the slope of the line will be the Weibull shape parameter β, and the characteristic life η will be the time at which 63.2% of the failures occurred.

10.6.1   WEIBULL ANALYSIS STEPS

The software makes Weibull analysis far more pleasant than it used to be. Conceptually, the software does the following:

- Groups the data in increasing order of time to failure, as in Section 9.2
- Obtains the *median rank* from tables for each time group (The median rank is explained in Section 10.6.3, and the median rank table for up to 12 samples is provided in Appendix A.2.)
- Plots on Weibull probability paper the median rank versus failure time of each observation

The April failures, for example, will have a median rank of 44.83% (Appendix 2), meaning that roughly 44.83% of them occurred up to and including April. The result is a plot such as in Figure 10.9. You can see from the graph that the shape parameter beta (β) is 2.04, and the characteristic life ETA (η) is 5.17. Furthermore, the mean life is 4.58 months (Table 10.1).

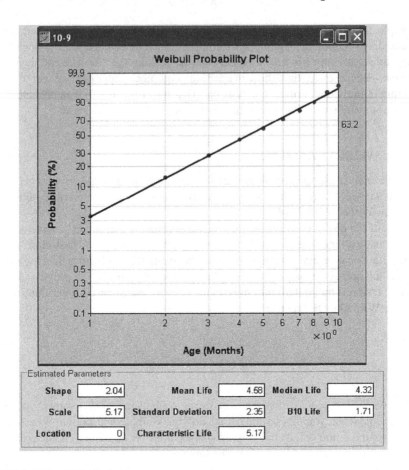

FIGURE 10.9   Weibull plot.

TABLE 10.2

Failures and Median Ranks by Munn

Month	Jan	Feb	Mar	Apr	May	Jun	Jul	Aug	Sep	Oct
Failures	2	5	7	8	7	6	5	4	3	1
Med Rank	3.47	13.80	28.28	44.83	59.51	71.72	82.06	90.34	96.53	98.57

10.6.2   ADVANTAGES

Much can be learned from the plot itself, even how the data deviate from the straight line – for example:

- Whether and how closely the data follows the Weibull distribution
- The type of failure (infant mortality, random, or wear-out)
- The component's B_n life (the time at which $n\%$ of a population will have failed)

- Whether there may be competing failures (e.g., from fatigue and abrasion occurring simultaneously)
- Whether some other distribution, such as log-normal, is a better fit
- Whether there may have been predelivery shelf-life degradation
- Whether there is an initial failure-free period that needs to be accounted for
- Forecasts for failures at a given time or during a future given period
- Whether there are batch or lot manufacturing defects

For more information about these deductions, see Abernethy.[2]

As you will see, surprisingly, few data are required to draw accurate and useful conclusions. Even with inadequate data, engineers trained to read Weibull plots can learn a lot from them. The horizontal scale measures life or aging such as start/stop cycles, mileage, operating time, take-off, or mission cycles. The vertical scale is the cumulative percentage failed. The two defining parameters of the Weibull line are the slope, β, and the characteristic life, η. (See Figure 10.3.)

The characteristic life, η, also called the $B_{63.2}$ life, is the age when 63.2% of the units will have failed. Similarly, where the Weibull line and the 10% CDF horizontal intersect is the age at which 10% of the population fails or the B_{10} life. The B1.0, B0.1, or B0.01 lives are readily obtained from the Weibull plot for more serious or catastrophic failures.

The slope of the Weibull line, beta, shows which failure class is occurring. This could be infant mortality (decreasing hazard function, $\beta < 1$), random (constant hazard function, $\beta = 1$), or wear-out (increasing hazard function, $\beta > 1$). η, or the $B_{63.2}$ life, is approximately equal to MTTF. (They are equal when $\beta = 1$. That is when Weibull is equivalent to the exponential distribution.)

A significant advantage of the Weibull plot is that even if you don't immediately get a straight line (e.g., in Figure 10.10) when plotting the data, you can still learn something quite useful. If the data points are curved downward (Figure 10.11a), it could mean the time origin is not at zero. This could imply that the component had degraded on the shelf or suffered an extended burn-in time after it was made but before delivery. On the other hand, it could show that it was physically impossible for the item to fail early on. Any of these reasons could justify shifting the origin and replotting.

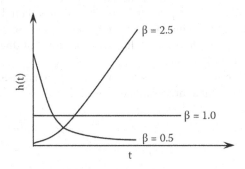

FIGURE 10.10   Hazard function for burn-in, random, and wear-out failures

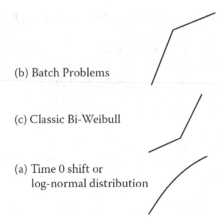

(b) Batch Problems

(c) Classic Bi-Weibull

(a) Time 0 shift or
log-normal distribution

FIGURE 10.11   Various Weibull plotted data.

Alternatively, the concave downward curve could be saying that the data really fit a log-normal distribution. You can, using software, quickly and conveniently test these hypotheses, replot with a time origin shift, or transform the scale for log-normal probability paper. Once you apply the appropriate origin shift correction (adding or subtracting a value t_0 to each data point time), the resulting plot will be straight, and you will know a lot more about the failure process.

Further, if the plotted data forms a dog leg downward (Figure 10.11b), it could mean that something changed when the failed part was manufactured. The first, steeper sloped leg reflects low-time failures. The second, lower sloped line indicates failures from batches without the defect. When there are dual failure modes like this, it's known as a batch effect. When there are many suspensions (parts removed for reasons other than failure, or parts that haven't failed at the end of the sampling period), it can be a clue that a batch problem exists. Failure serial numbers clustered closely along a leg support the theory that there was some manufacturing process change before the specific units were produced. Scheduled maintenance can also produce batch effects.[2]

An upward-pointing dog leg bend (Figure 10.11c) indicates multiple failure modes. Investigate the failed parts. If the different failures are plotted separately, treating the other failure as a suspension, two straight lines should be observed. This is the classic bi-Weibull, showing the need for a root-cause failure (RCF) analysis to eliminate, for example, an infant mortality problem. Several dog leg bends distinguish various multiple failure modes.[2]

When the Weibull plot curves concave downward, you may need to use an origin shift, t_0, equivalent to using a three-parameter Weibull[2]:

$$F(t) = 1 - e^{-\left(\frac{t-t_0}{\eta-t_0}\right)^{\beta}}$$

There should be a physical explanation of why failures cannot occur before t_0. For example, bearings, even in an infant mortality state, will continue to operate for some time before failing. Use a large sample size, at least 20 failures. If you know, from earlier Weibull analyses, that the third parameter is appropriate, you might use a smaller sample size, say, 8 to 10.[2]

For Weibull and other statistical modeling methods, the data requirements are straightforward. Maintenance personnel should be able to collect it during routine activities. There are three criteria for good data:

- The time origin must be clear.
- The scale for measuring time must be agreed upon.
- The meaning of failure must be clear.

Although all modern computerized maintenance management systems (CMMSs) and enterprise asset management (EAM) systems can provide this level of data collection, unfortunately, they have been mostly underused. To be fair, writing out failure and repair details isn't part of the tradesperson's job description. Many in the field don't yet see that adding meticulous information to the maintenance system makes organization assets function more reliably. Training in this area could yield untapped benefits.

Although the technicians performing the work provide the most useful information, collecting it is the responsibility of everyone in maintenance. Data must be continually monitored for both quality and relevance. Once the data flow is developed, it can be used in several areas, such as developing preventive maintenance programs, predictive maintenance, warranty administration, tracking vendor performance, and improved decision-making.[5] Data will improve as Weibull analysis and other reliability methods become prevalent through advanced software. The result? Management and staff will recognize the potential of good data to sharpen their company's competitive edge.

10.6.3 MEDIAN RANKS

Suppose that five components fail at 67, 120, 130, 220, and 290 hours. To plot these data on Weibull probability paper, you need the CDF's corresponding estimates. You must estimate the fraction that is failing before each of the failure ages. You can't simply say that the percentage that failed at 120 hours is 2/5 because that would imply that the cumulative probability at 290 hours, a random variable, is 100%. This small sample size doesn't justify such a definitive statement. Taken to the absurd, from a sample size of 1, you certainly couldn't conclude that the time of the single failure reflects 100% of total failures. The most popular approach to estimating the y-axis plotting positions is the median rank. Obtain the CDF plotting values from the median ranks table in Appendix A.2 or from a reasonable estimate of the median rank, Benard's formula:

$$\text{Median Rank} = \frac{i - 0.3}{n + 0.4}$$

TABLE 10.3
Modified Orders and Median Ranks for Samples with Suspensions

Hours	Event	Order	Modified Order	Median Rank
67	F	1	1	0.13
120	S	2		
130	F	3	2.25	0.36
220	F	4	3.5	0.59
290	F	5	4.75	0.82

Determined by either method, this item's cumulative failure probabilities are 0.13, 0.31, 0.5, 0.69, and 0.87, respectively, for the first, second, third, fourth, and fifth ordered failure observations. When you use reliability software, you do not have to look up the median ranks in tables or perform manual calculations. The program automatically calculates and applies the median ranks to each observation.

10.6.4 CENSORED DATA OR SUSPENSIONS

It is an unavoidable data analysis problem that, when you are observing and analyzing, not all the units will have failed. You know the age of the unfailed units and that they are still working properly, but not when they will fail. Also, some units may have been replaced preventively. In those cases, you don't know when they would have failed. These units are said to be *suspended* or *right-censored*. Left-censored observations are failures whose starting times are unavailable. While not ideal, statistically, you can still use censored data since you know that the units lasted *at least* this long. Ignoring censored data would seriously underestimate the item's reliability.

Account for censored data by modifying the order numbers before plotting and determining the CDF values (from Benard's formula or the median ranks table). Assuming item 2 was removed without having failed, the order numbers 1, 2, 3, 4, and 5 in Section 10.6.3 would become 1, 2.5, 3.75, and 5.625. The formula to calculate these modified orders is in Appendix A.3. Consequently, the modified median ranks would become those shown in Table 10.3.

Now you can plot the observations on Weibull probability paper normally.

10.6.5 THE THREE-PARAMETER WEIBULL

For the reasons described in Section 10.6.1, including physical considerations, the time of the observations may need to be shifted for the Weibull data to plot to a straight line. By changing the origin, you activate another parameter in the Weibull equation. To make the model work, you must estimate three parameters rather than two. See Appendix A.4 for the procedure to estimate the third parameter, also called the location parameter, the guaranteed life, or the minimum life parameter.[7] A two-parameter Weibull model is a special case of the more general three-parameter model, with the location parameter equal to zero.

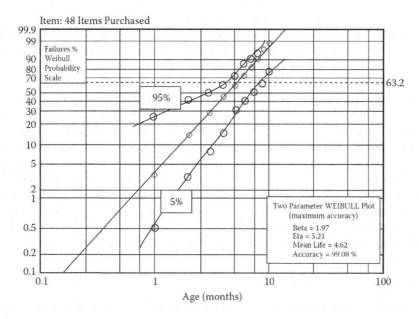

FIGURE 10.12 Weibull plot showing 5% and 95% confidence limits.

TABLE 10.4
Cumulative Probability Distribution, Including 95% and 5% Ranks

Month	Jan	Feb	Mar	Apr	May	Jun	Jul	Aug	Sep	Oct
Failures	2	5	7	8	7	6	5	4	3	1
Med rank	3.47	13.80	28.28	44.83	59.31	71.72	82.06	90.34	96.53	98.57
95% rank	9.51	23.17	39.57	56.60	70.41	81.43	89.85	95.91	29.26	99.89
5% rank	6.75	7.05	18.57	33.43	47.52	60.43	71.93	81.94	90.49	13.95

10.6.6 CONFIDENCE INTERVALS

You need to be confident that any actions you take based on your statistical analysis and observed data modeling will be successful. In Section 10.6 and Figure 10.9, we plotted failure observation times of 48 failures over a period of 10 months against the CDF values estimated by the median rank. The median rank estimates that 50% of the time, the true percentage of failures lies above and below it.

Similarly, you can estimate the CDF according to another set of tables, the 95% rank tables. This allows that 95% of the time, the percentage of failures will be below this value. The same applies to a 5% rank, where 95% of the time, the percentage will be above this value. Table 10.4 includes rows for the 5% and 95% rank values into our original example.

The 5% and 95% rank tables are given in Appendix A.5. Figure 10.12 shows the data from Table 10.4 on the Weibull plot. From the Weibull plot of the median, between 95%, and 5% rank lines, you can conclude that the distribution function at the end of May will have a value between 69% and 49%, with a confidence of 90%. This means that, after five months, in 90% of the tests you conduct, between 47% and 17% of the items will fail. Or, the reliability (which is 100% minus the CDF value) of this type of item surviving five months is between 30% and 53% for 90% of the time.

The vertical distance between the 5% and 95% lines represents a 90% confidence interval. By plotting all three lines, you get a confidence interval showing that the cumulative probability of failure will range from a to b with 90% confidence (90% of the time). As a second example, an item's failure probability, up to and including June, is between 60.43% and 81.43% for 90% of the time. Or, knowing that $R(t) = 1 - F(t)$, the item's reliability that it will survive to the end of six months is between 18.57% and 39.57% with 90% confidence. See Appendix A.5 for another example of how this methodology achieves a confidence interval for an item's reliability.

10.6.7 GOODNESS OF FIT

You would, quite naturally, like to have a quantitative measure of how well your model fits the data. How good is the fit? Goodness-of-fit testing provides the answer. Methods such as least squares or maximum likelihood (described in Appendix A.6) are used to fit an assumed distribution model to the data and estimate the parameters of its distribution function. That information, and the confidence intervals discussed in the previous section, will help you judge the validity of your model choice (be it Weibull, log-normal, negative exponential, or another) and your estimation method. One of the methods commonly used is the Kolmogorov-Smirnov test, described in Appendix A.7.

PERMISSIONS

* Chapter adapted from pages 229–249, *Asset Management Excellence: Optimizing Equipment Life-Cycle Decisions*, 2nd Edition by Editor(s), John D. Campbell, Andrew K. S. Jardine, Joel McGlynn Copyright (2011) by Imprint. Reproduced by permission of Taylor & Francis Group.

REFERENCES

1. Frankel, E.G., *Systems Reliability and Risk Analysis*, 2nd ed., Norwell, MA, Kluwer Academic, 1988.
2. Abernethy, R.B., *The New Weibull Handbook*, 5th ed., 2006, https://www.amazon.com
3. The Aladon Network, ACTOR software platform https://www.aladon.com
4. ReliaSoft, https://www.reliasoft.com/Weibull
5. Jardine, A.K.S., and A.H.C. Tsang, *Maintenance, Replacement, and Reliability: Theory and Applications*, 3rd ed., Boca Raton, CRC Press, 2022.
6. Fulton Findings, http://www.weibullnews.com
7. Kapur, K.C., and L.R. Lamberson, *Reliability in Engineering Design*, New York, John Wiley & Sons, 1991.

11 Maintenance Optimization Models*

Andrew K. S. Jardine

Maintenance optimization is about *getting the best result, given one or more assumptions*. In this chapter, we introduce the concept of optimization through a well-understood traveling problem: identifying the best mode of travel depending on different requirements. We also examine the importance of building mathematical models of maintenance-decision problems to help arrive at the best decision.

We look at key maintenance-decision areas: component replacement, capital equipment replacement, inspection procedures, and resource requirements. We use optimization models to find the best possible solution for several problem situations.

11.1   WHAT IS OPTIMIZATION ALL ABOUT?

Optimal means the most desirable outcome possible under restricted circumstances. For example, following a reliability-centered maintenance (RCM) study, you could conduct condition-monitoring tactics, time-based maintenance, or discard specific machine or system parts. In this chapter, we introduce maintenance-decision optimization. In the next chapter, we discuss detailed models for asset maintenance and replacement decision-making.

To understand the concept of optimization, consider this travel routing problem: you have to take an airplane trip, with three stops, before returning home to Chicago. The first destination is London, followed by Moscow, and then Hawaii. Before purchasing a ticket, consider several options, including airlines, fares, and schedules. You'd make your decision based on factors such as economy, speed, safety, and extras:

- If *economy* is most important, you will choose the airline with the cheapest ticket. That would be the optimal solution.
- If *speed* were it, you'd consider only the schedules and disregard the other criteria.
- If you wanted to optimize *safety*, you'd avoid airlines with a dubious safety record and pick only a well-regarded carrier.
- If you wanted a free hotel room (an *extra*) for three nights in Hawaii, you'd opt for the airline that would provide that benefit.

This list illustrates the concept of optimization. When you optimize in one area – economy, for example – you almost always get a less desirable (suboptimal) result in one or more of the other criteria.

Sometimes, you have to make a trade-off between two criteria. For example, though speed may be most important to you, the cost of traveling on the fastest schedule could be unacceptable. The solution is somewhere in the middle, providing an acceptable cost (but not the very lowest) and speed (but not the very fastest).

In any optimization situation, including maintenance-decision optimization, you should do the following:

- *Think* about optimization when making maintenance decisions.
- Consider *what* maintenance decision you want to optimize.
- Explore *how* you can do this.

11.2   THINKING OPTIMIZATION

Thinking about optimization means considering trade-offs: the pros and cons. Optimization always involves getting the best result where it counts most while consciously accepting less than that elsewhere.

The vice president asked a customer service manager of marketing what he thought his main mission was. His answer: To get every order for every customer delivered without fail on the day the customer specified, 100% of the time. To achieve this goal, the inventory of ready-to-ship goods would have to include every color, size, and style in sufficient quantities to ensure it could be shipped no matter what was called for. In spite of unusually big orders, a large number of customers randomly wanting the same thing at the same time, or machinery failure, the manager would have to deliver. His inventory would have had an unacceptably high cost.

The manager failed to realize that a delivery performance of just slightly less, say 95%, would be better. It would be a profit-optimization strategy, the best trade-off between inventory cost and an acceptable and competitive customer satisfaction level.

11.3   WHAT TO OPTIMIZE

As in other areas, you can optimize maintenance for different criteria, including cost, availability, safety, and profit.

Lowest-cost optimization is often the maintenance goal. The cost of the component or asset, labor, lost production, and perhaps even customer dissatisfaction from delayed deliveries are all considered. Where equipment or component wear-out is a factor, the lowest possible cost is usually achieved by replacing machine parts late enough to get good service out of them but early enough for an acceptable rate of on-the-job failures (to attain a "zero" rate, you'd probably have to replace parts every day).

Availability can be another optimization goal: getting the right balance between taking equipment out of service for preventive maintenance and suffering outages due to breakdowns. If *safety* is most important, you might optimize for the safest possible solution but with an acceptable impact on cost. If you optimize for *profit*, you will consider cost and the effect on revenues through greater customer satisfaction (better profits) or delayed deliveries (lower profits).

11.4 HOW TO OPTIMIZE

One of the main tools in the scientific approach to management decision-making is building an evaluative model, usually mathematical, to assess various alternative decisions. Any model is simply a representation of the system under study. We frequently use a symbolic model when applying quantitative techniques to management problems. The system's relationships are represented by symbols and properties described by mathematical equations.

To understand this model-building approach, examine the following maintenance stores problem. Although simplified, it illustrates two important aspects of model use: constructing the studied problem and its solution.

A Stores Problem

A store controller wants to know how much to order each time an item's stock level reaches zero. The system is illustrated in Figure 11.1.

The conflict is that the more items are ordered at any time, the more ordering costs will decrease, but holding costs will increase since more stock is kept on hand. These conflicting features are illustrated in Figure 11.2. The stores' controller wants to determine which order quantity will minimize the total cost. This total cost can be plotted, as shown in Figure 11.2, and used to solve the problem.

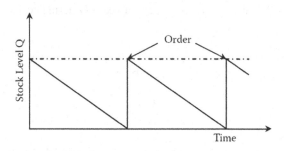

FIGURE 11.1   An inventory problem.

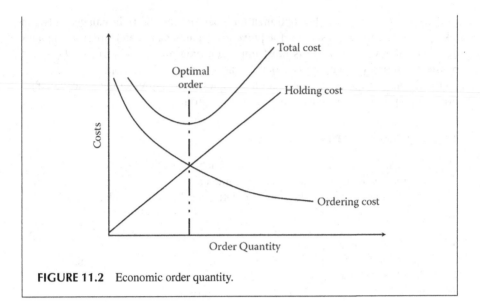

FIGURE 11.2   Economic order quantity.

However, a much more rapid solution is to construct a mathematical model of the decision situation. The following parameters can be defined:

D = total annual demand
Q = order quantity
C_o = ordering cost per order
C_h = stockholding cost per item per year

Total cost per year of ordering and holding stock = Ordering cost per year + Stockholding cost per year
Now,

$$\text{Ordering} \frac{\text{cost}}{\text{year}} = \text{Number of orders placed per year} \times \text{Ordering cost per order} = \frac{D}{Q}C_o$$

Stockholding cost/year = Average number of items in stock per year (assuming linear decrease of stock) × Stockholding cost per item

$$\frac{Q}{2}C_h$$

Therefore, the total cost per year, which is a function of the order quantity and is denoted $C(Q)$, is

$$C(Q) = \frac{D}{Q}C_o + \frac{Q}{2}C_h \tag{11.1}$$

Equation 11.1 is a mathematical model of the problem relating order quantity, Q, to total cost, $C(Q)$. The stores' controller wants to know the number of items to order to minimize the total cost, that is, the right-hand side of Equation 11.1. The answer is obtained by differentiating the equation with respect to Q, the order quantity, and equating the answer to zero as follows:

$$\frac{dC(Q)}{dQ} = -\frac{D}{Q^2}C_o + \frac{C_h}{2} = 0$$

Therefore,

$$\frac{D}{Q^2}C_o = \frac{C_h}{2}$$

$$Q = \sqrt{\frac{2DC_o}{C_h}} \qquad (11.2)$$

Since the values of D, C_o, and C_h are known, substituting them into Equation 11.2 gives the value of the order quantity Q. You can check that the value of Q obtained from Equation 11.2 is a minimum and not a maximum by taking the derivative of $C(Q)$ and noting the positive result. This confirms that Q is optimal.

Example

Let $D = 1{,}000$ items, $C_o = \$5$, $C_h = \$0.25$

$$Q = \sqrt{\frac{2 \times 1000 \times 5}{0.25}} = 200\,\text{items}$$

Each time the stock level reaches zero, the stores' controller should order 200 items to minimize the total cost per year of holding and ordering stock.

Note that the various assumptions made in the inventory model may not be realistic. For example, no consideration has been given to quantity discounts, the possible lead time between placing an order and its receipt, and the fact that demand may not be linear or known for certain. The purpose of the previous model is to illustrate constructing a model and attaining a solution for a particular problem. There is abundant literature about stock control problems without many of these limitations. If you are interested in stock control aspects of maintenance stores, see Nahmias.[1]

It's clear from the previous inventory control example that we need the right kind of data, properly organized. Most organizations have a computerized maintenance management system (CMMS) or an enterprise asset management

(EAM) system. The vast amount of data they store makes optimization analyses possible.

Instead of building mathematical models of maintenance-decision problems, the software is available to help you make optimal maintenance decisions. This is covered in Chapter 12.

11.5   KEY MAINTENANCE MANAGEMENT DECISION AREAS

Maintenance managers must address four key decision areas to optimize their organization's human and physical resources. These areas are depicted in Figure 11.3. The first column deals with component replacement, the second with inspection decisions, including condition-based maintenance, and the third with establishing the economic life of capital equipment. The final column addresses decisions concerning resources required for maintenance and their location.

To build strong maintenance optimization, you need an appropriate source, or sources, of data. The foundation for this, as shown in Figure 11.3, is the CMMS/ EAM system/enterprise resource planning (ERP) system.

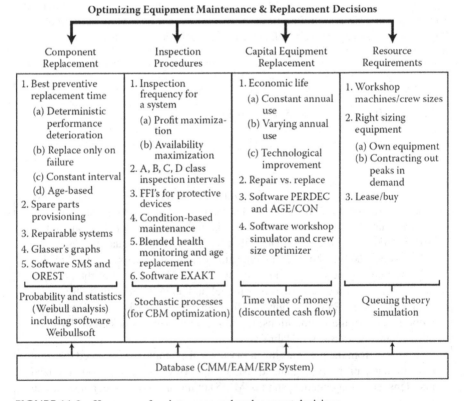

FIGURE 11.3   Key areas of maintenance and replacement decisions.

Chapter 12 discusses optimizing key maintenance decisions of component replacement, inspection procedures, and capital equipment replacement (Columns 1, 2, and 3 of Figure 11.3). The framework, foundation, or database is discussed in detail in Chapter 5.

Extensive development and discussion of models, including case studies, are provided in Jardine and Tsang.[2]

PERMISSIONS

* *Asset Management Excellence: Optimizing Equipment Life-Cycle Decisions*, 2nd Edition by Editor(s), John D. Campbell, Andrew K. S. Jardine, Joel McGlynn Copyright (2011) by Imprint. Reproduced by permission of Taylor & Francis Group.

REFERENCES

1. Nahmias, S., *Production and Operations Analysis*, 3rd ed., New York, Irwin/McGraw-Hill, 1997.
2. Jardine, A.K.S., and A.H.C. Tsang, *Maintenance, Replacement, and Reliability: Theory and Applications*, 3rd ed., Boca Raton, CRC Press, 2022.

12 Optimizing Maintenance and Replacement Decisions*

Andrew K. S. Jardine
Previous contributions from Andrew K. S. Jardine and Murray Wiseman

12.1 CHAPTER OVERVIEW

This chapter explores the strategies and tools you need to make the best maintenance and replacement decisions. In particular, you need to know the optimal replacement time for critical system components (line-replaceable units, or LRUs) and capital equipment.

At the LRU level, we examine age- and block-replacement strategies. You'll learn how OREST[1] software can help you optimize LRU maintenance decisions to keep costs under control and increase equipment availability.

With capital equipment, it's critical to establish economic viability. We examine how to do this in two operating environments:

- Constant use, year-by-year
- Declining use, year-by-year (older equipment is used less)

In this section, you'll discover how to extend the life of capital equipment through a significant repair or rebuild and when that is more economical than replacing it with new equipment. The optimal decision is the one that minimizes the long-run equivalent annual cost (EAC), the life-cycle cost of owning, using, and disposing of the asset.

Our study of capital equipment replacement includes AGE/CON[1] and PERDEC[1] software, which simplifies the job of managing assets.

Finally, we look at maintenance resources: what resources should be, where they should be located, who should own them, and how they should be used. The role that the theory of queues and simulation can play will be highlighted.

12.2 INTRODUCTION: ENHANCING RELIABILITY THROUGH PREVENTIVE REPLACEMENT

Generally, preventing a maintenance problem is always preferred to fixing it after the fact. You'll increase your equipment reliability by learning to replace critical

DOI: 10.1201/9781032679600-15

components optimally before a breakdown occurs. When is the best time? That depends on your overall objective. Do you most want to minimize costs or maximize availability? Sometimes the best preventive replacement time accomplishes both objectives, but not necessarily.

Before you start, you need to obtain and analyze data to identify the best preventive replacement time. Later in this section, we present some maintenance policy models of fixed interval and age-based replacements to help you do that.

But, for the preventive replacement to work, these two conditions must first be present:

- If cost is most important, the total cost of a replacement must be more significant after failure than before. If reducing total downtime is essential, the total downtime of a failure replacement must be greater than a preventive replacement. In practice, this usually happens.
- The risk of a component failing must increase as it ages. How can you know? Check that the Weibull shape parameter associated with the component's failure times is greater than 1. Often, it is assumed that this condition is met, but be sure that is the case. See Chapter 10 for a detailed description of Weibull analysis.

12.2.1 BLOCK REPLACEMENT POLICY

The block replacement policy is sometimes called the group or constant interval policy since preventive replacement occurs at fixed times and failure replacements whenever necessary. The policy is illustrated in Figure 12.1. C_p is the total cost of a preventive replacement, C_f is the total cost of a failure replacement, and t_p is the interval between preventive replacements.

You can see that there isn't a failure for the first cycle of t_p, while there are two in the second cycle and none in the third and fourth. As the interval between preventive replacements decreases, fewer failures will occur. You want to obtain the best balance between the investment in preventive replacements and the economic consequences of failure replacements. This conflict is illustrated in Figure 12.2.

$C(t_p)$ is the total cost per week of preventive replacements occurring at intervals of length t_p, with failure replacements occurring whenever necessary. See Appendix A12 for the total cost curve equation.

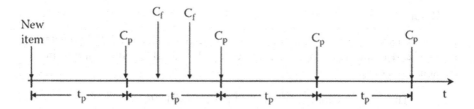

FIGURE 12.1   Constant interval replacement policy.

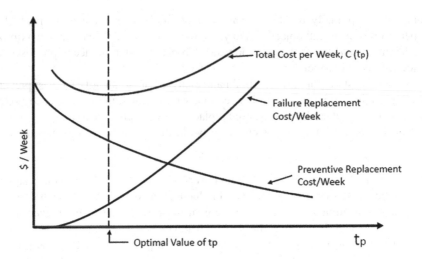

FIGURE 12.2   Constant interval policy: optimal replacement time.

The following problem is solved using OREST[1] software, incorporating the cost model and Weibull analysis, to establish the best preventive replacement interval.

PROBLEM

There has been a bearing failure in the blower used in diesel engines. The failure has been established according to a Weibull distribution, with a mean life of 10,000 km and a standard deviation of 4,500 km. The bearing failure is expensive, costing ten times as much to replace than if it had been done as a preventive measure. Determine:

- The optimal preventive replacement interval (or block policy) to minimize total cost per kilometer
- The expected cost saving of the optimal policy over a run-to-failure replacement policy
- Given that the cost of a failure is $2,000, the cost per km. of the optimal policy

Figure 12.3 shows a screen capture from OREST:

RESULT

You can see that the optimal preventive replacement time is 3,873 kilometers (3,872.91 in the chart). Also, the figure provides valuable additional information. For example:

- The cost per kilometer of the best policy is: $0.09/km
- The cost saving compared to a run-to-failure policy is: $ 0.11/km (55%)

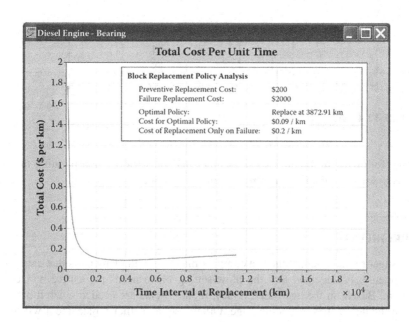

FIGURE 12.3   OREST output: block replacement.

12.2.2 AGE-BASED REPLACEMENT POLICY

In the age-based policy, the preventive replacement time depends upon the component's age. If a failure replacement occurs, the component's time clock is reset to zero, unlike the block replacement policy, where preventive replacements occur at fixed intervals regardless of the operating component's age. In this case, the component is only replaced once it reaches the specified age.

Figure 12.4 illustrates the age-based policy. C_p and C_f represent the block replacement policy and t_p represent the component age when preventive replacement occurs.

In Figure 12.4, you can see that there aren't any failures in the first cycle, which is t_p. After the first preventive replacement, the component must have failed before reaching its next planned preventive replacement age. The clock is set to zero after the first failure replacement, and the next preventive replacement is scheduled. However, before it's reached, the component is again replaced due to failure. After this second failure replacement, the component survives to the planned preventive replacement age of t_p. Similarly, the next replacement cycle shows the component has reached its planned preventive age.

The conflicting cost consequences are identical to those in Figure 12.2, except that the x-axis measures the item's actual age (or utilization) rather than a fixed time interval. See Appendix A for the mathematical model depicting this preventive replacement policy.

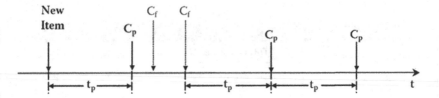

FIGURE 12.4   Age-based replacement policy.

The following problem is solved using OREST software incorporating the cost model to establish the best preventive replacement age.

PROBLEM

A sugar refinery centrifuge is a complex machine with many parts that can fail suddenly. It's decided that the plow-setting blade needs preventive replacement. Based on the age-based policy, replacements are needed when the setting blade reaches a specified age. Otherwise, a costlier replacement will be needed when the part fails. Consider:

- The optimal policy to minimize the total cost per hour associated with preventive and failure replacements
- To solve the problem, you have the following data:
- The labor and material cost of a preventive or failure replacement is $2,000
- The value of production losses for a preventive replacement is $1,000, and $7,000 for a failure replacement
- The failure distribution of the setting blade can be described adequately by a Weibull distribution with a mean life of 152 hours and a standard deviation of 30 hours.

RESULT

Figure 12.5 shows a screen capture from OREST. As you can see, the optimal preventive replacement age is 112 hours (111.49 hours in figure), and there's additional key information that you can use. For example, the preventive replacement policy costs 44.8% of run-to-failure ([(59.21–32.66)/59.21]), making the benefits very clear. Also, the total cost curve is relatively flat in the 90 to 125 hours region, providing a flexible planning schedule for preventive replacements

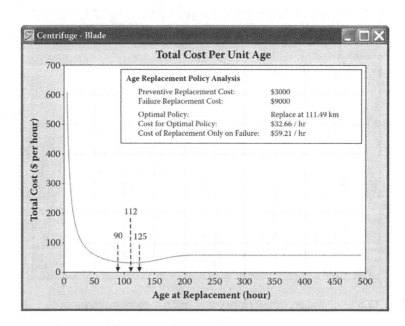

FIGURE 12.5 OREST output: Age-based replacement.

12.2.2.1 When to Use Block Replacement

On the face of it, age replacement seems to be the only sensible choice. Why replace a recently installed component that is still working properly? In age replacement, the component remains in service until its scheduled preventive replacement age.

To implement an age-based replacement policy, you must keep an ongoing record of the component's current age and change the planned replacement time if it fails. Clearly, this cost is justified for expensive components, but for an inexpensive one, the easily implemented block replacement policy is often more cost-efficient.

12.2.2.2 Safety Constraints

With block and age replacement, the objective is to establish the best time to preventively replace a component, to minimize the total cost of preventive and failure replacements.

If you want to ensure that the failure probability doesn't exceed a particular value, say 5%, without cost considerations being taken into account, you can determine when to schedule a preventive replacement from the cumulative failure distribution. This is illustrated in Figure 12.6. You can see that the preventive replacement should be planned once the item is at 3,000 km.

Alternatively, you can preventively replace a critical component so that the risk, or hazard, doesn't exceed a specified value, such as 5×10^{-5} failures per kilometer. In this case, you must use the hazard plot to identify the preventive replacement age. This is illustrated in Figure 12.7, which shows the component's appropriate preventive replacement age is 4,000 km.

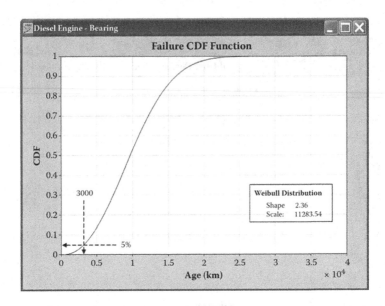

FIGURE 12.6   Optimal preventive replacement age: risk-based maintenance.

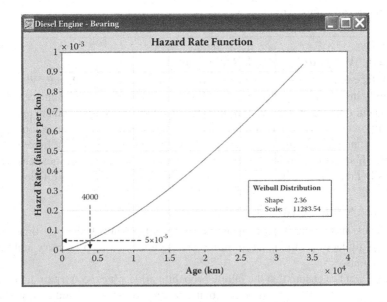

FIGURE 12.7   Optimal preventive replacement age: Hazard limit.

12.2.2.3   Minimizing Cost and Maximizing Availability

In block and age replacement, the objective is minimizing cost. Maximizing availability requires replacing the models' total preventive and failure replacement costs with their total downtime. Minimizing total downtime is then equivalent to maximizing availability. In Appendix A, you'll find total downtime minimization models for block and age replacement.

12.3   DEALING WITH REPAIRABLE SYSTEMS

Rather than completely replace a failed unit, you may be able to get it operating with minor corrective action. This is what's known as a minimal and general repair. Models for addressing this case are discussed in Jardine and Tsang.[2]

12.4   ENHANCING RELIABILITY THROUGH INSPECTION

A big part of the maintenance mandate, of course, is to ensure that the system is reliable. One way to achieve this is to identify critical components that are likely to fail and preventively replace them. Section 12.2 covered methods to identify which components (line-replaceable units) are candidates for preventive replacement and how to decide when.

An alternative is to consider the system as a whole and make regular inspections to identify problem situations. Then, perform minor maintenance to prevent system failure, such as changing a component or topping up the gearbox with oil. You need to know, though, the best frequency of inspection.

Yet another approach is to monitor the system's health through predictive maintenance and only act when you get a signal that a defect is about to happen, which, if not corrected, will create a system failure. This second approach is covered in detail in Section 12.4.

12.4.1   ESTABLISHING THE OPTIMAL INSPECTION FREQUENCY

Figure 12.8 shows a system composed of five components, each having its failure distribution. (In Reliability Centered Maintenance terminology, these components are different system failure modes.)

Because the Weibull failure pattern is so flexible, all the components can likely be described as failing but with different shape parameters. Depending on the component, the risk of it failing as it ages can either increase, remain constant, or decrease. The overall effect is that there will be system failures from any number of causes. If you analyze the overall system failure data, a Weibull would fit again. The shape parameter almost certainly would equal 1.0, indicating that system failures occur strictly randomly. This is what you should expect. The superposition of numerous failure distributions creates an exponential failure pattern, the same as a Weibull with a shape parameter of 1.0 (Drenick[3]).

To reduce these system failures, you can inject inspections, with minor maintenance work, into the system, as shown in Figure 12.8. The question is then: How frequently should inspections occur?

While you may not know the individual risk curves of system components, you can determine the overall system failure by examining the maintenance records. The pattern will almost certainly look like Figure 12.9, where the system failure rate is constant but can be reduced by increasing the inspection frequency.

The risk curve may not be constant if system failures emanate from one leading cause. In this case, the system failure distribution will be identical to the component's failure pattern.

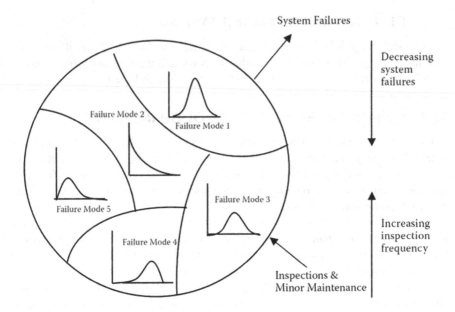

FIGURE 12.8   System failures.

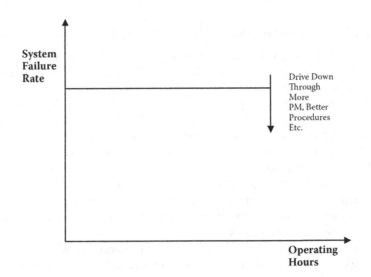

FIGURE 12.9   System failure rate.

Another way of viewing the situation is to consider the system's mean time to failure (MTTF). As the system failures decline, the MTTF will increase. This is illustrated in the probability density functions in Figure 12.10.

If the optimal inspection frequency minimizes the total downtime of the system, you get the conflicting curves of Figure 12.11. See Appendix A for the underlying mathematical model that establishes the optimal frequency.

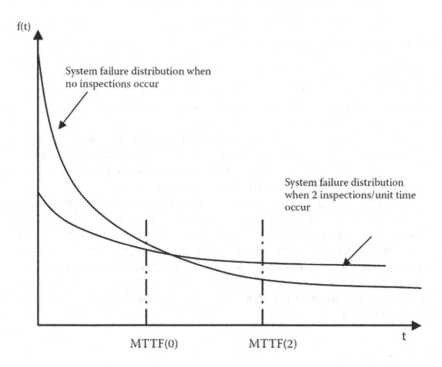

FIGURE 12.10 Inspection frequency versus MTTF.

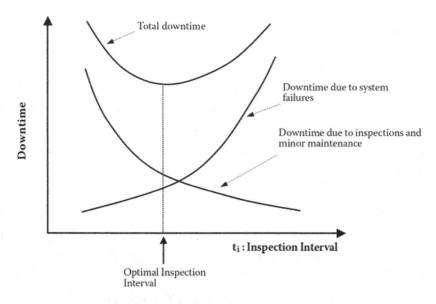

FIGURE 12.11 Optimal inspection frequency.

Case Study

An urban transit fleet had a policy of inspecting the buses every 5,000 kilometers. An A, B, C, or D check occurred at each inspection. An A check is a minor maintenance inspection, while a D check is the most detailed. The policy is illustrated in Table 12.1.

Since the 5,000 km. inspection policy was not followed precisely, some inspections took place before 5,000 km, and others were delayed. As a result, Figure 12.12 shows that the average distance traveled by bus between failures decreased as the interval between the checks was increased.

Knowing the average time a bus was out of service due to repair and inspection established the optimal inspection interval at 8,000 km, as Figure 12.13 shows. Note that the total downtime curve is very flat around the optimum, so the current policy of making inspections at the easily implemented 5,000 km interval might be best. Before the analysis, though, whether the current policy was appropriate or should be modified wasn't known. As you can see, the data-driven analysis revealed the answer. The Jardine and Hassounah study provides full details.[4]

TABLE 12.1
A, B, C, and D Inspection Policy

Transit Commission's Bus Inspection Policy Inspection Type

Km (1,000)	"A"	"B"	"C"	"D"	
5	X				
10		X			
15	X				
20			X		
25	X				
30		X			
35	X				
40			X		
45	X				
50		X			
55	X				
60			X		
65	X				
70		X			
75	X				
80				X	
Total	8	4	3	1	$\Sigma = 16$
Ri	0.5	0.25	0.1875	0.0625	$\Sigma = 1.0$
Where					

R_i = No of type I inspections/Total No. of inspections, I = A, B, C, or D

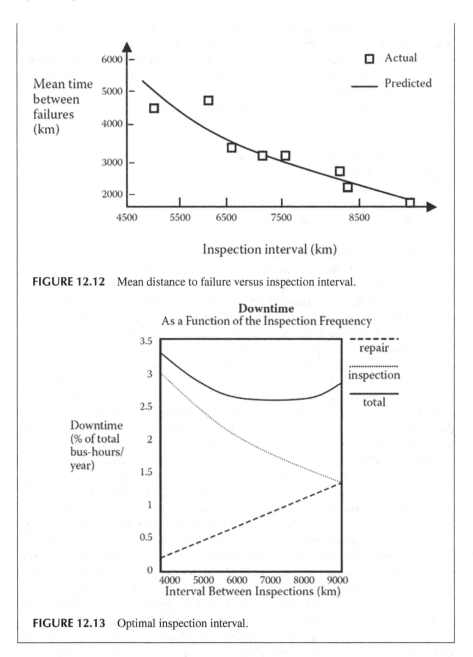

FIGURE 12.12   Mean distance to failure versus inspection interval.

FIGURE 12.13   Optimal inspection interval.

12.5   ENHANCING RELIABILITY THROUGH INSPECTION: OPTIMIZING CONDITION-BASED MAINTENANCE

12.5.1   INTRODUCTION

Condition-Based Maintenance (CBM) is a good idea. Of course, you want the maximum useful life from each physical asset before taking it out of service for preventive maintenance. But translating this into an effective monitoring program and timely maintenance isn't necessarily easy. It can be difficult to select the monitoring parameters most likely to indicate the machine's state of health and then to interpret the influence of those measured values on the machinery's remaining useful life (RUL). We address these problems in this chapter.

The essential questions to pose when implementing a CBM program are:

Why monitor?
What equipment components to monitor?
What monitoring technologies to use?
How (what signals) to monitor?
When (how often) to monitor?
How to interpret and act upon the condition monitoring results?

Reliability-Centered Maintenance (RCM), described in Chapter 8, helps to find the correct answers to the first three questions. However, additional optimizing methods are required to handle the remainder. In Chapters 10 and 11, you learned that the way to approach these problems is to build a model describing the factors surrounding maintaining or replacing equipment. In Chapter 10, we dealt with the lifetime of components considered independent random variables, meaning that no additional information, other than equipment age, is used to schedule preventive maintenance.

CBM, however, introduces new information, called condition data, to determine more precisely the most advantageous moment to make a repair or replacement. We extend the models in Chapter 10, Section 10.2, to include the influence of condition data on the RUL of machinery and its components. The extended modeling method we introduce in this chapter, which considers measured data, is known as Proportional Hazards Modeling (PHM), and the measured condition data are referred to as covariates.

Since D.R. Cox's (1972) pioneering paper[5] on PHM, it has been used primarily to analyze medical survival data. Since 1985, it has been used extensively, including applications to marine gas turbines, nuclear reactors, aircraft engines, and disk brakes on high-speed trains. In 1995 A.K.S. Jardine and V. Makis at the University of Toronto initiated the CBM Laboratory[6] to develop general-purpose software applying proportional hazards models to available maintenance data. The software was designed to be integrated into the plant's maintenance information system to optimize its CBM activities. The result, in 1997, was a program called EXAKT, now in its fourth version and rapidly earning attention as a CBM optimizing methodology. We produced the examples in this chapter, with their graphs and calculations, using the EXAKT program. For a demonstration version of EXAKT, contact OMDEC.[7]

Industry has adopted various monitoring methods that produce a signal when a failure is about to occur. The most common are vibration monitoring and oil analysis. Moubray[8] provides an overview of condition monitoring techniques, including the following:

Vibration analysis
Ultrasonic analysis
Ferrography

Magnetic chip detection
Atomic emission spectroscopy
Infrared spectroscopy
Gas chromatography
Moisture monitoring
Liquid dye penetrants
Magnetic particle inspection
Power signature analysis

As Pottinger and Sutton[9] said, "Much condition-monitoring information tells us something is not quite right, but it does not necessarily inform us what margins remain for exploitation before real risk, particularly real commercial risk, occurs." In this chapter, you will learn how to accurately estimate equipment health using condition monitoring. The goal is to make the optimal maintenance decision, blending economic considerations with estimated risk.

Early work on estimating equipment risk dealt with jet engines on aircraft.[10] Oil analysis was conducted weekly, and if unacceptable metal levels were found in the samples, the engine was removed before its scheduled removal time of 15,000 flying hours. A PHM was constructed by statistically analyzing the condition data and the corresponding age of the engines that functioned for the duration and those removed due to failure (in this case, operating outside tolerance specifications). Three key factors, from a possible 20, emerged for estimating the risk that the engine would fail:

Age of engine
Parts per million iron (Fe) in the oil sample at the time of inspection
Parts per million chrome (Cr) in the oil sample at the time of inspection

The PHM also identifies the weighting for each risk factor. The complete equation used to estimate the risk of the jet engine failing was:

$$\text{Risk at time of inspection} = \frac{4.47}{24,100}\left(\frac{t}{24,100}\right)^{3.47} e^{0.41\text{Fe}+0.98Cr}$$

Where the contribution of the engine age toward the overall failure risk is

$$\frac{4.47}{24,100}\left(\frac{t}{24,100}\right)^{3.47}$$

(this is termed a Weibull baseline failure rate) and the contribution to overall risk from the risk factors from the oil analysis is $e^{0.41\text{Fe} + 0.98Cr}$.

The constants in the age contribution portion of the risk model 4.47, 24,100, and 3.47 are obtained from the data and will change depending on the equipment. They may even differ in the same equipment operating in a different environment. The key iron and chrome risk factors are also equipment and operating environment specific.

In this case, the absolute values of iron and chrome were used. In other cases, rates of change or cumulative values may be more meaningful for risk estimation. By carefully analyzing condition monitoring data and information about the age and reason for equipment replacement, you can construct an excellent risk model.

Optimizing maintenance decisions requires that more than just the risk of failure is taken into account. You may want to maximize the operating profit and/or equipment availability or minimize the total cost. In this section, we assume your objective is to minimize the total long-term preventive and failure maintenance cost. Besides determining the risk curve, you must get cost estimates for prevention and failure replacement and failure consequences.

Detecting failure modes, which gradually lessen functional performance, can dramatically impact overall costs. This, therefore, is the first level of defense in the RCM task planning logic in Figure 12.14.

The logic diagram in Figure 12.14 shows that CBM is preferred if the impending failure can be "easily" detected in ample time. This proactive intervention is illustrated in the P-F interval[11] in Figure 12.15. Investigating the relationship between past condition surveillance and past failure data helps develop future maintenance management policies and specific maintenance decisions. Modern, flexible maintenance information management systems also compile and report performance, cost, repair, and condition data in numerous ways.

Maintenance engineers, planners, and managers perform CBM by collecting data reflecting equipment or component health. These condition indicators (or covariates)

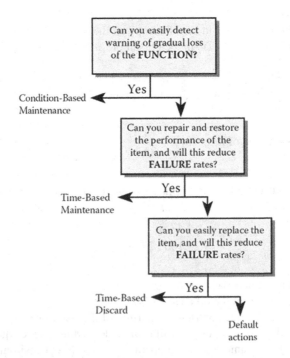

FIGURE 12.14   RCM logic diagram.

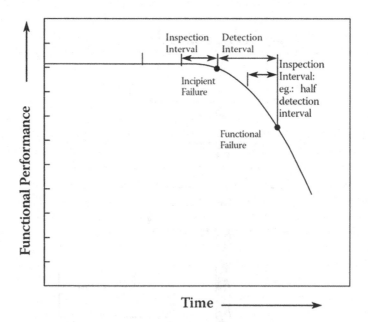

FIGURE 12.15   The P-F interval.

can take various forms. They may be continuous, such as operational temperature or feed rate of raw materials. They may be discrete, such as vibration or oil analysis measurements.[13] They may be mathematical combinations or transformations of the measured data, such as measurement change rates, rolling averages, and ratios. The condition indicator choices are endless since it's hard to know exactly why failures occur. Without a systematic means of judging and rejecting superfluous data, you will find CBM far less useful than it should be. Proportional hazard modeling is an effective approach to information overload because it can decipher the equipment's current condition and make an optimal recommendation based on substantial historic conditions and corresponding failure age data.

In this section, we describe, with examples, the key steps of the PHM process, as provided in Figure 12.16. An integral part of the process is the statistical testing of various hypotheses. This helps avoid the trap of mindlessly following a method without adequately verifying whether the model suits the situation and data.

12.5.1.1 Step 1: Data Preparation

No matter what tools or computer programs are available, you should always examine the data in several ways.[14] Many data sets can have the same mean and standard deviation and differ. That can be of critical significance. Also, instrument calibration and transcription mistakes have likely caused some data errors. You may have to search archives for significant data that are missing. To develop accurate decision models, you must be fully immersed in the operating and maintenance context. You must know the failure and repair work order process. Properly collected and validated subjective data reflecting all currently known about a problem is invaluable,

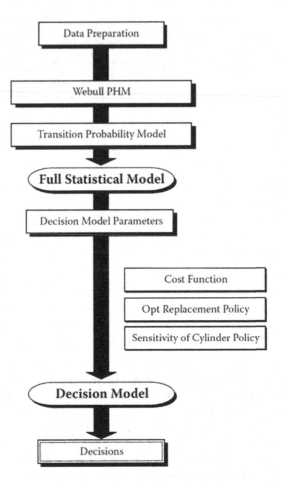

FIGURE 12.16   PHM flow diagram.

but it's not enough. You must also collect valuable objective data about the problem or process for a complete analysis and to confirm subjective opinions.[15]

Generally, the maintenance engineer or technologist starts with an underlying model based on the data type, where the observations came from, and previous experience. After obtaining or converting the data into well-structured database tables, you must verify the model in stages before plunging ahead. This is where the tools of descriptive statistics – graphics (frequency histograms, cumulative frequency curves) and numbers (mean, median, variance, skewness) – are very useful. Some of these EXAKT™ methods are described in the following sections.

Although robust computerized maintenance management systems (CMMS) and enterprise asset management (EAM) systems are growing, too little attention has been paid to data collection. Existing maintenance information management database systems are under-used. A clear relationship between accurately recorded component age data and effective maintenance decisions has been lacking. Tradespeople need to be educated about the value of such data, so they meticulously record it when replacing failed components. Consistent with the principles of Total Productive Maintenance described in Chapter 7, maintenance and operational staff are the true custodians of the data and models it creates.

12.5.1.1.1   Events and Inspections Data

Unlike the simple Weibull analysis described in Chapter 11 and applied in Section 12.2, PHM requires two types of equally important information: *Event* data and *Inspection* data. Three types of events, at a minimum, define a component's lifetime or history:

> The Beginning (B) of the component life (the time of installation)
> The Ending by Failure (EF)
> The Ending by Suspension (ES), i.e., a preventive replacement

The model should include additional events if they directly influence the measured data. One such event is an oil change. Into the model, you should input that, at each oil change, some covariates – such as the wear metals – are expected to be reset. Periodic tightening or re-calibrating the machinery may have similar effects on measured values and should be accounted for in the model.

EXAMPLE 1

In a food processing company, shear pump bearings are monitored for vibration. In this example, 21 vibration covariates from the shear pump's inboard bearing – represented in Figure 12.17 – are reduced to only the three significant ones shown in Table 12.2.

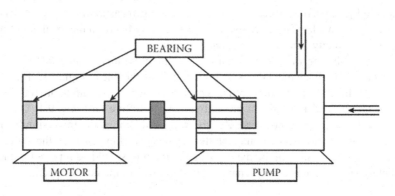

FIGURE 12.17   Shear pump.

TABLE 12.2
Significant Covariates

Indent	Date	Working Age	VEL_1A	VEL_2A	VEL_1V
B1	29-Sep-94	1	0.09	0.017	0.066
B1	08-Nov-94	41	0.203	0.018	0.113
B1	24-Nove-94	57	0.142	0.021	0.09
B1	25-Nov-94	58	0.37	0.054	0.074
B1	26-Nov-94	59	2.519	0.395	0.081
B2	11-Jan-95	46	0.635	0.668	0.05
B2	12-Jan-95	47	1.536	0.055	0.0078
B3	19-Apr-95	97	0.211	0.057	0.144
B3	04-May-95	112	0.088	0.022	0.079
B3	06-May-95	114	0.129	0.014	0.087
B3	29-May-95	137	0.225	0.021	0.023
B3	05-Jun-95	144	0.05	0.017	0.04
B3	20-Jul-95	189	1.088	0.211	0.318
B4	21-Jul-95	1	0.862	0.073	0.102
B4	22-Jul-95	2	0.148	0.153	0.038
B4	23-Jul-95	3	0.12	0.015	0.035
B4	24-Jul-95	4	0.065	0.021	0.018
B4	21-Aug-95	33	0.939	0.1	0.3

The bearing measurements were taken in three directions: axial, horizontal, and vertical. In each direction, the fast Fourier transform of the velocity vector was taken in five frequency bands. The overall velocity and acceleration were also measured. This provided a total of 21 vibration measurements from a single bearing. Example 1 analyzes these 21 signals using EXAKT• software, concluding that only three vibration monitoring measurements are key risk factors to consider. There are two different velocity bands in the axial direction and one velocity band in the vertical direction.

By combining the proportional hazard model with economic factors, you can devise a replacement decision policy, such as the one in Figure 12.18. The cost of a failure repair, compared to a preventive repair, was input into the decision model, and the ratio of preventive cost to failure cost was 9:1. On the graph, the composite covariate Z – the weighted sum of the three significant influencing factors, are points plotted against working age. If the current inspection point falls in the dark (red) area, the unit should be repaired immediately since there is a good chance it will fail before the next inspection.

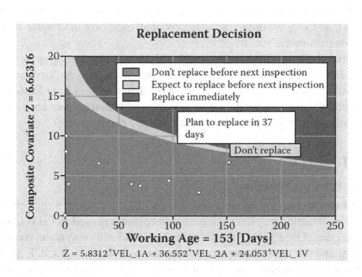

FIGURE 12.18   Optimal condition-based replacement policy.

Suppose the current point is in the medium gray (green) area. In that case, the optimal decision to minimize the long-run cost is to continue operating the equipment and inspect it at the next scheduled inspection. If the point is in the light (yellow) zone, it is optimal to keep operating it but preventively replace the component before the next scheduled inspection. The chart also indicates how much longer (remaining useful life) the equipment should run before being repaired or replaced.

Using the Figure 12.18 decision chart, total maintenance cost was reduced from $59.46/day to $26.83/day – an impressive 55% saving. Also, by following the recommendation on the chart, the mean time between bearing replacement is expected to increase by 10.2%.

12.5.1.2   Step 2: Building the PHM

Examining proportional hazards, you can see that they're an extension of the Weibull hazard function described in Chapter 10 and applied in Section 12.2.

$$h(t,Z) = \frac{\beta}{\eta}\left(\frac{t}{\eta}\right)^{\beta-1} e^{\gamma_1 Z_1(t)+\gamma_2 Z_2(t)+\dots+\gamma_n Z_n(t)} \tag{12.1}$$

The new part factors in (as an exponential expression) the covariates $Z_i(t)$. These are the measured signals at a given time t of, for example, the parts per million of iron or other wear metals in the oil sample. The *covariate parameters* γ_i specify each covariate's relative impact on the hazard function. A very low value for γ_i indicates that the covariate isn't worth measuring. You'll find that software programs provide valuable criteria to omit unimportant covariates.

To fit the proportional hazards model to the data, you must estimate not only the parameters β and η, as we did in the simple Weibull examples in Chapter 10, but also the covariate parameters γ_i.

Remember that "condition monitoring" isn't monitoring the equipment's condition per se but variables you *think* are related. Those variables or "covariates" influence the failure probability shown by the hazard function, $h(t)$, in Equation 12.1. From the model you construct, based on past data, you want to learn each covariate's degree of influence (namely the size of the covariate parameters γ_i).

12.5.2 THE OPTIMAL DECISION

12.5.2.1 The Cost Function

Chapter 10 taught you how models optimize an objective, such as the overall long-run cost of maintaining a system. An analogous process is used in PHM to optimize the cost function. Once again, we compare the costs for a preventive versus failure replacement and ask the software to calculate the cost function graph, Figure 12.19, which shows the minimum cost associated with an optimal replacement policy.

The cost function is the sum of the costs due to preventive replacements (upper part) and failure replacements (lower). When the risk level increases, the cost of failure replacements also increases. When the risk level rises to infinity, the cost function increases to the cost of the failure replacement policy indicated by the value at the right-hand edge of the graph. You want to select the lowest possible risk level (horizontal axis) without increasing the total maintenance cost per hour. Naturally, zero risk would entail infinite cost. An infinite risk is equivalent to a "run-to-failure" policy whose cost is indicated by the dashed line.

12.5.2.2 The Optimal Replacement Decision Graph

The replacement decision graph, Figure 12.20, reflects the entire modeling exercise to date. It combines the proportional hazards model results, the transition probability

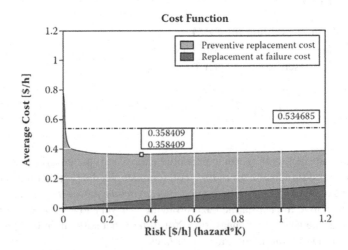

FIGURE 12.19   Cost function.

Replacement Decision

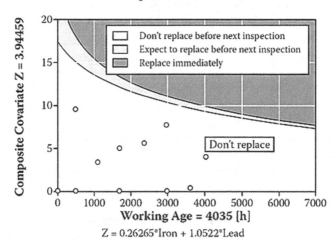

FIGURE 12.20   The optimal replacement decision graph.

model, and the cost function into the best decision policy for the component or system in question.

The ordinate is the composite covariate, Z – a balanced sum of covariates statistically influencing failure probability. Each covariate's contribution is weighted by its influence on the failure risk in the next inspection interval.

A significant advantage of this system is that a single graph combines the information you need to make a replacement decision. The alternative is to examine the trend graphs of perhaps dozens of parameters and "guess" whether to replace the component immediately or a little later. The Optimal Replacement Decision Graph recommendation is the most effective guide to minimize maintenance costs *in the long run*.

12.5.3   SENSITIVITY ANALYSIS

Considering your plant's ever-changing operations, how do you know that the optimum replacement decision graph constitutes the best policy? Are the replacement assumptions you used still valid? If not, What will be the effect of those changes? Is your decision still optimal? These questions are addressed by sensitivity analysis.

The assumption you made in building the cost function model centered around the relative costs of a planned replacement versus those of a sudden failure. That cost ratio may have changed. If your accounting methods don't provide exact repair costs, you have to estimate them when building the cost portion of the decision model. In either case, these uncertain costs can doubt whether the Optimal Replacement Decision Graph policy is well founded.

The sensitivity analysis allays unwarranted fears and indicates how to obtain more accurate cost data.

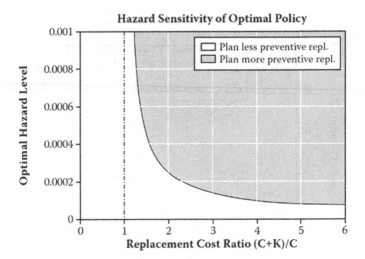

FIGURE 12.21   Sensitivity of optimal policy to cost ratio.

Figure 12.21, the Hazard Sensitivity of Optimal Policy graph, shows the relationship between the optimal hazard or risk level and the cost ratio. If the cost ratio is low, less than 3, the optimal hazard level will increase exponentially. We need, then, to track costs very closely to substantiate the benefits calculated by the model. On the other hand, if the cost ratio is in the 4 to 6 range, the curve is relatively flat, and the optimal replacement decision graph is accurate.

The Cost Sensitivity of the Optimal Policy graph, Figure 12.21, has two lines.

Solid line: If the actual $(C + K)/C$ (the cost of a failure replacement divided by the cost of a preventive replacement) differs from that specified when you built the model, it means that the current policy (as dictated by the Optimal Replacement Graph) is no longer optimal. The solid line tells how much % more you're paying above the optimal cost/unit time initially calculated.

Dashed line: Again, assume the actual cost ratio has strayed, and you want to rebuild the model using the new $(C + K)/C$. The dashed line tells how much your new optimal cost would change if you follow the new policy. The sensitivity graphs assume that only Cf (failure repair cost) changes and Cr (planned repair cost) remains the same.

12.5.4 CONCLUSION

This section explores a new approach to presenting, processing, and interpreting condition data. You have seen the benefits of applying proportional hazards models to condition monitoring and equipment performance data in several industries, including petrochemical, mining, food processing, and mass transit. Table 12.3 summarizes the advantages of CBM optimization by proportional hazard modeling in 4 companies. Good data and the increasing use of the software will fuel even more significant maintenance progress in the coming years.

TABLE 12.3

Summary of Recorded Benefits

Industry	Data Reduction of Key Condition Indicators	Cost Savings Over Run-to-Failure Policy or Simple Age-Based Maintenance Policy	Average Extension In Replacement Life
Mining	21 oil analysis measurements, 3 were found to be significant.	25%	13%
Mass transit	A single-color observation was used.	55%	More frequent maintenance inspections but less extensive repairs are required.
Food processing	21 vibration signals, 3 were found to be significant.	5%	
Petrochemical	12 vibration signals, 2 found to be significant.	42%	

12.6 ENHANCING RELIABILITY THROUGH ASSET REPLACEMENT

Eventually, replacing an aging asset with a new one becomes economically justifiable. Since it's usually years between replacements rather than weeks or months, as it often is for component preventive replacement, you must consider that money changes in value over time. This is known as discounted cash flow analysis. Figure 12.22 shows the different cash flows when replacing an asset on a 1, 2, and 3-year cycle.

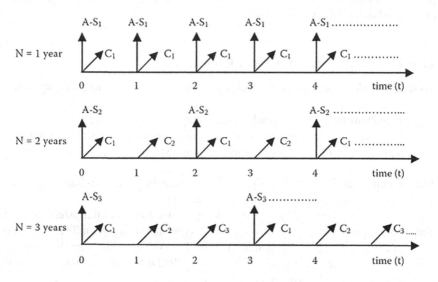

FIGURE 12.22 Asset replacement cycles.

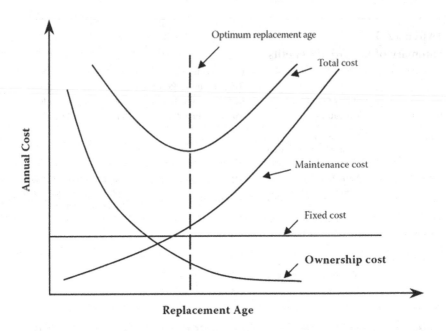

FIGURE 12.23 Economic life problem.

To decide which of the three alternatives would be best, you must compare them fairly. You can do this by converting all cash flows associated with each cycle to today's prices or their current value. This is the process known as discounting cash flows. Also, you must compare all possible cycles in the alternatives over an infinite period. While this may seem unrealistic, it isn't. It keeps the analysis straightforward and is used in the following section.

12.6.1 ECONOMIC LIFE OF CAPITAL EQUIPMENT

There are two key conflicts in establishing the economic life of capital equipment:

- the increasing operations and maintenance costs of the aging asset
- the declining ownership cost in keeping the asset in service since the initial capital cost is being written off over a more extended period.

These conflicts are illustrated in Figure 12.23, where the horizontal line also depicts fixed costs (such as operator and insurance charges):

The asset's economic life is the time when the total cost is minimized. You will find the total cost curve equation, along with its derivation, in Appendix A15. Rather than directly relying on the economic life model, you can use software that incorporates it. The following problem is solved using PERDEC software, which contains the economic life model provided in Appendix A16.

PROBLEM

To minimize total discounted cost, Canmade Inc. wants to determine the optimal replacement age for its materials handling turret side loaders. Historical data analysis has produced the information in Table 12.4 (all costs are in present-day prices). The cost of a new turret side loader is $150,000, and the interest rate, for discounting purposes, is 12%.

SOLUTION

Using PERDEC produces Figure 12.24, which shows that the economic life of a side loader is 3 years, with an annual cost of $79,973. This amount would be sufficient to buy, operate, and sell-side loaders on a 3-year cycle. This is the optimal decision. Note that the annual budget to fund replacements would be calculated based on the number and age of side loaders and a three-year replacement cycle.

TABLE 12.4
Maintenance Cost and Resale Value

Year	Average Operating and Maintenance Cost ($/year)	Resale Value at The End Of The Year ($)
1	16,000	100,000
2	28,000	60,000
3	46,000	50,000
4	70,000	20,000

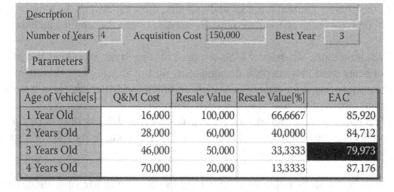

Description				
Number of Years 4	Acquisition Cost 150,000		Best Year 3	
Parameters				

Age of Vehicle[s]	Q&M Cost	Resale Value	Resale Value[%]	EAC
1 Year Old	16,000	100,000	66,6667	85,920
2 Years Old	28,000	60,000	40,0000	84,712
3 Years Old	46,000	50,000	33,3333	79,973
4 Years Old	70,000	20,000	13,3333	87,176

FIGURE 12.24 PERDEC optimal replacement age.

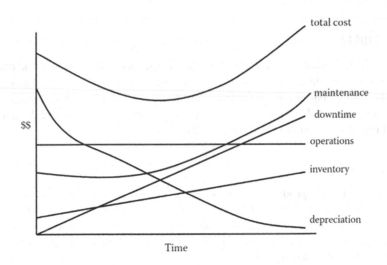

FIGURE 12.25   APWA economic life model: establishing equipment's economic life where utilization varies during its life.

While the above example has only considered operating and maintenance (O&M) costs and resale value, you must include all relevant costs. Figure 12.25, taken from an American Public Works Association publication[16] shows that, for example, the cost of holding spares (inventory) and downtime is included in the analysis.

The calculation of this section has used the "classic" economic life model that assumes equipment is being replaced with similar equipment. It also assumes that the equipment is steadily used year-by-year. You may find this useful for some analyses, such as forklift truck replacement if the equipment's design isn't impacted significantly and its use is constant.

Some equipment isn't used steadily, year to year. You might use new equipment frequently and older equipment only to meet peak demands. In this case, you have to modify the classic economic model and examine the total cost of the similar equipment group rather than individual units. Examples, where this applies include:

- Machine tools, where new tools are used to meet basic workload and older tools are used to meet peak demands, say during annual plant shutdowns.
- Materials handling equipment, such as older forklift trucks in a bottling plant, are kept to meet seasonal peak demands.
- Trucking fleets undertake both long-distance and local deliveries. New trucks are used on long-haul routes initially, and then, as they age, they are relegated to local deliveries.

The following is an example of establishing the economic life of a small fleet of delivery vehicles using AGE/CON software:

PROBLEM

A company has 8 vehicles to deliver their products to customers. The company uses its newest vehicles during regular demand periods and older ones to meet peak demands.

In total, the fleet travels 100,000 miles per year, and these miles are distributed, on average, among the 8 vehicles as follows:

Vehicle number 1 travels 23,300 miles/year
Vehicle number 2 travels 19,234 miles/year
Vehicle number 3 travels 15,876 miles/year
Vehicle number 4 travels 15,134 miles/year
Vehicle number 5 travels 12,689 miles/year
Vehicle number 6 travels 8,756 miles/year
Vehicle number 7 travels 3,422 miles/year
Vehicle number 8 travels 1,589 miles/year

Determine the optimal replacement age for this class of delivery vehicle.

SOLUTION

You must first establish how often the vehicle is used as it ages. (See Appendix A for the underlying mathematical model of when to replace aging equipment. It features a case study that establishes the economic replacement policy for a large fleet of urban buses.)

The utilization data will look like Figure 12.26. The equation of a straight line can describe the trend of Figure 12.26:

$$Y = a - bX$$

Where "Y" is the "miles/year" figure and "X" is the "vehicle number," vehicle number 1 is the one most utilized. Vehicle number 8 is the least utilized.

Using the actual figures given above, you can establish from AGE/CON (or by plotting the data on graph paper or using a trend-fitting software package) that the equation, in this case, would read:

$$Y = 26,152 - 3,034X$$

Next, establish the operating and maintenance (O&M) costs trend.

For *Vehicle 1* (your newest vehicle and the one used the most), you need the following information:

Miles traveled last year (already given)	23,300
O&M cost last year	$3,150
Cumulative miles on the odometer to the mid-point of last year	32,000

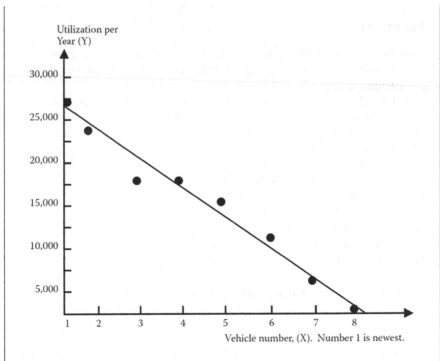

FIGURE 12.26   Vehicle utilization trend.

As you can see, the O&M cost/mile is $ 0.14 for Vehicle 1. Do the same for all 8 vehicles.

Vehicle 8 (the oldest and the one least used) may have data that looks like this:

Miles traveled last year (already given)	1,589
O&M cost last year	$765
Cumulative miles on the odometer to the mid-point of last year	120,000

In this case, the O&M cost/mile is $0.48 for Vehicle 8. A plot of the O&M cost trend would look like Figure 12.27. Each vehicle's O&M cost is represented by a "dot" on the graph. Vehicles 1 and 8 are identified in the diagram. The straight line is the trend that has been fitted to the "dots."

The equation you get this time is:

$$Z = 0.0164 + 0.00000394\,T$$

where

Z = $/mile,
T = Cumulative miles traveled

The two trend lines are both used as input to AGE/CON

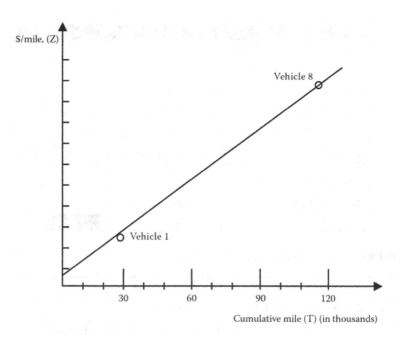

FIGURE 12.27   Trend in O&M cost

Note

In both cases, a straight (linear) relationship existed for $Y(X)$ and $Z(T)$ so that the fitted lines read $Y = a - bX$ and $Z = c + dT$. A polynomial equation often gives a better "fit" to a particular data series. These polynomial equations can be generated using a standard statistical package such as Minitab or SPSS.

To solve the problem, you require additional information:

Assume that a new delivery vehicle costs $40,000. The resale values for this particular type of vehicle are:

$$Z = 0.0164 + 0.00000394\,T$$

The interest rate for discounting purposes is 13%. Figure 12.28 from AGE/CON shows that the optimal replacement age of a delivery vehicle is 4 years with an associated equivalent annual cost (EAC) of $14,235.

To implement this recommendation, you'd likely need to replace a quarter of the fleet each year, so the same number of vehicles would be replaced yearly. All vehicles would be replaced at the end of their fourth year of life.

Age of Vehicle[s]	Q&M Costs	Resale Value	Resale Rate[%]	EAC
1 Year Old	1,432	28,000	70,0000	18,818
2 Years Old	2,405	20,000	50,0000	16,724
3 Years Old	2,156	13,000	32,5000	15,348
4 Years Old	1,118	6,000	15,0000	14,236

AGE/CON - Main

File Edit View Parameters Help

Description Delivery Vehicle

Number of Years 4   Acquisition Cost 40,000   Best Year 4

Parameters

FIGURE 12.28 AGE/CON optimal replacement age.

12.6.2 Before and After-Tax Calculations

In most cases, you conduct economic life calculations on a before-tax basis, and this is always the case in the public sector, where tax considerations are not applicable. Your financial group can help decide the best course of action in the private sector. In many cases, the after-tax calculation doesn't alter the decision, although the EAC is reduced when the result is in after-tax dollars.

Figure 12.29 illustrates what happened when a Feller Buncher was replaced in the forestry industry. The data used is provided in Tables 12.5 and 12.6.

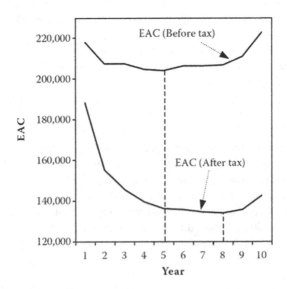

FIGURE 12.29 Economic life: before- and after-tax calculation (Feller Buncher data).

TABLE 12.5
Feller Buncher Data

The acquisition cost of Feller Buncher	$ 526,000
Discount rate	10%
Capital Cost Allowance (CCA)	30%
Corporation Tax (CT)	40%

TABLE 12.6
Feller Buncher: Annual Data

	O&M Cost	Resale Value
1	8,332	368,200
2	60,097	275,139
3	107,259	212,423
4	116,189	169,939
5	113,958	104,189
6	182,516	95,085
7	173,631	85,981
8	183,883	83,958
9	224,899	73,842
10	330,375	40,462

You can see from Figure 12.29 that the general trend of the EAC curve remains the same, but it veers downwards when the tax implications are included in the economic life model. In before-tax dollars, the economic life of the Feller Buncher is five years, while in after-tax dollars, the minimum is eight years. Note that, in both cases, the total cost curve is flat around the minimum. In this example, a replacement age between five and eight years would be good for a before-tax or after-tax calculation.

If you make calculations after tax, ensure all relevant taxes and current rules are incorporated into the model. One example is the But Buttimore and Lim report16, which deals with the cost of replacing shovels in the mining industry on an after-tax basis. It includes not only corporation tax and capital cost allowances but federal and provincial taxes applicable to the mining industry at the time.

12.6.3 THE REPAIR VERSUS REPLACEMENT DECISION

You may face a sudden significant maintenance expenditure for equipment, perhaps due to an accident. Or, you may be able to extend the life of an asset through a significant overhaul. In either case, you have to decide whether to make the maintenance expenditure or dispose of the asset and replace it with a new one.

Here's an approach that can help you make the most cost-effective decision:

PROBLEM

A significant piece of mobile equipment, a front-end loader used in open pit mining, is 8 years old. It can remain operational for another 3 years, with a rebuild costing $390,000. The alternative is to purchase a new unit, costing $1,083,233. Which is the better alternative?

SOLUTION

You need additional data to decide to minimize the long-run equivalent annual cost (EAC). Review historical maintenance records for the equipment; to help forecast the future O&M costs after rebuild and obtain O&M cost estimates and trade-in values from the supplier of the potential new purchase.

Figure 12.30 depicts the cash flows from acquiring new equipment at time T where:

- R is the cost of the rebuild
- $C_{p,i}$ is the estimated O&M cost of using the present equipment after rebuild in year i, $i = 1, 2, ..., T$
- A is the cost of acquiring and installing new equipment
- T is when the change-over occurs from the present equipment to a new one. $T = 0, 1, 2, 3$
- $S_{p,T}$ is the trade-in value of the present equipment at the change-over time, T
- $C_{t,j}$ is the estimated O&M cost associated with using the new equipment in year j, $j = 1, 2, ..., n$
- S_n is the trade-in value of the new equipment at age n years
- n is the economic replacement age of the new equipment

The necessary data for the current equipment is provided in Table 12.7, and The necessary data for the new equipment is provided in Table 12.8. The interest rate for discounting is 11%, and the new equipment's purchase price is $A = $1,083,233$. Evaluating the data gives the new equipment an economic life of 11 years, with an associated EAC of $494,073.

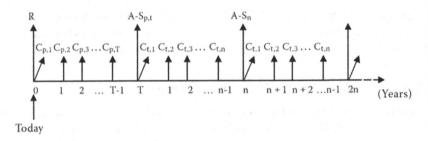

FIGURE 12.30   Cash flows associated with acquiring new equipment at time T.

TABLE 12.7
Cost Data: Current Equipment

$C_{p,1}$ = \$138,592	$S_{p,0}$ = \$300,000
$C_{p,2}$ = \$238.033	$S_{p,1}$ = \$400,000
$C_{p,3}$ = \$282,033	$S_{p,2}$ = \$350,000
	$S_{p,3}$ = \$325,000

TABLE 12.8
Cost Data: New Equipment

$C_{t,1}$ = \$38,188	S_1 = \$742,500
$C_{t,2}$ = \$218,583	S_2 = \$624,000
$C_{t,3}$ = \$443,593	S_3 = \$588,000
$C_{t,4}$ = \$238,830	S_4 = \$450,000
Etc.	Etc.

TABLE 12.9
Optimal Change-Over Time

	Change-Over Time to New Loader, T			
	$T = 0$	$T = 1$	$T = 2$	$T = 3$
Overall	449,074	456,744	444,334	435,237
EAC (\$)				↑
				Minimum

SOLUTION

To decide whether or not to rebuild, calculate the EAC associated with not rebuilding (i.e., $T = 0$), replacing immediately, rebuilding then replacing after 1 year (i.e., $T = 1$), and rebuilding then replacing after 2 years (i.e., $T = 2$), rebuilding and replacing after 3 years (i.e., $T = 3$), and so on. The result of these various options is shown in Table 12.9. The solution is to rebuild and then plan to replace the equipment in 3 years at a minimum equivalent annual cost of \$435,237.

NOTE

In a full-blown study, the rebuilt equipment might be kept longer. See Appendix A19 for the model used to conduct the above analysis. The same model is used in the following section.

12.6.4 Technological Improvement

If a new, more technically advanced model of equipment you are using becomes available, you will have to weigh the costs and benefits of upgrading. See Appendix A19 for a basic model to evaluate whether or not to switch. The Buttimore and Lim case study dealing with shovel replacement in an open pit mining operation[16] shows that a better technical design improved productivity. (This is an extension of the model in Appendix A19.)

12.6.5 Life-cycle Costing

Life-cycle costing (LCC) analysis considers all costs associated with an asset's life cycle, which may include:

- Research and development
- Manufacturing and installation
- Operation and maintenance
- System retirement and phase out

When making decisions about capital equipment, be it for replacement or new acquisition, reflect on all associated costs. Figure 12.31 is a good illustration of what can be involved.

The iceberg (Blanchard[17]) in Figure 12.31 shows that, while costs like upfront acquisition are apparent, the total cost can be many times greater. In the airline

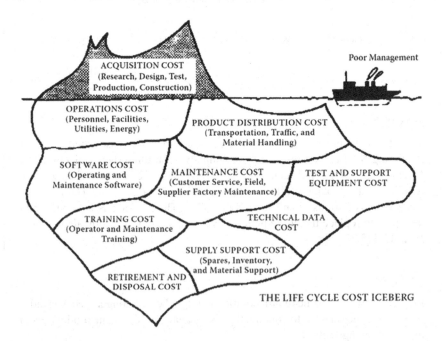

FIGURE 12.31 The life-cycle cost iceberg.

industry, the life-cycle cost of an aircraft can be five times its initial purchase. A Compaq Computer Corp. ad states that 85% of computer costs are usually hidden, going to administration (14%), operations (15%), and the bulk – about 56% – to asset management and service and support costs.

In the economic life examples covered in this section, we take an LCC approach, including costs for:

* Purchase price
* Operations and maintenance cost
* Disposal value

Of course, if necessary, you may include other costs contributing to the LCC, such as spare parts inventory and software maintenance.

12.7   RESOURCE REQUIREMENTS

When it comes to maintenance resource requirements, you must decide what resources should be, where they should be located, who should own them, and how they should be used.

If sufficient resources aren't available, your maintenance customers will be dissatisfied. Having too many resources, though, isn't economical. Your challenge is to balance spending on maintenance resources such as equipment, spares, and staff with an appropriate return for that investment.

12.7.1   ROLE OF QUEUING THEORY TO ESTABLISH RESOURCE REQUIREMENTS

The branch of mathematics known as queuing theory, or waiting line theory, is valuable when bottlenecks occur. You can explore the consequences of alternative resource levels to identify the best option. Figure 12.32 illustrates the benefit of using queuing theory to establish the optimal number of lathes for a workshop.

In this example, the objective is to minimize the total cost of owning and operating the lathes and tying up jobs in the workshop.

12.7.2   OPTIMIZING MAINTENANCE SCHEDULES

In deciding maintenance resource requirements, you must also consider how to use resources efficiently. An important consideration is scheduling jobs through a workshop. If there is restricted workshop capacity and jobs cannot be contracted out, you must decide which job should be done first when a workshop machine becomes available.

Sriskandarajah et al.[18] present a unique and highly challenging maintenance overhaul scheduling problem at the Hong Kong Mass Transit Railway Corporation (MTRC). In this case, preventive maintenance keeps trains in a specified condition, considering the maintenance cost and equipment failure consequences. Since maintenance tasks are either "too early" or "too late" and can be costly, how maintenance activities were scheduled was important.

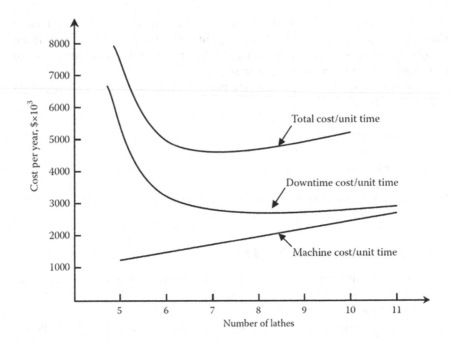

FIGURE 12.32   Optimal number of machines in a workshop.

Smart scheduling reduces the overall maintenance budget. Establishing a schedule, it's essential to acknowledge constraints, such as the number of equipment/machines that can be maintained simultaneously and economic, reliability, and technological concerns. Most scheduling problems are a combination of factors. Because of this complexity, a heuristic technique known as genetic algorithms (GAs) was used to arrive at the global optimum.

The algorithm's performance compared well with manual schedules established by the Hong Kong Mass Transit Railway Corporation. For example, total costs were reduced by about 25%. The study supports the view that GAs can provide reasonable solutions to complex maintenance scheduling problems.

12.7.3   OPTIMAL USE OF CONTRACTORS (ALTERNATIVE SERVICE DELIVERY PROVIDERS)

The above decision process assumed that all the maintenance work had to be done within the railway company's maintenance facility. If you can contract out the maintenance tasks, you must decide, for instance, whether to contract out all, some, only during peak demand periods or none of them. The conflicts in making such decisions are illustrated in Figure 12.33.

You must contract out all the work if your organization lacks maintenance resources. The only cost will be paying for the alternate delivery service. However, if your organization has a maintenance division, there will be two cost components. One is a fixed cost for the facility, shown in Figure 12.33, to increase linearly as the

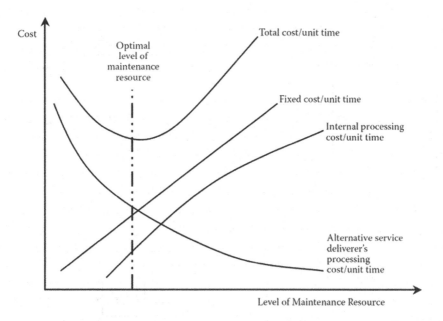

FIGURE 12.33   Optimal contracting-out decision.

size expands. The second cost is variable, increasing as more work is handled internally but leveling off when over-capacity exists.

The optimal decision is a balance between internal resources and contracting out. Of course, this isn't always the best solution. You must assess the demand pattern and internal and external maintenance costs. If it is cheaper not to have an internal maintenance facility, the optimal solution would be to contract all the maintenance work. Alternatively, the optimal solution could be to gear up your organization to do all the work internally. See Appendix A11 for a model of this decision process.

12.7.4   ROLE OF SIMULATION IN MAINTENANCE OPTIMIZATION

Common maintenance management concerns about resources include:

- How large should the maintenance crews be?
- What mix of machines should there be in a workshop?
- What rules should be used to schedule work through the workshop?
- What skill sets should we have in the maintenance teams?

Some of these questions can be answered by using a mathematical model. However, a simulation model of the decision situation is often built for complex cases. You can use the simulation to evaluate various alternatives, then choose the best. There are many simulation software packages available that require minimal programming.

Many resource decisions are complex. Take, for example, a situation where equipment in a petrochemical plant requires attention. If the maintenance crew is limited

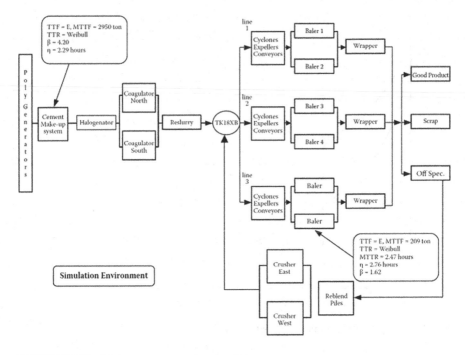

FIGURE 12.34   Maintenance crew size simulation.

by size or attending to other tasks, the new job may be delayed while another emergency crew is called in. Of course, this wouldn't have occurred if a large crew had been in place initially. Since there can be many competing demands on the maintenance crew, it's challenging to establish the best crew size.

This is where simulation is valuable. Figure 12.34 illustrates a case where simulation was used to establish a petrochemical plant's optimal maintenance crew size and shift pattern. For illustration purposes, statistics are provided for the failure pattern of select machines, along with repair time information. The study was undertaken to increase the plant's output. The initial plan was to add a fourth finishing line to reduce the bottleneck at the large mixing tank (TK18XB) just before the finishing lines. The throughput data showed that once a machine failed, there was often a long wait until the maintenance crew arrived. Rather than construct another finishing line, it was decided to increase maintenance resources. Using simulation, the plant's throughput was significantly increased with little additional cost. The crew size increase was a much cheaper alternative and achieved the required increased throughput.

12.7.5   Software That Optimizes Maintenance and Replacement Decisions

This chapter has referred to several software packages: OREST for component replacement decisions, EXAKT for CBM optimization, AGE/CON for mobile equipment replacement, and PERDEC for fixed equipment replacement. There are other good software tools on the market, and more are becoming available all the time, all of which you can find on the Web.

ACKNOWLEDGMENTS

Much of this chapter is based on sections from Chapters 3, 4, and 5 in the book *Maintenance, Replacement and Reliability: Theory and Applications*, Third Edition, Andrew K.S. Jardine and Albert H.C. Tsang, CRC Press/Taylor & Francis Group, 2022.

Section 12.5 on CBM is heavily based on the chapter in CBM written by Murray Wiseman in the first edition of this book.

PERMISSIONS

* Pages 259–299 #, *Asset Management Excellence: Optimizing Equipment Life-Cycle Decisions*, 2nd Edition by Editor(s), John D. Campbell, Andrew K. S. Jardine, Joel McGlynn Copyright (2011) by Imprint. Reproduced by permission of Taylor & Francis Group.

REFERENCES

1. OREST – Details at The Aladon Network, ACTOR software platform https://www.aladon.com
2. Jardine, A.K.S. and Tsang, A.H.C., *Maintenance, Replacement, and Reliability: Theory and Applications*, Third Edition, CRC Press, Taylor & Frances, 2022.
3. Drenick, R.F., "The Failure Law of Complex Equipment," *Journal Society Industrial Applied Math*, Vol. 8, pp. 680–690, 1960.
4. Jardine, A.K.S. and Hassounah, M.I., "An Optimal Vehicle-Fleet Inspection Schedule," *Journal of Operations Research Society*, Vol. 41, No. 9, pp. 791–799, 1990.
5. Cox, D.R., Regression Models, and Life Tables (with discussion). *Journal of the Royal Statistical Society. Series B*, 34, 187–220, 1972.
6. https://cmore.mie.utoronto.ca
7. www.omdec.com
8. RCM II 2.2 second edition, John Moubray, ISBN 0 7506 3358 1 Butterworth-Heinemann, 1999.
9. Pottinger, K. and Sutton, A., *Maintenance Management: An Art or Science*, Maintenance Management International, 1983.
10. Anderson. M, Jardine, A.K.S., and Higgins, R.T., The Use of Concomitant Variables in Reliability Estimation, *Modeling, and Simulation*, Vol. 13, pp. 73–81, 1982.
11. Nowlan, F.S. and Heap, H., *Reliability-centered Maintenance*. Springfield Virginia, National Technical Information Service, U.S Department of Commerce, 1978.
12. "Proportional Hazards Modelling, a new weapon in the CBM arsenal," Murray Wiseman and A.K.S. Jardine, *Proceedings Condition Monitoring '99 Conference*, University of Wales, Swansea 12–15 April 1999.
13. Statistical Methods in Reliability Theory and Practice, Brian D. Bunday, Ellis Horwood Limited, 1991.
14. On the Application of Mathematical Models in Maintenance. *Philip A. Scarf European Journal of Operational Research*, 1997.
15. Green, H. and Knorr, R.E., "Managing Public Equipment," *The APWA Research Foundation*, pp. 105–129, 1989.
16. Buttimore, B. and Lim, A., "Noranda Equipment Replacement System," *Applied Systems and Cybernetics*, Vol. 2, pp. 1069–1073, 1981.
17. Blanchard, B.S., *Logistics Engineering and Management*, Fifth Edition, Prentice Hall, 1998.
18. Sriskandarajah, C., Jardine, A.K.S. and Chan, C.K., Maintenance Scheduling of Rolling Stock Using a Genetic Algorithm, *Journal of the Operational Research Society*, Vol. 49, pp. 1130–1145, 1998.

13 A Maintenance Assessment Case Study*

Don M. Barry
and Previous contributions from Don M. Barry

The maintenance excellence assessment described in the previous chapter and detailed in the book's Appendix has been applied to hundreds of maintenance operations looking for areas to improve and prioritize their identified action items.

This chapter will review a modified "actual" example of a maintenance self-assessment report that a third-party consulting company completed and how this report can help maintenance and operations experts identify actions to improve their operation. The case study has been adapted from an actual report. The recommendations described later in this chapter come from maintenance and operations experts who understand leading practices' elements from the ten categories of the Maintenance Excellence pyramid. (See Figure 13.1 for a picture of the Asset Management Excellence pyramid.)

The ABC Consulting Company was engaged to perform a Site Assessment of UTO Company's maintenance management practices at their Base Valley operations and compare current UTO practices with best practices. UTO Base Valley operations include three production plants (Falcon, West End, and Argile), a mining site (Lake Operations), Central Services and Utilities, Materials and Logistics, and Railroad.

13.1 SITE OVERVIEW

The Base Valley operations produce several bulk chemical products. These are commodity products with little or no differentiation from competing products and compete primarily on price. Due to its remote location, shipping costs are a significant component of the price to their customers. This location places considerable pressure on keeping manufacturing costs down.

The remote location and small local community affect the site's ability to hire locally, although employees tend to be long-term once hired. The remote location also affects the local supplier support, as the nearest major center is 100 miles away.

The plant's location in California imposes scrutiny from the State on environmental issues, notably air pollution. This impacts the operation of coal-fired steam boilers in Utility operations.

Bill Jackson and Associates has financed the purchase of several companies through bank loans. Bank assessment would focus on expected corporate cash flow first, then asset value. Banks would place restrictive covenants to ensure no significant flow or asset value change without their consent. The expected outcome is a corporate

 DOI: 10.1201/9781032679600-16

FIGURE 13.1 Maintenance Excellence pyramid elements fishbone.

focus on cash flow conservation, with tight capital spending and a short-term focus on spending.

13.2 METHODOLOGY

The methodology used involved:

- Participation by each site in a self-assessment using a standard questionnaire
- On-site walkthroughs, mapping, and interviews by ABC Consulting to determine what practices and processes were being used at each site
- Analyzing and presenting the results of the "As Is" assessment data collected
- Comparison of current UTO practices with leading practices
- Recommendations
- Presentation of the findings to the site
- Report on findings

The first three steps of the methodology are illustrated in Figure 13.2.

The areas are assessed at each site corresponding with the following ten elements of maintenance best practices:

- Maintenance Strategy
- Maintenance Tactics
- Risk and Reliability Analysis/Engineering
- Key Performance Metrics/Benchmarking
- IT Systems Support

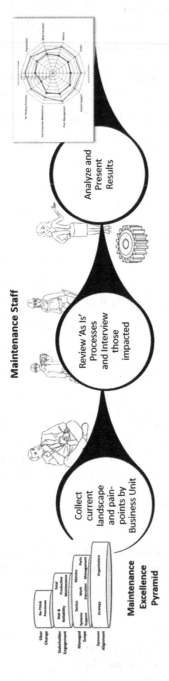

FIGURE 13.2   The first three steps of the maintenance Assessment process.

- Work Execution (Planning and Scheduling)
- Parts Management
- Maintenance Processes Re-Think
- Organization and Human Resources
- Total Productive Maintenance

Process mappings were carried out for the whole Valley and corresponded to actual practices at UTO. Base Valley operations. The mappings reflect what and how the processes are executed. Process maps do not necessarily exist for all strategic elements examined. Most maps pertain to planning and scheduling (work management) and materials management.

In this process, UTO would expect to receive a report which outlines the following:

- UTO practices based on ABC Consulting observations and mapping
- UTO self-assessment results
- A description of the best practices
- A set of recommendations for implementation

What follows is a list of findings that were derived from the Maintenance Excellence questionnaire and the follow-on interviews of the participants:

13.3 FINDINGS

13.3.1 MAINTENANCE STRATEGY

Current Situation:

- Their developed vision is communicated.
- A Maintenance council has been formed, and its mission statement has been defined and communicated. The goal of the Maintenance council is to establish a World-Class Maintenance and Reliability organization.
- UTO uses a combination of centralized services with area-based workforce crews
- UTO is IS0-9002 certified
- Some sites (Utilities, Lake) share a common vision and values between operations and maintenance.
- The maintenance strategy is not clearly defined.
- Long-term objectives relate only to cost and headcount, not long-term improvement initiatives.
- There is a lack of a clear and consistent vision for asset management throughout the Valley.
- Key-value drivers are defined but not effectively communicated or understood at all levels.
- Maintenance and capital expenditures appear inadequate, with unclear definitions and objectives.
- Production complains that maintenance does not view them as customers and does not provide good customer service.

13.3.2 ORGANIZATION

Current Situation:

- Good area organization design with some central support
- Some centralized Computerized Maintenance Management System (CMMS) support and maintenance policies (council)
- The trades to supervisors (crew size) ratio is very low at 6 to 8:1
- There are many planners, 1:6 to 10, and there is a lack of maintenance engineers
- The number of levels within maintenance is high in some areas at 4
- Backlog not measured by person-hours or crew weeks
- Evidence suggests a lack of some technical support/PdM (predictive maintenance)
- Some maintenance supervisors would like a pool of mechanics they could pull from for shutdowns, vacations, etc. No backlog planning

13.3.3 HUMAN RESOURCES

Current Situation:

- The workforce is not unionized, except for railroad employees.
- Training program for trades is in the early stage of development.
- Leadership training for supervisors is well underway; classes are held once a month for all supervisory staff.
- The formal policy on training is 60 hours/year but is not strictly applied; however, most trades indicated that when training was requested, it was usually granted.
- A performance bonus is in place for the salaried employees.
- Industrial relations policy appears positive, but supervisors have raised some concerns about their open-door policy.
- Safety is a high priority.
- The maintenance staffing level is adequate, capable, and experienced.
- Overall, overtime represents approximately 10% of total person-hours and is well distributed amongst trades and areas.
- Contractors are used for special projects and are held accountable for their work on the same basis as UTO employees.
- Horizontal communications are generally effective but sometimes variable.
- The morale of the maintenance workforce is currently low, mainly due to the rumors of outsourcing maintenance processes.
- No skill or performance-based incentives for trades.
- Technical documentation is missing or outdated; there is a lack of discipline in the documentation.
- Shifts vary from 10 hours, 4 days/week to 8 hours, 5 days/week, depending on the department and the service. Therefore, the workforce is reduced on

Mondays and Fridays, and maintenance and operations shift hours do not match.
- Many trades indicated they would like to have the apprenticeship program reinstate.
- Contractors are used to replace UTO employees who leave (retirements, long-term disabilities, etc.).
- Vertical communications are often complex due to a large number of hierarchical levels.

13.3.4 Maintenance Tactics

Current Situation:

- The Preventive Maintenance (PM) program is widespread and comprises up to 45% of the work.
- Condition-based maintenance is well started.
- XXX 2120 portable devices are used to take readings for vibration analysis, and Master Trend (XXX Software) is used for data analysis. Most people indicated they were delighted with the vibration analysis services they received. However, some people indicated they would like more feedback from the VA crew.
- Maintenance performs thermography on electrical equipment.
- Fundamental oil analysis is performed in-house at Argile and otherwise sent to contractors.
- Many employees were trained to perform laser alignment. However, not all sites have laser alignment equipment.
- Monthly equipment availability for Boron is approximately 96–99%.
- Emergencies represent 10–30% of work.
- Schedule compliance is approximately 80%.
- There is no formal reliability-based program for determining the correct PM routines. PMs are defined mainly based on vendor recommendations and intuition and, to a lesser extent, equipment history in Revere (CMMS System).
- There is little practical analysis of tactics, systematic approach, or formal and systematic use of equipment histories.
- The lubrication team at the WE site is used primarily as helpers for maintenance, so lubrication is not always done; failures due to lack of lubrication were reported.
- Central Services PM their equipment (i.e., vehicles, jib cranes, fire ext., grinders, electric boxes, etc.).
- PMs represent, on average, approximately 30% of total person-hours. The AC/R team indicated they didn't have time to perform PM tasks in the summer.
- At Argile, maintenance crews spend much time performing demand maintenance on the fluid bed dryers at the Bi-Carb.

13.3.5 PERFORMANCE METRICS/BENCHMARKING

Current Situation:

- Maintenance performance is measured for inputs (i.e., costs), with some process measures (i.e., labor distribution, schedule compliance) and output measures (i.e., availability, production).
- Internal comparisons of best practices by the plant were started. Five years prior, the Maintenance Best Practice Team was created with members from various Jackson companies. This team aimed to identify maintenance strengths and weaknesses and, by sharing this knowledge, improve the overall effectiveness of maintenance within the Jackson group.
- Labor and material costs are accumulated and reported against critical systems and equipment but are challenging to access in Revere (CMMS System).
- Key measures/value drivers are trended but not effectively communicated. Key-value drivers tracked include safety (number of incidents), % availability, % emergency, %P M, % compliance, % schedule efficiency, % OT, % machines in alarm, % expenses ($actual/$budget).
- The workforce does not sufficiently understand value drivers.
- External Best Practice Benchmarking is not formally done.
- Goals are often not set nor well communicated.

13.3.6 IT SYSTEMS SUPPORT

Current Situation:

- Revere (CMMS system) is fully functional, although not always easy to use. It is an older and less user-friendly system than the newer one.
- Most equipment is recorded in Revere.
- A project management system (MS Project) is used to plan and schedule shutdowns.
- Many supervisors indicated that they would like to look at the equipment history, but it is lengthy and complicated.
- Data analysis of inventory using the current system is limited.
- CMMS and Financial systems are not fully integrated.
- Labor times must be entered twice: once in Revere to report time against Memos (WO) and a second time in the payroll system.

13.3.7 WORK EXECUTION (PLANNING AND SCHEDULING)

Current Situation:

- Most of the work is covered by a WR, then made into a Memo (WO) within Revere.
- The approval process is rapid and effective. Some approvals are done electronically.
- Weekly planning meetings are held between maintenance and production.

- Major shutdowns are planned well in advance through monthly meetings, which become weekly as the shutdown approaches. Utility managers, maintenance and production superintendents, engineering, mechanical and electrical and instrumentation supervisors are usually present at these planning meetings.
- In most cases, work estimates are done quickly (without really assessing the needs) and are sometimes not done.
- WO priorities are well-defined but often abused.
- Work backlog is not formally tracked or tracked in meaningful units (most people could tell us how many pages of the backlog they had, but they did not know how much time it represented).
- Work management is overdone (trades often must report to more than one person).
- Different sites plan their work and often conflict in utilizing shared resources, especially cranes.

13.3.8 PARTS MANAGEMENT

Current Situation:

- Stores management has recently been contracted out to a third-party company; they are transitioning.
- Stock Keeping Unit (SKU): approximately 30,000 line items in stock for a total value of $7.7M. The objective is to eliminate $700,000 of obsolete stock to get a stock value of $7M and eventually $5M by decreasing ISA division and OSVS division stocks.
- Stock rotation is approximately 1.6.
- Service level is evaluated to 97% by Stores.
- The accuracy of inventory is high.
- ABC analysis is performed regularly; Items are categorized as A, B, C, or D. A items are counted monthly, B items are counted quarterly, C items are counted every six months, and D items are counted yearly.
- Procurement credit cards are available to some individuals to accelerate the direct purchase process.
- Most departments depend on parts stocked in their area to have parts readily available, especially when some distance from Stores (i.e., West End); this material is not tracked once it has left Stores and increases material holding costs.
- Order quantities are not based on EOQ (economic order quantity).
- Complaints were formulated about part descriptions. Parts are often named using inconsistent naming techniques, making them difficult to locate in the system. The resulting descriptions are not complete enough to uniquely identify the part.
- Standardization of parts and equipment is not formally done.
- Material purchases done by Central Services must be approved by the client first. Some Supervisors complain that the work hasn't been done yet, only to realize they forgot to approve the PR (Purchase Requisition).

- 5% of material is verified when received at Stores. 95% is verified by the customer who has to return material to Stores when there is a problem.
- A central tool crib is available, but discipline is lacking when returning the tools. They are often stored in individual lockers.

13.3.9 TOTAL PRODUCTIVE MAINTENANCE

Current Situation:

- Partnering is exhibited in some areas only (i.e., back shifts).
- The "Helping-hand" concept is used only in specific operations or under certain conditions (i.e., Lake).
- Operators are willing to perform simple maintenance tasks, and maintenance doesn't mind transferring these tasks.
- Although maintenance is not represented on every shift, maintenance trades quickly respond to call-outs after hours and with minimum effort from operations.
- Self-directed work teams are not used.
- Little opportunity for autonomy.
- Management communication is often one way only.
- No formal multi-skilling or cross-skilling was done.
- Some Maintenance trades said they would like to receive training on operations to evaluate equipment criticality.
- When maintenance work is required after hours, operations must decide what support is needed. There is no on-shift maintainer to take care of this task.
- Argile operators said that they often operate different production lines, which makes it more difficult for them to develop a sense of ownership.
- Argile operators also reported that some operators change the operation parameters of their equipment, which increases the production on their shift but negatively affects the production capacity of the next shift. Because of the strong production focus of the company, these operators are praised instead of reprimanded. Some operators feel it would be better to produce at a constant rate without changing the parameters. There is no consistency in operating strategy or philosophy, which indicates some areas where UTO operation is lacking and its results.

13.3.10 RISK AND RELIABILITY-CENTERED MAINTENANCE

Current Situation:

- Availability is tracked (i.e., total duration of failures).
- Equipment histories are kept by saving completed WOs but are not used systematically for reliability analysis.
- Under the Quality program, root cause analysis is performed for significant failures.

- Some equipment redesign is done to improve reliability in specific areas (i.e., Pump Crew, Utilities, Lake).
- PM tasks are mainly based on the manufacturer's recommendations and experience.
- Change Order Process: to get as many people as possible involved in the decision process when a change is considered, engineering, maintenance, production, safety, and environment representatives have to sign the Change Order Process Form, which describes the changes considered. However, there were many complaints regarding new or revised installations and equipment, even with this procedure.
- Routes for lubrication and vibration analysis are established by the crews responsible for these tasks and are kept on the computer for the most part.
- Reliability (MTBF, i.e., failure frequency) and maintainability (MTTR, i.e., repair time) are not tracked.
- Operations and maintenance indicated that they seldom have sufficient input into engineering projects. They feel that communication between maintenance, operations, and engineering could be improved.
- There are only a few maintenance engineers assigned to maintenance departments.
- The maintenance department was criticized for not solving the problem at its root causes. Instead, the tendency is to fix the problems that arise quickly, often cutting corners to save production time.

13.3.11 MAINTENANCE PROCESS RE-THINKING

Current Situation:

- Most maintenance and materials processes have been documented but are not always followed.
- No formal process exists to review and revise current processes to eliminate non-value-added activities.
- Activity costs are not measured.

13.4 ADDRESSING THE FINDINGS

This company ultimately reviewed these findings, along with leading practices, developed a set of opportunities, and recommended actions from their combined findings. Along with the recommended actions (or initiatives), they defined their success criteria so that they would recognize when the project was started and when they achieved their goals. They then worked to prioritize their findings into categories such as "benefit" and "cost to implement" and developed a roadmap for managed business transformation going forward. This approach often helps executives appreciate that the whole set of business issues is more likely to be addressed. As such, the management and staff will feel that their issues were, at the least, considered, and this contributes to managing change acceptance when the final roadmap is communicated.

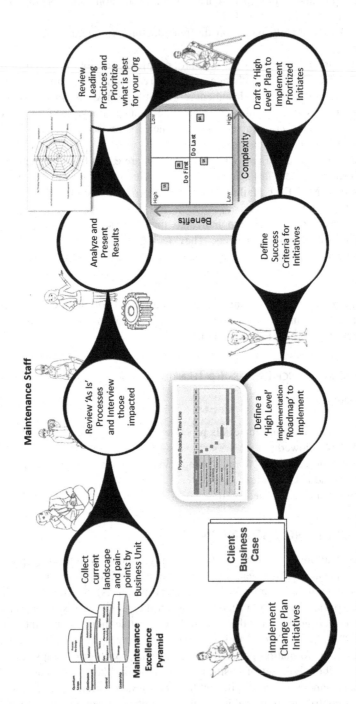

FIGURE 13.3   The complete maintenance assessment process steps.

Without revealing their exact business strategy or going into the cultural dynamics that will affect every business, we gave these findings to maintenance experts. We asked them what they would derive as recommendations given the above report. As a result, three or four recommendations were identified for each assessment section for this case demonstration.

They observed that the findings were not all tidy and wrapped in a way that clarified the recommendations. In other words, if you were looking for planning and scheduling issues and opportunities, you needed to review the whole report to determine if you captured all the issues identified from the interviews. Although this is common sense, sometimes, ideas arise from stakeholders that impact other asset management excellence areas.

13.4.1 Strategy

The first request of the maintenance experts was: What should the primary goal of this maintenance assessment be for this business?

With only ten to fifteen minutes to answer, the thoughts and hopes that were returned included:

- Holistic analysis (SWOT) of Strengths, Weaknesses, Opportunities, and Threats
- An understanding of how to bridge financial and manufacturing constraints
- An understanding of the company's viability
- An understanding of how this company can affect production optimization and equipment influence on availability, costs, and safety or environment
- Understanding what the go-forward Production Strategy should be and how it affects productivity or employee satisfaction

These are lofty goals. The maintenance assessment will help you get part of the way there. However, you will, no doubt, also require an understanding of the company's change history and the commitment of the management team to understand how some of the recommendations that will arise can be accepted and implemented.

The team did establish a hypothetical set of "Strengths, Weaknesses, Opportunities and Threats" charts to call out what they thought would summarize the most apparent health of UTO.

To align with the observed business imperatives, the maintenance strategy should support the business requirement to improve operations and maintenance delivery, focusing on cash flow and asset value. The team developed the following strategic recommendations:

- Develop a clear vision for Asset Management and a defined PM program
- Identify how asset management (and critical assets) contribute to cash flow
- Identify pain points that affect production, sustainment, and short vs. long-term goals
- Communicate the asset management strategy to management and staff
- Develop craft skills for future needs, including cross-training, so that flexibility can be leveraged

TABLE 13.1
Example of SWOT Analysis That Could Apply to the Use Case

Strengths	Weaknesses
• Competent maintenance staff • Safety is a priority • Training is a priority	• CMMS usability • Operations and maintenance not working together on production issues • No tech support
Opportunities	**Threats**
• Gains to be made by sharing a common vision across the company sites • Develop maintenance strategies that align with the local operating context • Integrate an enhanced CMMS solution with their financial system	• Lack of available resources (supplies and skilled personnel) • Employee morale • Metrics to track activity costs needed to support cash flow concerns

Another discussion point to consider is looking at their "go-to-market" strategy and understanding how the cost of distribution and manufacturing contributes to their competitive pricing model. What are the value drivers of their success in the market?

13.4.2 ORGANIZATION

Their maintenance management had a few opportunities identified by the team.
There is a defined desire to review the maintenance team mix, including:

- The ratio of Supervisors vs. crafts (maybe too many Supervisors vs. workers)
- Consider adding Schedulers and reducing the number of Planner roles

To develop a more substantial resource pool in their organization:

- Consider cross-training the crafts to create the ability to leverage existing staff
- Create a Craft pool to help manage workload balance and backlog
- Review the need and benefits of an apprenticeship program
- Entertain the concept of an "on-call" maintenance supervisor for after-hours maintenance coordination

Create an environment of communication and visibility of key business processes by:

- Leverage the CMMS across the company to maintain maintenance standards
- Publishing issues and planned change actions to keep all stakeholders informed and improve employee morale
- Promoting and celebrating the perception that they had strong maintenance staff and are well trained

13.4.3 Maintenance Tactics

Generally, maintenance tactics did not draw many recommendations from the group. They acknowledged a displayed use of preventive and predictive maintenance and that mobile workforce tools were used to capture some of the equipment readings. They suggested that a root cause failure analysis could be done to focus on high-end emergencies (30% of total maintenance). It was also suggested that tightening PM execution across the company and understanding why some PMs were missed were short-term activities.

13.4.4 Performance Metrics/Benchmarking

Plan to educate the staff and communicate the current focus's key measures/value drivers. Key-value drivers tracked include safety (number of incidents), % availability, % emergency, % PM, % compliance, % schedule efficiency, % OT, % machines in alarm, and % expenses ($actual/$budget).

13.4.5 IT Systems Support

With an understanding of leading practices and efficiencies gained from managing data, the team recommended replacing or upgrading the existing CMMS system and integrating it with their financial system. In addition, maintenance history data and resource management improvements will help them assess what needs to be looked at to initiate a work order and manage the resource backlog.

13.4.6 Work Execution (Planning and Scheduling)

Schedule compliance is 80%, so, on the surface, they appear to have a healthy planning and scheduling maintenance operation. However, they lack long-term planning, which could come with successful short-term execution that can be managed in the new CMMS solution. There is a need to understand why operations often overrule work orders with well-defined priorities. This situation could be a smoldering fuse to a real problem or, in any event, a process that needs to be corrected.

Short-term quick hits could include:

- Reassigning a Planner to a Scheduler role and reviewing backlog, and coordinating crane availability;
- Establish roles and responsibility rules and lines of authority between Planners and Schedulers, Maintenance and Operations;
- Set up daily operations meetings for daily schedule alignment;
- Planners to attend shutdown meetings;
- Engineers to be assigned to support Planners; and
- Begin measuring backlog by craft hours.

13.4.7 Parts Management

- Establish visibility of parts availability across the company
- Reduce material carrying costs

- Ensure content management completes on SKU identification
- Develop a "service level agreement" with the parts outsourcing company
- Notify supervisor on purchase requisitions not approved

13.4.8 TOTAL PRODUCTIVE MAINTENANCE

- Maintenance Control Reporting System
- Engage operations, maintenance, and engineering in PM validation reviews
- Support Total Productive Maintenance (TPM) with a maintenance strategy education program

13.4.9 RISK AND RELIABILITY-CENTERED MAINTENANCE

- Operating context-specific maintenance task assignment
- Identify critical assets and do a Reliability Centered Maintenance (RCM) analysis.

13.4.10 MAINTENANCE PROCESS RE-THINKING

- Consider routes for maintenance personnel once RCM analyzed tasks and frequencies have been assigned.
- Train crafts on more than one discipline for maintenance delivery flexibility

PERMISSIONS

Section IV

Optimizing Maintenance and Replacement Decisions

14 Real Estate, Facilities, and Construction*

Brett Barlow and Atif Sheikh
Original by Andrew Carey and Joe Potter

14.1 INTRODUCTION

Expecting more financial value in facilities and property than in all other asset classes combined should be no surprise. Therefore, getting the management of property assets right will go a long way in supporting the bottom line of an asset-intensive enterprise.

Property is often among the most significant items on an organization's profit and loss account and balance sheet. By moving from typical to best practice (or leading practice), businesses can improve value for money and reduce their real estate and facilities costs by up to 20% while still improving services and consistency of performance. Property activities also frequently encompass complex construction projects that, if managed effectively, can be delivered more quickly and at a lower cost. Building construction activities also critically influence the lifecycle of assets and operations. In addition, Real Estate and Facilities Management (REF) change can be used to leverage business transformation, thus enhancing agility, improving customer service, and contributing to staff retention.

This chapter focuses on "leading practice" approaches in the management of all aspects of corporate real estate and facilities management, notably including:

Asset Strategy, Process, and Organization
Lifecycle Delivery considerations
Information and Communication Technology Systems
Optimization.

14.2 OVERVIEW

14.2.1 Definitions

Real Estate: This text refers to real estate, facilities, and property interchangeably. Real Estate is defined as the buildings, land, and associated ancillary assets (i.e., roads, car parks, etc.) that an entity may own or lease to run its enterprise.

Facilities Management: Facilities Management (FM) includes all support services for the built environment and the management of their impact on the organization, people, and the workplace. Therefore, FM applies equally to all types of facilities, including office, industrial, retail, and residential accommodation.

DOI: 10.1201/9781032679600-18

Facilities Management services are often categorized as either "Hard" or "Soft," with the majority of suppliers having grown from either a technical "Hard" or a services-oriented "Soft" base.

The primary "Hard" categories include:

- Building Fabric Maintenance
- Mechanical and Electrical Maintenance
- Minor Projects

Soft Facilities Management services include:

Cleaning
Pest Control
Catering
Manned Security
Office Services
Waste Disposal

Procuring Soft and Hard Facilities Management from separate providers is often appropriate. Nevertheless, the overall approach to service delivery, including such considerations as performance and environmental management, can be common to all services and contracts.

Construction: Construction involves building new property assets or extending and refurbishing existing property assets. It is often referred to as "capital projects": however, this title infers the traditional finance approach for such projects and can, at times, be misleading. This chapter treats construction as a subset of Real Estate (for new build) or Facilities Management (extensions and refurbishments).

14.2.2   Facilities Management Business Drivers

There is a strong synergy between Real Estate, Facilities Management, and Construction disciplines. Businesses that understand the relationships between these three activities use holistic REF strategies to support their business drivers.

Whether a business is a retailer, an office-based company, or a manufacturer, the relationship between its operational and REF strategies can be critical to its success. For example, the world of investment banking may appear far removed from FM. Still, the link becomes immediately apparent if the cooling and ventilation fail at the data center serving the trading floor.

Major businesses are increasingly locating their operations in the most economically advantageous parts. As a result, it is common to hear firms relocating their manufacturing capability to Eastern Europe or China, while Bangalore has become synonymous with outsourced customer service centers.

Real Estate and Facilities (REF) services are often procured and managed predominantly locally despite this evident mobility in core business areas and other support services. Individual facilities service categories are commonly sourced separately, often at a site level, with the inherent lack of leverage and increased administrative workload that such decentralization entails. Service levels vary widely and can be difficult to compare. Data on estates and the associated services and spend is often poor and only held locally, if at all. Therefore, benchmarking between sites and suppliers is unrealistic, and implementing demand management discipline is complicated.

This inefficient state of affairs has not been helped by the historic inability of much of the supply market to deliver consistent multi-site or multi-national service. Moreover, client managers have often struggled to see benefits from developing a consistent company-wide approach to REF services procurement and management.

Many REF services have a robust local delivery component. For example, it is not currently feasible to clean or repaint a building remotely. However, the management of such services can increasingly be procured on a more aggregated level, leading to significant economies of scale.

REF service providers in the United States and Western Europe extend their global reach. Therefore, there are opportunities for significantly more efficient estate operating models for organizations that are both aware of them and willing to embrace the necessary change.

14.3 ASSET STRATEGY, PROCESSES, AND ORGANIZATION

14.3.1 OVERVIEW

An organization's objectives drive operational strategies. These strategies, to work, require a structured approach where the right performance matrices are defined to drive the correct behavior. These matrices, providing a line-of-sight alignment with the organization's financial, operational, and growth objectives, are enabled through i) an integrated set of cross-enterprise processes and ii) an organization structure ensuring the right combination of competencies and collaboration.

14.3.2 ASSET STRATEGY

Asset strategy is about translating the overall organizational strategy into a real-estate/physical infrastructure management strategy that will support the aims of the business. It can be described as a process to ensure the development of physical asset strategies, policies, and portfolio management plans which support the core business and help optimize performance while providing an appropriate level of flexibility.

Many vital questions are to be considered when assessing a facility's asset strategy:

- How does it support the core business?
- How can the asset's lifecycle value be optimized?
- What policies are there to guide REF decisions and activities? How are the decisions made?
- How is the REF function organized internally and in relation to the rest of the business?

	Stage 1	Stage 2	Stage 3	Stage 4
	Review Business Environment	Develop Physical asset strategy & operating model	Set Physical asset policies & standards	Plan & budget for physical assets
Inputs	• Business objectives • Business opportunities • Business constraints imposed by assets	• Business strategy • IT & HR strategies • Flexibility/ responsiveness • Statutory requirements • Required performance versus current	• Physical asset strategy • Operating model • Required performance changes • Business requirements	• Current asset portfolio • Future business needs • Forecast changes in service volumes or standards • New markets

Interventions	• Business strategy process
	• Define role of assets in meeting objectives
	• Document business strategy
		• Organizational design principles
		• Identify key processes
		• Key roles:
		  • Asset ownership
		  • Asset management
		  • Operations & maintenance
			• Identify key policy levers to drive change
			• Draft policies
			• Design input to draft standards
			• Consult on drafts
				• Portfolio planning
				• Asset Lifecycle cost analysis
				• Capital project planning
				• Facilities planning
				• Location studies
				• Scenario analysis (Twinning)
				• Risk assessment

Outputs	• Business strategy • Description of role of assets in strategy (high-level)	• Physical asset strategy • Operating model • Outsourcing scope • Performance management metrics • IT/systems linkages • HR implications • Change plan	• Policies (i.e., asset charging, business case appraisal) • Standards (i.e., asset allocation, fit-out)	• Portfolio/asset plan • Capital program • Facilities program • Budgets • Risk management plan

FIGURE 14.1 Stages of strategy development.

- Does the current structure encourage sufficient demand challenge or require a robust assessment of supply options?
- How does it align with the company's sustainability goals?
- How is flexibility addressed?

14.3.3 Stages of Asset Strategy Development

Below, the chart illustrates four critical stages of asset strategy development:

Each stage depicts contributing inputs, processes, interventions, and the resulting artifacts and outputs.

14.3.3.1 Strategy Development Stage 1: Review Business Environment

The first stage to consider is the business environment.

There are many aspects to ponder when considering how physical assets support an organization. External influences (political, economic, social, and technological) must be considered potential drivers of existing and future strategies and processes. These drivers will also influence business objectives and identify constraints the REF assets impose.

One key driver is how much flexibility the business needs. Most corporations must remain agile enough to seize new opportunities and respond to competitive threats. It should be recognized that planning horizons are considerably shorter than they used to be. Flexibility is, therefore, an ongoing requirement.

In understanding the business environment, shareholder value is essential, as it is the basis on which many major companies are run. A new asset strategy can impact the operations and financing of a company and impact shareholder value by affecting future cash flows and the future share price of an organization. For example, decisions on whether to lease or own property and in what proportion can significantly impact the market's view of the net worth of a corporation.

The Public Sector has different goals, yet parallel with shareholder value. Rather than maximizing the share price, the aim is to deliver public goods and services at the best value for money (VfM). Therefore, a new strategy can help achieve departmental objectives by impacting several of the same drivers available to companies seeking to influence shareholder value. For example, reducing asset operating costs and utilization levels may improve VfM.

A typical high-level REF strategy response to a business strategy is shown in Table 14.1. This strategy also needs to consider how REF will be operated and funded.

14.3.3.2 Strategy Development Stage 2: Develop Physical Asset Strategy and Operating Model

The next area to consider is the development of the Physical Asset Strategy and Operating Model.

A good strategy will address the following challenges:

- What is the required business contribution of the asset class?
- How will we fund our assets? What are the affordability criteria?

TABLE 14.1
Typical High-Level Asset Strategy

Responsive:
- Anticipates potential future changes and provides appropriate flexibility.
- Customizes services to fit customer needs, adding value.
- Aggregates data, turning it into useful information.
- Enables its people to make rapid, well-informed decisions.

Variable:
- Shifts from a predominately fixed cost structure to a predominately variable one.
- Builds for average capacity and supplements internal capabilities externally.
- Outsources selected functions entirely to achieve optimum variability and performance.

Resilient
- Knows own exposure to operational, market, and other risks in real-time.
- Effectively distributes risk with strategic partners/proactively manages the remainder.
- Build robust, "self-healing" capabilities (asset, technology, and processes).
- Recovers quickly from external disruptions to operations.

Focused:
- Concentrates resources on activities that add value.
- Leverages scale efficiencies through partners.
- Benefits from competitive advantage by insourcing superior capabilities.

- What level of risk transfer is preferred?
- What are the required high-level specifications or capabilities?
- What asset volumes or capacity do we need?
- In what broad locations do we need the assets? What performance and reliability standards do we need?
- What asset-related services and service standards are required?
- What flexibility are we likely to need to change any of the above?
- How does this compare to the existing asset base and performance?
- Tools and techniques are available to develop a new strategy:

Table 14.2 includes a non-exhaustive list of tools and approaches that can help understand a business environment and, thus, develop a new strategy.

Careful consideration must be given to the degree of consultation required with the business. If, as usual, they need to "own" the finished strategy, then activities like workshops and structured interview programs are invaluable in helping to build this.

Developing the asset management organization's operating model is an integral part of the strategy, as it will impact many aspects of asset performance.

TABLE 14.2
Typical Tools and Approaches for Strategy Development

Maturity profiles	Voice of the customer analysis (interviews, questionnaires)	Away days
Competitor analysis "Porter 5 forces" analysis"	PEST analysis (Political, Economic, Sociocultural, and Technological)	Workshops
Shareholder value analysis (private sector)	SWOT analysis (Strengths, Weaknesses, Opportunities, Threats)	Supplier discussions

14.3.3.3 Strategy Development Stage 3: Set Physical Asset Policies and Standards

The next area of consideration is the policies and standards that should apply to physical assets.

Below is a list of areas that guide decision-making and should be covered when setting policies and standards for physical assets:

- Investment and funding policies
- Appraisal policies
- Asset Planning and Procurement policies
- Asset Management Human Resources and Competency Management policies
- Asset management and maintenance execution policies
- Internal charging policies
- Service levels required
- Risk management policies
- Regularity Compliance
- Business-specific standards
- Level of flexibility required
- Environment and Sustainably

The diagram below illustrates an example of a high-level property policy framework.

TABLE 14.3
Typical Policy Framework

	Demand Policies	Supply Policies
Acquisition	• How the business units articulate demand for new accommodations	• How group property (GP) considers flexibility and property risks • How GP appraises property supply options • How GP fits client accommodation
Management	• How business units articulate demand for FM services and set service levels • How GP procures services • How business units occupy the accommodation efficiency • How internal charging operates	• How GP plans property supply • How business units budget for property • How GP manages relationships with business units • How GP manages external suppliers • How GP manages business-critical buildings • How GP manages multi-occupied buildings
Change	• How business units articulate changes in their requirements	• How GP manages moves/churn • How GP manages capital expenditure projects
Disposals	• How business units identify surplus accommodation	• How GP manages surplus properties

14.3.3.4   Strategy Development Stage 4. Plan and Budget for Physical Assets

The final area to consider is the planning and budgeting for physical assets.

Real Estate and physical infrastructure plans need to be prepared in an agreed strategy and policy framework; the diagram below shows how planning relates to asset strategy while also illustrating how the different elements of planning inter-relate.

14.3.4   PLANNING FRAMEWORK

The planning process is iterative and will usually consist of three core elements.

14.3.4.1   Portfolio Planning

A portfolio plan establishes which individual properties or other physical facilities are required to meet the business's needs. It should also describe how flexibility should be provided to meet changes in business needs and avoid unnecessary legacy costs should existing or planned facilities become surplus to requirements. Finally, the plan changes the existing Real Estate or physical infrastructure portfolio to meet current and future business needs.

14.3.4.2   Facilities Management Planning

A facilities management (FM) plan usually flows from the Portfolio Plan. It describes the services and service levels required to maintain and operate the portfolio of properties and other physical facilities in line with business needs. The FM plan describes how these FM services should be provided and includes a facilities management budget.

The critical stages of an FM plan are:

- Define FM Specifications and Service Levels
- Define FM Sourcing Strategy
- Prepare FM Plan

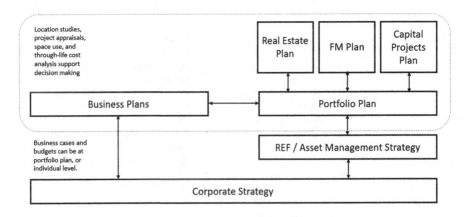

FIGURE 14.2   Typical planning framework.

14.3.4.3 Capital Program Planning

The Capital Program (CP) also flows from the Portfolio Plan. It prioritizes and phases the program of the capital projects required to meet agreed business needs.

The critical stages of a CP plan are:

- Agree to CP Aims and Objectives
- Define CP and Individual capital projects, including Capital Strategy and timescales
- Prepare cost plan and resources requirements for each project
- Assess CP risks
- Finalize CP and project plans

14.3.4.4 Real Estate Planning

The consolidated output from these three elements is a Real Estate and Physical Infrastructure Plan. This needs to be signed off by key business stakeholders.

The critical stages of a Real Estate and Infrastructure plan are:

- Summarize the Portfolio Plan, FM Plan, and CP plan elements
- Determine the contract strategy
- Review draft plan with key business stakeholders
- Finalize Real Estate and Infrastructure Plan

14.3.5 ASSET STRATEGY AND CONSIDERATIONS FOR OPERATING MODEL

Before diving into the asset value delivery lifecycle details, some basic operating model "design principles" should be established. The organization design must explicitly respond to the critical business drivers (i.e., shareholder value, performance improvement, and cost reduction). It will also provide additional areas typically considered at this design stage.

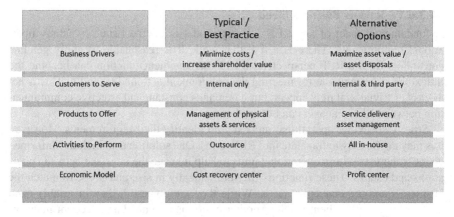

	Typical / Best Practice	Alternative Options
Business Drivers	Minimize costs / increase shareholder value	Maximize asset value / asset disposals
Customers to Serve	Internal only	Internal & third party
Products to Offer	Management of physical assets & services	Service delivery asset management
Activities to Perform	Outsource	All in-house
Economic Model	Cost recovery center	Profit center

FIGURE 14.3 Typical operating model decision matrix.

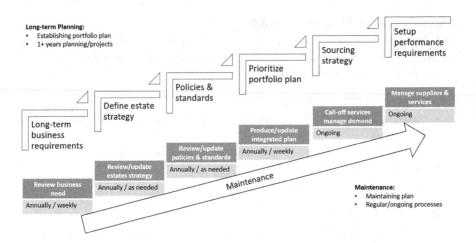

FIGURE 14.4　REF delivery model.

Usually, the design principles adopted in each area will focus on supporting the core business as "the customer" rather than serving external customers. For example, the "product to offer" is typically the provision of REF services to support the business, not the exploitation of physical assets as a source of profit in their own right (although this is a consideration for some corporations, i.e., in selling-off surplus land).

14.3.6　DELIVERY FRAMEWORK

Figure 14.4 highlights a delivery framework that can be used to develop more detailed processes. The model distinguishes between initial "long-term planning" tasks at an establishment and ongoing management of delivery ("maintaining").

The establishment tasks identified in the top row need to be undertaken as part of any project associated with the REF Process and Organization design, while the ongoing activities in the bottom row are described in more detail in later sections.

14.3.6.1　Review Business Need

A fundamental tenet of any REF delivery model is for the Estate to satisfy operational requirements. There is a risk that while it is reasonable to apply genuine constraints from a REF perspective, a REF management function can become too inflexible. A prerequisite for meeting the requirement is to understand it. This is not as easy or trivial as it may appear, but it demands constant and effective communication between REF managers and their clients.

Strictly speaking, operational units should identify their requirements. However, this may not be easy to translate into REF terms. One solution is to include "informed client" functions within operational units to help manage the interface between operations and estates. These functions vary dramatically in size and scope to match the business's and the concept's needs. While they are powerful, they invariably need tailoring to specific client needs. Another option is to embed an "account management" function within the REF organizational units.

14.3.6.2   Review/Update REF Strategy

A REF Strategy forms a vital part of the interface between the REF organization and the rest of the business. It should extract the key themes from the business strategy, develop a vision, set priorities for the Estate, and define the delivery model to achieve the vision according to clear priorities. It should also link to Information and Communication Technology (ICT) infrastructure, risk management, and financing strategies.

14.3.6.3   Review/Update Policy and Standards

The REF function must direct high-level policies (i.e., tenures and group-wide issues, such as sustainable development, health, safety and environment). They must also provide standards to be used as technical design requirements (i.e., specifications for sprinkler systems). Policies and standards should be derived from both business needs and external policy.

14.3.6.4   Produce/Update Integrated Plan

The REF organization must ensure that business requirements are captured and translated into coordinated and fully budgeted project plans.

14.3.6.5   Call-Off Services/Manage Demand

Call-off services from existing suppliers must be managed while ensuring demand for those services is controlled according to agreed processes, policies, and standards.

14.3.6.6   Manage Suppliers and Services

This activity entails maintaining and developing working relationships with suppliers to deliver planned work and providing an appropriate level of service to fulfill policy standards and meet budgets (i.e., developing and implementing procurement strategies) while monitoring and challenging performance where necessary.

14.3.7   CONNECTING DEMAND WITH SUPPLY

At its highest level, the provision of REF services is no different from delivering any other service or product, i.e., demand satisfaction through supply. Figure 14.5 shows the basis for any relationship between end-users (i.e., staff in the business unit who generate demand for accommodation and services) and the suppliers of those services (i.e., the internal or external contractors working for the internal REF management organization).

This conceptual model separates supply (i.e., the REF organization and its third-party suppliers) from demand (i.e., the operational business units and REF services' ultimate customers). The upper portion of the diagram focuses on customer requirements. At the same time, the lower part (i.e., the estate management organization and its supply chain) is structured toward managing supply and the organization's suppliers.

The interface between the supply and demand sides must be designed in line with the business operating model. This may mean that the "Informed Client" role resides within the supply organization as a form of "account" or "service manager," or it may

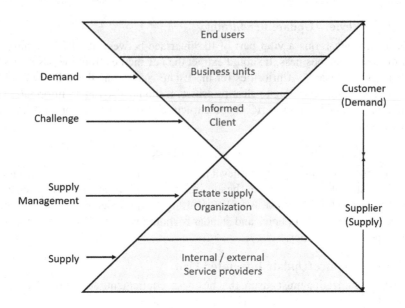

FIGURE 14.5   Customer/supplier separation, supply/demand model.

mean that it lies within the client business as indicated above. However, irrespective of location, the Informed Client must "own" and support its role of communicating the REF requirements of the business to the REF organization.

Demand challenge concerns ensuring that the requested services genuinely are required and that the client's needs cannot be satisfied in an alternative, more efficient way. A simple example is the cleaning of offices: the offices need to be maintained with an appropriate level of hygiene, but this does not necessarily mean that they must be cleaned every night to maintain that level. Once the REF organization understands its clients' genuine (often implicit) needs, it can procure services to satisfy them and those of its other clients cost-effectively.

Experience has shown that the most significant overall savings in cost often come from exercising effective demand control. However, those savings in other areas, including supplier unit costs, may have a far lower relative value.

14.3.8   ORGANIZATION DESIGN METHODOLOGY

Figure 14.6 shows a typical high-level approach to organization design. Once the scope and design principles (including the desired operating model discussed above) have been established, the design team steps around the wheel clockwise, performing the activities indicated. Going around the wheel more than once may be necessary since this process has an iterative element. This is mainly when this exercise is carried out parallel with procurement and ICT workstreams.

The skill base of the existing organization, and its ability to recruit, are aspects that should be considered with particular care. For example, it is pointless to create roles that only upper decile staff could perform, or if the current staff does not meet this criterion, salary structures will not attract such staff through recruitment.

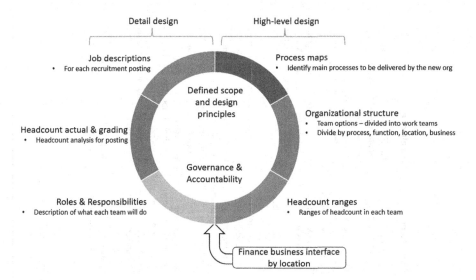

FIGURE 14.6   Organization design delivery model.

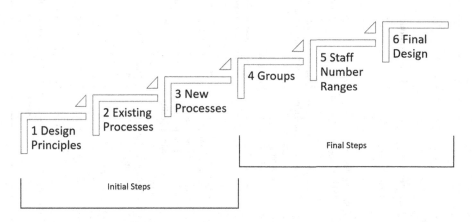

FIGURE 14.7   Organization design stages (initial and final stages).

The high-level delivery model depicted in Figure 14.7 can be broken down into individual stages (or steps), as depicted below.

14.3.9   FACILITY MANAGEMENT MATURITY MATRIX

Maturity profiling compares an organization's current practices with recognized "leading practice" asset management processes. For example, a sample maturity profile in Table 14.4 relates to Facility Management.

An organization, or its constituent parts, is mapped by type of process (i.e., Planning processes, Service delivery, etc., in the example below), and its "maturity" is each determined on the continuum of "Innocence" to "Excellence." Comparing

TABLE 14.4
Facility Management Maturity Matrix

	Innocent	Awareness	Understanding	Competence	Excellence
FM strategy linked to operational strategy	Link not considered	Ad hoc discussions	Occasional process	Straightforward process – infrequent application	The FM strategy is aligned to support overall operational objectives
Policies and standards	Ad hoc policy development: low levels of enforcement	Some policies are in place; HSE and manufacturer led	Key policies further developed; moderate levels of enforcement	Many policies are in place; high enforcement levels	Value engineering policies in place to deliver service levels – enforced
Planning processes	Local budgets – often stopped or reduced	Local planning for FM: not coordinated with planning & scheduling (P&S) or business need	Planning linked to business need; some demand challenge	Strong link with business need and P&S; demand challenge in place	Periodic planning round integrated with business needs and FM policies; efficient intra-charging
Commercial processes	Local appointments	Some bundling and grouping	Centralized purchasing in critical areas; contract development in-house	Central purchasing; contract structure further developed	Appropriate contract structures in place; suppliers actively managed
Service management	In-house – detached from the rest of the organization	In-house – better integration – few formalized contract management processes	Active management of service provisions	Formalized management structures and reporting	Contracted services delivered to contract; in-house services effectively managed; key performance indicator regime
Service delivery	Unproductive and high cost + extensive use of specialists	Some adjustments to cost of operation – output quality varied	Right-sized, market-rate workforce; output quality varied	Market standard organization; output quality addressed	Well-trained, efficient, and motivated labor force and supervision; market cost
Organization	Local organizations with many variations	Standardized local operations	"Designed in" regionalization and centralization	An organization built around end-to-end processes	Centralization and outsourcing as appropriate
Management information (MI) and supporting information technology systems -FM related	Inconsistent and ad hoc use of MI across the organization	MI is supported by ad hoc Information Technology (IT) plus the use of some standard systems	Common MI across some functions; not fully aligned with IT	Common MI supported by networked IT	Integrated centralized and networked systems designed to support MI strategy

Facility Management Maturity Matrix

Managing Processes Overview

The maturity profile for an organization shows pockets of "leading practice" and significant opportunities for improvement.

where the business feels it should be in each process area builds the momentum for change. It focuses on where existing processes must be improved or expanded in the new organization.

14.4 REF LIFECYCLE DELIVERY MANAGEMENT

14.4.1 OVERVIEW

Asset strategies and operating models are applied to REF assets to optimize asset value as assets are operated, maintained, and refurbished throughout their useful life. Asset strategies are also applied to dispose of end-of-life assets responsibly and sometimes create revenues by selling them at their residual values.

Asset management strategies and operating models may differ based on asset use context. However, key themes focusing on lifecycle value optimization remain the same. These themes include the following principal considerations:

- Early design considerations and strategic procurement – While competitive purchasing remains the basic principle, the overall life cycle delivery cost must be the focus, not just the first-time PO $ value.
- Asset strategies must address the operations and maintenance (O&M) phase of asset lifecycle delivery. This is where the most significant opportunities for value optimization lie. How assets are operated, maintained, and utilized as a part of an integrated operations plan determines their service continuity and, in turn, their return on investment.
- O&M management is perpetually faced with challenges related to the change. As the organization responds to internal and external changes, the only way to continually deliver value is to embed flexibility in the asset operating model. This flexibility should allow for timely course correction in strategic direction to support both short-term and long-term objectives alignment.

14.4.2 PLANNING AND PROCUREMENT

Typical drivers for organizations to undertake a program of procurement activity (i.e., outside the typical churn of periodic contract renewal) include cost-reduction initiatives, process improvement, the establishment of shared services centers, and consolidation following merger and acquisition activity.

Procurement activity typically follows six key steps:

- Strategy
- Sourcing
- Specification
- Tendering and solicitation
- Mobilization
- Contract award.

As they apply to procure REF services, these generic steps are explained at a high level below:

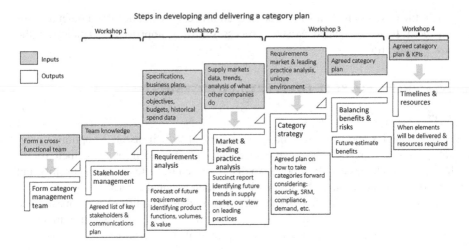

FIGURE 14.8   Typical steps in delivering a category plan.

14.4.2.1   Sourcing Strategy

Implementing a REF procurement strategy can significantly maximize the VfM achieved from its REF assets and associated activities. Category Management can provide structure and focus to this critical stage.

Figure 14.8 outlines a typical Category Management approach.

14.4.2.2   Tender Process

The procurement functions of businesses worldwide operate many subtly different variants of a broadly generic tender process. Typically, the steps shown in the figure below are followed, although some activities may occur concurrently to reduce the overall timescale. Similar processes apply for one-off projects and the formation of framework contracts that facilitate the subsequent call-off of services during the contract's life.

During the evaluation stages, complexity is added to the process, with qualitative and quantitative measures needing to be evaluated to select suppliers. In addition, scenario modeling is often required to assess the best VfM and facilitate negotiations with bidders.

14.4.2.3   Specification

The Specification needs to translate business requirements into user requirements that a supplier can understand and deliver for an agreed price.

Traditional "input-based" specifications in all three REF areas of Estates Management, Facilities Management, and Construction tend to be replaced by "output-based" specifications. These output-based specifications allow the supplier greater flexibility to deliver their solutions, albeit with a generally higher level of risk transfer. Provided that responsibilities are carefully and adequately defined, these typically allow the provider to achieve significant efficiency savings, which can be passed on to the client.

14.4.2.4 Mobilization

A good mobilization phase can significantly contribute to a successful contractual relationship. This phase includes due diligence, site set-up, and document review. It can also transfer staff from the client organization or incumbent suppliers to new providers. With service contracts (i.e., Facilities Management), this phase ensures that service continuity risk is adequately managed. This phase can ensure that all site procedures (i.e., health and safety) are fully operational with delivery contracts.

14.4.2.5 Contract Award

Ultimately, the contract should be able to inform both parties of their and others' responsibilities concerning their mutual relationship. While the REF industry has tried to reduce some inherent complexity by producing standard documentation, the practical reality is that significant contracts often require significant periods to reach a financial close.

One consideration often missed in contracts is that the individuals involved in day-to-day contract management from both the client and supplier sides have rarely been involved in the initial negotiations. There is, therefore, a risk that the original intent of particular contract clauses can be lost. To mitigate the intent risk, involving such parties early in the process can be beneficial to ensure that a detailed handover is affected.

14.4.3 OPERATIONS AND MAINTENANCE

As assets and facilities are procured and commissioned, they become part of the organization's annual operational expenses. These expenses typically comprise operating costs (costs directly and indirectly associated with running normal operations, including human resources) and maintenance costs (directly and indirectly associated with asset upkeep and regular upgrades). Since the O&M phase constitutes the most significant share of an asset's total lifecycle cost, it also manifests significant opportunities for asset lifecycle value optimization, efficiency gains, and long-term cost savings.

O&M strategies play a crucial role in realizing these opportunities and must include the following considerations for Facilities Management:

- Line-of-sight alignment with the asset Cost of Ownership (total Lifecycle cost)
- Focus on demand organization, clarity of expectations, and market/business environment trends when planning new investments
- Asset maintainability and useability when planning asset design and procurement
- Focus on strategies that help identify and prioritize maintenance optimization opportunities versus new grassroots investments
- Maximize utilization of facilities assets to reduce long-term per capita cost – focus on user experience
- Integrate operations and maintenance planning to minimize net asset outage – minimize failure rates and mean times to repair
- Safety Systems health monitoring and inspections program

- Continuous improvement plans for iterative O&M optimizations: small but consistent and valuable measures to maximize asset utilization and asset availability
- Ensure that all O&M activities are aligned with, and promote the organization's environment and sustainability objectives

14.4.4 Retiring and Disposals

As facilities assets reach their end-of-useful life, they should be planned for replacement or decommissioning, as the case may be. Replacement or decommissioning usually constitutes a significant cost and must be included in long-term asset planning and forecast. The following are the key considerations when planning a facilities asset retirement and disposal:

- Minimize user disruption and arrange alternate operations as applicable
- Disposal or Reallocation of resources (staff, energy, utilities, amenities, etc.) to alternate operations, as applicable
- Community and Environment obligations, as applicable
- Retirement of ancillary or auxiliary assets and spare parts
- Alignment with the Accounting department for accurate settlement of final depreciated cost. Some companies allow the sale of assets resulting in positive cash flow and net credit to the company's account.

14.4.5 Governance and Standards

Governing standards provide the North Star with structured improvements in the management of assets and asset portfolios. These standards provide a practical roadmap to an organized, competency-based continuous improvement journey. While the ISO5000 series of asset management standards lay down a clear framework for establishing a management system for value optimization of your tangible and nontangible assets, the ISO 41000 series of standards focus primarily on concepts and context of Facilities Management.

ISO 41000 family of standards, developed between 2017 and 2018, is the only leading global framework providing detailed guidelines on the development, maintenance, and upkeep of an effective management system for Facilities Management.

Built upon the common theme of the PDCA (Plan, Do, Check, Act) cycle, the standard covers the following critical areas in Facility Management:

- Demand Planning
- Organization leadership
- Workforce Productivity and well being
- Supporting resources and competencies
- Operations and asset care
- Performance Management, Assurance, and Conformance
- Monitoring and continuous improvement

14.4.6 CHANGE MANAGEMENT

The success of any change program depends on having three factors in place. Firstly, a clear rationale for change must be developed to generate the necessary momentum to embark on what can otherwise be a very challenging exercise lasting many months. Secondly, a clear direction must be provided in a broad strategic outline.

REF, particularly Facilities Management delivery, can affect people more personally than most other services since it directly influences an individual's working environment: for example, their desk, chair, ambient lighting, and temperature, as well as collective and personal hygiene. As a result, any change in responsibilities, particularly where outsourcing is involved, will be perceived as a threat to established roles and power bases. Change projects in this field are thus especially vulnerable to resistance and require rigorous governance processes.

Senior management should first be convinced of the need for any change: It should ensure adequate program resources are available and that more junior stakeholders in all geographies actively contribute to its success. The co-option of key personnel at both core and country levels to shape and implement the change strategy will help obtain their buy-in and turn them into advocates for the program.

Whether the drivers are regional, national, or international, experience has demonstrated that the following are the most common reasons for the change:

- Reducing operating costs as part of an organization-wide cost-saving initiative
- Improving service quality, often as a reaction to changes to the business that means its operations no longer meet the business need
- Increasing service consistency as the support services should operate similarly to the other business functions
- Increasing focus on the core business, thus avoiding management distraction on support and non-core activities
- Mergers and acquisitions with the consequent drive for rapid and effective rationalization
- Sustainability, with the need to reduce carbon footprint (i.e., greenhouse gases – GHG)

The typical benefits of any change program are summarized in Figure 14.9.

14.4.7 PERFORMANCE MANAGEMENT

As a critical phase in the Plan-Do-Check-Act cycle, Performance monitoring is pivotal to understanding the impact of asset strategies and applications of various O&M operating models and business processes. Several tools like Key Performance Indices (KPIs) and Balance Score Cards are used to quantify, calculate and report various aspects of a facility's performance: Financial, Occupancy Rates, Utilities Cost per unit of space utilization, Maintenance Performance, Asset Reliability, Planning efficiency, etc.

Benefits to a REF change program

REF Strategy linked to Operational Strategy	• Integration and clarity of strategic direction
Policies and Standards	• Positive impact on service levels & customer satisfaction
Planning Process	• Integration of business need & REF policies to avoid overspecification • Efficient cross-charging
Commercial Processes	• Ensures optimal leverage
Service Management	• Improves service quality & performance monitoring
Service Delivery	• Improved competence & motivation of staff at reduced cost
Organization	• Generation of cost benefits; fully integrated & centralized organization
MI & supporting IT systems	• Improved management information & reporting; increased compliance

FIGURE 14.9   Typical benefits of a REF change program.

KPIs are monitored and reported to ensure a line-of-sight alignment with meeting organizational objectives. These also benchmark overall performance targets across business units and industries. In the short to medium term, these KPI benchmarks become a "North Star" to guide organization priorities and behavior.

When implementing the right KPIs and other performance measures, care should be taken to ensure the following:

- Performance measures should be applied in a cascaded taxonomy, with the lowest level being the most detailed and hence directly relevant to and actionable by the tactical level in an organization.
- KPIs should aggregate as the levels move up, getting limited to the most strategic few at C Suite Dashboard, driving strategic decisions and objectives.
- Consider unified calculations and formulae – preferably standardized across the industry for normalized benchmarking.
- Single data source, with explicit assumptions and exceptions, clearly called out where required.

14.5   TECHNOLOGY IN FACILITIES ASSET MANAGEMENT

14.5.1   OVERVIEW

In the early 2000s, most organizations used Computerized Maintenance Management Systemss (CMMS) or EAM solutions to manage the general FM tasks and processes. However, they had limited functionality or solutions specialized to address REF processes (e.g., leasing, space planning, etc.). Software vendors then started to develop complete real estate-specific package solutions providing functionality to support the various tasks and processes needed by REF professionals. Gartner coined the term Integrated Workplace Management Solutions (IWMS) to cover these new solution platforms. These solutions have continued evolving and added new features to meet sustainability and data analytics needs.

14.5.2 CURRENT CHALLENGES

The introduction to this chapter highlighted that organizations are increasingly centralizing the management of their property assets and are thus treating them as strategic rather than local assets. Furthermore, they are also increasingly outsourcing the delivery of their REF services.

The former trend means operating on the traditional basis of all asset knowledge held locally is no longer feasible. Consequently, corporate-level systems must be employed to ensure consistent and effective management of property assets.

At the same time, the trend toward outsourced service delivery means that an organization no longer holds comprehensive information about its portfolio's performance. Instead, the service provider manages this data partially and is limited to the extent of services provided. The situation becomes more complex when multiple service providers have their data island produced and used without standardization or consolidation. This fragmentation of data systems and ownership increasingly drives organizations to consider the benefits of using common systems for all stakeholders.

14.5.3 BIG DATA ANALYTICS AND DATA STANDARDS

Big Data is a term coined to represent a vast volume of diverse data an organization produces during its operations and its ability to capture, mine, transform, standardize, and curate that data for insightful information that facilitates critical and strategic decision-making.

With decreasing cost of data mining and transformation (due to technological advances in internet of things (IOT) and sensor manufacturing), advanced analytical tools can now be deployed for a better understanding of data correlations and past trends, as well as for determining future opportunities in business transformation, otherwise not possible ten to fifteen years ago. Big Data and Analytics space has matured to be considered a strategic asset for an FM organization.

A critical consideration, often overlooked, is the need to define (or adopt) and implement common data standards and taxonomy. For example, suppose different smart hardware vendors or building management solution vendors have proprietary data standards. In that case, anyone building out solutions that need to aggregate and normalize this data into actionable insights will need to perform the translation or transformations during the development of these analytical solutions.

Key considerations for advanced data analytics outcomes in the FM industry include:

- Connected assets – a well-thought-of network of cloud-based applications and telematic devices
- Simplified dashboards driving actionable insights through feedback, performance monitoring, and benchmarking
- Enablement of asset lifecycle optimization and lower cost of ownership through:
  - Failure/Incident Prevention and service continuity

- Enhanced security and risk analysis
- Energy efficiency via real-time consumption analysis and timely course adjustment
- Regulatory compliance with sustainability and green initiatives
- Facilities' user (tenant) experience enhancement
- Intelligent buildings and infrastructure for achieving sustainability goals

14.5.4 AI IN FACILITIES OPERATIONS MANAGEMENT

AI-based capabilities are only starting to take hold in EAM and CMMS solutions focused on core asset maintenance. These capabilities include asset health monitoring, predictive analytics, failure forecasting, and prescriptive maintenance, assisting with failure diagnosis and corrective actions. All are based on artificial intelligence (AI) built upon structured and unstructured data from asset maintenance history, technical manuals, drawings, and other documentation that considers historical root cause analysis data.

IWMS and other point solutions have also started incorporating AI capabilities into their REF-focused applications. Some examples include using visual recognition to track actual space usage versus planned. Many IWMS solutions provided the capability to "Reserve" or book space for the day, but the difficulty for space planners trying to review actual usage against the REF strategy had difficulty confirming if the space was used. This input is critical to determine if the REF strategy meets its target goals. Combining AI capabilities with core IWMS functions is starting to take hold across the IWMS ecosystem.

14.5.5 TECHNOLOGY PLATFORMS IN FACILITIES ASSET MANAGEMENT

The advent of the IWMS platforms has provided ERP-like capability specific to real Estate. And like the evolution of ERP solutions, where new features and functions are added over time, the IWMS solutions continue to evolve. Many have added capabilities around environmental management and sustainability. Not all IWMS solutions (similar to ERPs) have the same capabilities, and organizations need to review those against their organizations' REF strategies and goals. Even though real Estate is the second or third highest cost on an organization's balance sheet, many companies underestimate the time and budget required to implement these solutions successfully.

14.5.6 SYSTEMS IN FACILITIES ASSET MANAGEMENT

There are other solutions supporting core REF capabilities other than IWMS solutions. These include a variety of Building Management Solutions (BMS) with a wide range of functional capabilities. These solutions will have facility-specific functions or modules for heating, lighting, water, sewage, etc. but can also be a platform to collect data from other point-specific solutions like elevators or fire suppression. With more of these systems, including intelligent features or data sensors, they become aggregators of huge volumes of data that can provide actionable insights, as discussed in the data and analytics section earlier.

14.5.7 System Type Summary

In ICT terms, two types of systems support REF functions: "Transactional" and "Property Performance Management" systems. The diagram of a typical REF series of functions, as represented by Table 14.5, is a useful starting point in developing any ICT strategy.

This diagram notably introduces the topic of environmental management. This has been historically included within either Estates Management or Facilities Management or sometimes treated separately altogether and managed together with the core functions of, for example, a manufacturing plant due to its business-critical importance.

TABLE 14.5
REF Functions

	Estate Management	Facilities Management	Environmental Management	Capital Projects
Property Performance Management	\Property Cost % of the overall budget, Total property cost per m²/FTE. Maintenance per Chartered Institute of Public Finance (CIPFA) category, carbon footprint per m². FTE/CIPFA category			
Functional Performance Management	• Property list • % Freehold/ Leasehold • Lease events • Rent paid on time • Service charges paid to time	• Asset list • School reporting • % Target Construction grade • Maintenance backlog %	• Energy consumption per m² /FTE/ CIPA • Category by month, year to date, etc. • Carbon footprint	• Project capacity • Impact • Number of accidents • Milestones • Performance to budget, time, and change impact
Operational Asset Management	• Portfolio details • Legal/ commercial • Lease events • Rents, rates • Service charges • Insurance • Legal obligations • Easements, RoW, etc. • Landlord/ tenant valuations	• Asset register • Surveys (condition, etc.) • Planned/reactive maintenance • HSE inspections • Tree management • Warranty management • SLAs • Reservation management • Accommodation management	• Energy management • Carbon management • Waste management • Building energy spend	• Design • Costing • Planning • Build management • Project progress • Change management • Project financials
	Staff, data, documents, finance, IT infrastructure, and systems management Service delivery performance management			

14.6 REF OPTIMIZATION

Facilities management, whether as a core business or as a supporting arm of a large manufacturing concern, remains a significant line item under O&M Expenses. In addition, unfortunately, it is not uncommon to see most REF projects marred by severe cost overruns and delays. Also, the post-Covid world of hybrid work practices creates an ever-growing demand for flexibility, remote access management, and compliance with health and hygiene protocols.

With the advent of transformational technologies and global standards, REF Management industry, however, now finds itself surrounded by islands of potent opportunities, from O&M cost reduction via asset lifecycle optimizations to enhanced efficiencies in space management and, from proactively supporting sustainability measures to successfully meeting regulatory compliance.

The application of AI in IWMS systems brings opportunities to transform the workspace culture, where space utilization and occupancy trends are monitored for timely asset availability and reliability assurance. In addition, real-time planning for extensions and risk-mitigated new real estate investments will also contribute to a facility-intensive organization's bottom line.

PERMISSIONS

15 IT Service Management Lifecycle*

Brian Helstrom
Previous contributions from Brian Helstrom and Ron Green

Chapter Summary

At their core, all organizations have the same objective: to achieve their business mission most effectively. And, while most institutions rely on technology tools to facilitate the achievement of this mission and its related business goals, the tools they rely on often prove to be more of a hindrance than a help.

As a result, many organizations have turned to process models to take a more structured approach to information technology (IT) management and IT asset management (ITAM). Two such models, Total Lifecycle Asset Management (TLAM) and the Information Technology Infrastructure Library (ITIL), created by Britain's Office of Government Commerce (OGC), provide the guidance that IT organizations crave.

In concert, TLAM and ITIL can help organizations reduce operational costs, improve transaction efficiencies, enhance customer experience, and better meet the business mission. Despite their complementary capabilities, only some businesses have merged these two methodologies. However, organization leaders are beginning to recognize that, with thoughtful deployment, TLAM and ITIL can be woven together to make IT assets even more effective in serving and achieving the business mission and goals.

IT Service Management consists of multiple parts, which all add up to providing support to the organization from IT. Over the last 30 years, the IT infrastructure has become as crucial to the operation of an organization as the other classes of assets addressed in this book. Managing IT assets has similarities to other assets; however, they often use different terminology.

IT people like to have their language and have adapted one that fits the nature of their delivery of services. However, when we take the time to analyze the IT languages and terms, it is remarkably similar to the maintenance strategies and services discussed throughout this book. This chapter will discuss some unique language used to describe IT Service Management while showing the parallelisms with other forms of maintenance and why it is part of this text.

15.1   INTRODUCTION

IT Service Management (ITSM) has become an icon of sorts as the global community encircles itself with ITIL and Control Objectives for Information Technology (COBIT), to name a few. The idyllic solution of optimal management of IT services has been the topic of many books, seminars, and actions. So, why consider ITSM as a chapter in a book on maintenance excellence?

In IT, like any other part of an asset-based organization, it requires management of the assets that deliver capabilities toward the goal of successful production. The focus of any asset's lifecycle, whether in IT or the large equipment in the manufacturing field, is keeping the investment operating efficiently to serve the business. In comparison, we understand that some large equipment pieces have a cost base where their obsolescence is generally much higher and can be extremely expensive to maintain. IT assets can have a relatively shorter lifespan in the organization but can be equally as expensive over time.

An organization's IT assets focus more on providing value to the company through the services they enable rather than on the equipment itself. However, the layered collection of interdependencies often makes for a challenge in assessing the actual cost of the individual piece of equipment. Further, some IT assets are not within the company's walls but require maintenance. That is, the assets act more as a collaborative form of associated services that become augmented by the people who support the technology in providing those services. This is the foundation of ITSM: Delivering and supporting the IT functions of the business through technology.

15.2   MAINTENANCE IN A SERVICES BUSINESS

Unlike a plant that produces products, whether oil, gas, cars, electricity, or whatever, IT is a services-based business. IT aims to provide enablement through the technology for the business it services. So, as we go about the effort to maintain IT assets, we are, in turn, maintaining and managing the services that support the business. The IT group has no function if the business has no use for information technology, so this idealizes what a services business is; its sole purpose is to provide services that enable it. The IT functions are there to provide assistance to support the enablement of the business through information technology (IT assets).

Understanding that a service business is only there to support its customers, we can better appreciate the focus on maintenance in this structure. As a service business, we perform maintenance to ensure that we can provide the services our customer requires. This is independent of whether the customer is internal to the organization. In the discussions here, the customer is the user of the IT service and may be part of the same organization or company. Likewise, an IT service business could be a department that manages IT for the same organization as easily as it could be a separate company.

In understanding this premise, we can quickly understand the level of importance in the services business. Maintenance in a services business must focus on providing service levels appropriate to the customer's demands. This may entail servicing assets and equipment that provide services and servicing the services themselves.

The types of investments that offer services, especially in IT, are, by their design, usually layered across many component pieces, such as in the network, which comprises routers, hubs, switches, multiplexers, authentication devices, firewalls, bridges, and miles of cable. Within IT, these services layer upon other services to form a complete offering to the customer. The measure for the service level to support the customer must be across all these components, which by definition is not additive but multiplicative.

As an illustration, when assessing just two dependent components A+B, where reliability is 80% of component A when serialized with 80% of component B, it does not equal 80% for A+B together, but instead, it equals 64% for A+B. This measurement causes the most significant problem in supporting services to a customer. We cannot just measure the maintenance of the single component as it is not the single component that offers the service. Measuring the complete service delivery as the business expects it is crucial.

To support this, you must ensure that your organization has the tools to help critical business aims and that these assets function efficiently at total capacity. However, securing smooth, always-up IT-asset functionality can prove challenging.

Add to this conundrum the global business environment that, by its very nature, requires distributed IT capabilities beyond the superstructures of the business. They must be capable of functioning across the enterprise and across the geographic expanses of that enterprise. As a result, keeping track of IT assets enterprise-wide – where these assets are, how they are used, and who is using them – can become an issue. This is further expounded by the fact that some assets are now pure services, such as Software-as-a-Service (SaaS) type assets. In keeping your organization's technology assets up-to-date and functional, focus can often drift away from the core mission to ensure IT assets work how they need to.

Whether organizations are based in the corporate world or within a government agency, several challenges regarding business technology assets must be addressed. First, most enterprise executives must confront a growing gap between the IT asset portfolio and what is needed to achieve the business mission and support related strategies. But bridging this gap requires careful thought and planning, as pressures to reduce costs seem to grow increasingly stringent.

Additionally, post-September 11, not only does it make good business sense for organizations to better account for and manage their assets and information, it is now often a legal requirement. An increasing number of national and international regulations and security directives must be complied with – for example, Basel II, the Sarbanes-Oxley, and the USA PATRIOT Acts. Couple all this with inadequate or skewed investment in staff and IT infrastructure, resulting in a misalignment between an organization's asset base and productive use of those assets.

15.3 UNDERSTANDING ITIL AS A BASELINE

Information Technology Infrastructure Library (ITIL) is the world's most widely accepted approach to ITSM. The library is a compilation of books developed for Britain's Office of Government Commerce (OGC) (now under license through AXELOS). It consists of a cohesive set of best practices for managing IT in any organization. ITIL resulted from years of effort by several public and private sector firms and consultancies to develop a comprehensive collection of processes, frameworks, techniques, and tools to measure and manage IT.

ITIL has continued its journey of continuous improvements from its original debut as the preferred method for managing IT services. In addition, ITIL, through its various derivations, has expanded its focus on needs in the IT services business.

In its latest derivations, ITIL has expanded to include 34 practices to accommodate the full range of delivery of IT across the business. These areas take the practices and processes and cross-map these through the following four dimensions: Organization and People, Information and Technology, Partners and Suppliers, and Value Streams and Processes. This provides a holistic view of all the functions and processes delivering IT through these four dimensions. This revised view in ITIL includes many of the newer aspects of IT, such as Artificial Intelligence and updated architectural concepts.

However, under it all, there are still the foundational aspects of maintaining the IT assets that we have been discussing in this book. The concept of Maintenance Excellence fits as a strategy for the tactical and continuous improvement of optimizing the delivery of IT to the business needs.

Using ITIL aligns well with integration within the business to COBIT and TLAM to deliver the best tools to fulfill the organization's needs to manage its IT assets and services to their optimal value. It reinforces the alignment with the goals of maintenance excellence in supporting IT assets: strategic, tactical, and continuous improvement across the organization while optimizing this within the information technology realm of the organization.

This chapter is not about the lengthy details of these ITIL processes and their framework. Instead, it will try to explain how these various concepts can be redefined or aligned with the terminology of the maintenance excellence realm.

Yes, the terminology is different between plant maintenance and IT. But what is most interesting to note is that the resulting high-level actions and activities are reasonably similar and align similarly. So, for example, if we view some terminology differences, we can see why IT and Plant maintenance are close.

Not all ITIL practices need to be discussed, but as shown, some of the practices inside ITIL align well with plant asset maintenance.

Incident Management in IT is similar to a repair notification in the plant environment, where we must address issues and restore productive operations as quickly as possible.

Problem Management in IT deals with problem analysis and root cause determination through problem resolution to correct deficiencies in IT operations. This varies slightly from the enterprise definition of root cause failure and corrective action, which addresses weaknesses in its equipment's operational characteristics. However, not too dissimilar as the goal is to assess a means to avoid its recurrence.

Release Management in IT deals with significant or collective changes to the IT environment. In contrast, we see the same level of complexity in what the plant would call a Turnaround or significant overhaul. These often consist

IT Terminology		Plant Terminology
Incident	⇔	Notification
Incident type	⇔	Failure Codes
Incident Assignment	⇔	Dispatching
Change management	⇔	Management of change (MOC)
Change request	⇔	Service request/Work order
Configuration item	⇔	Equipment/Assembly
Hardware	⇔	Equipment
Problem analysis	⇔	Root cause failure analysis
Problem resolution	⇔	Corrective action

FIGURE 15.1   ITIL terms versus industrial terms.

of wholesale changes in the environments with considerable project efforts introduced to make the changes or a collection of many changes that work in tandem to make a significant effort that incorporates into a one-time activity.

Through making these simple, high-level comparisons, we can illustrate that ITIL is similar to the traditional plant maintenance methods for optimizing support services to the organization. ITIL still aligns with the full scope of the Asset Lifecycle and fits the capability maturity model for excelling within its impact on the business. It becomes evident that ITIL is an extension of these principles of uptime to the organization but with a tighter focus on a more service-based approach to delivering IT services. Is that not what IT is: A service-based component of the business?

IT services provide business services to the rest of the company to be productive or deliver other services driving the core business functional operations. Just as plant maintenance ensures effective use of the equipment for core business functional operations. Both services represent a necessary component of enabling business (production) success.

15.4   APPLYING ITIL FIRST TO ITAM

It has been estimated that enterprises that begin an asset management program experience up to a 30% reduction in cost per asset in the first year – including people, process, and technology costs – and continued savings of 5–10% annually over the next five years. These savings can be found by recovering assets rather than buying new ones and eliminating unused assets with costs involved (such as maintenance costs for unused equipment).

ITIL offers a framework of processes that can be used to develop and support an enterprise IT asset management (ITAM) program. Here is a simple truth: **ITAM** is 80% process and 20% tools. This reinforces why developing process strategies around the framework provided by ITIL is the first step before implementing any ITAM solution or tool. You cannot implement one without the first having the other. ITIL provides the processes and identifies the people roles that should be defined so that you can develop a clearly defined and successful ITAM solution.

It is also essential that with all the various models of IT deployed, including service offerings, IT assets are not always tangible internal assets. This poses a challenge under the older models for tracking assets and is uniquely different from typical plant assets. But just the same, these system-as-a-service rollouts are assets that must be managed through contacts and service-level agreements. Tracking them inside an asset management framework is still necessary and fundamental to ensure the continuous measure of asset productivity under any maintenance excellence program.

To better understand some of the features that will drive a complete asset management solution for the organization can be done by leveraging a capability maturity model that measures and identifies the level of capability an organization has reached concerning their IT asset management practices.

How do we describe ITAM best practices and their relationship with maintenance excellence within the framework of ITIL? ITIL implicitly contains the maturity model concept in its presentation of best practices. The maturity of asset management practices is an underlying theme for many ITIL concepts. The ITIL functions defined below illustrate the interface between the asset management strategy and the business processes that support it. Each of these functions describes criteria for a successfully managed process and allows you to identify the level of maturity within a maturity model to which these functions apply.

15.4.1 CONFIGURATION MANAGEMENT

Configuration management identifies, controls, maintains, and verifies configuration items (CIs) in a configuration management database (CMDB). CIs include individual assets, business processes, aggregations, and virtually allocated resources.

The plant analogy may be to consider this as the Engineering design that details asset function for how all its components integrate into supporting business delivery.

Within the ITAM maturity model, an "aware" state requires the following developments within the organization:

- Instantiation of a CMDB;
- Definition of CIs for shared infrastructure;
- Conducting a discovery of shared infrastructure and entering results into CMDB; and
- Established control and verification of processes for the CMDB.

While the "capable" state requires the following:

- Creation of interfaces among existing CMDBs;
- Feeding a central CMDB with validated information; and
- Propagating identification, control, verification, and maintenance processes throughout the organization.

15.4.2 INCIDENT MANAGEMENT

Incident management is the effort to minimize the impact of events on services and the mission. Incidents can be equipment failures in managed assets, service outages, acquisitions, or the sudden discovery of undocumented assets. Incident management deals with effects and symptoms. A related function, problem management, is analyzing and resolving causes. Often, problem definition will arise from the root cause analysis of an incident.

Within the ITAM maturity model, an "aware" state requires the following to be in place:

- Proper maintenance service functions for the asset management system as well as assets within the mission dependency chain;
- Incident reporting triggers for repair and maintenance processes; and
- Incident records that remain available to verify maintenance efforts.

While the "capable" state requires that incident management is closely tied to change and configuration management to track every asset's supplier, maintenance history and methods, licensing, and usage.

15.4.3 CHANGE MANAGEMENT

Change management in the context of IT assets regulates and oversees requests for change and change processes. This applies to any changes to assets, operations, or organization. The "aware" state is implied in having a change process. Still, more importantly, the "capable" form requires that the implemented methods are such that any change automatically results in inventory updates and triggers configuration management.

15.4.4 FINANCIAL MANAGEMENT

Financial management is managing and reporting the costs, funding sources, availability of funds, budgets, and return on investment for assets. The maturity model that identifies the "aware" state requires that the financial reporting processes and supporting infrastructure exist such that groups and individuals charged with finance and accounting responsibilities can access all relevant data and that the asset costs, maintenance costs, and replacement costs are tracked with CI's in the CMDB. The "capable" state requires that the financial management policies are explicitly part of all asset management processes and that the Incidents and changes trigger economic effects and reporting without additional intervention.

15.5 MOVING INTO ADDITIONAL ASSET CLASSES

Asset and Service Management is a set of processes and practices used to manage the optimal performance of all critical assets in alignment with the requirements and expectations of the organization. This can be accomplished through an effort in asset classification that delivers a common repository for assets and services within a standard business model. We enable a view of all assets from the perspective of the physical relationship, the service it performs, and the visible part it plays in the business.

Classifying assets allows us to understand the criticality of a unique piece of equipment as it applies to the schema of the business rather than the value of the individual asset. The ongoing innovation of assets and external forces (regulatory governance) is causing businesses and organizations to implement tighter controls around assets. Three areas of focus can be identified that drive the demand for how we classify assets and enable us to manage the risks implicated by these changes.

- The **interdependency** that exists between assets is on the rise; therefore, it is no longer practical to manage assets independently, which is especially true with assets-as-a-service type models;
- The **boundaries** between asset classes are fading; therefore, it is necessary to treat equipment, buildings, and IT assets as an ecosystem where operational assets not only have embedded IT assets, but they are dependent on IT assets to function well and likewise are IT assets dependent on active operational support such as a heating, ventilation, and air conditioning (HVAC) solution and power conditioners to function correctly; and
- Influencing **pressures**, such as globalization and regulation, have created the need for more transparency of assets throughout the asset life cycle.

15.6   THE EVOLUTION OF TOOLS

Businesses need to use and reuse existing technology more efficiently. But simultaneously, they must find a way to deliver services even more effectively. Total Lifecycle Asset Management (TLAM) is one avenue companies can use to streamline efficiencies and cut costs. And TLAM practices provide a framework that can assist with meeting these goals by helping IT staff better optimize and align IT investments that support the enterprise's overall mission and strategies.

The TLAM methodology takes a holistic asset management approach and calls for the review of virtually everything IT-related. TLAM also looks at operation, maintenance, modification, and disposal across the enterprise, beginning with IT strategy and planning, evaluation and design, acquisition and build.

Using TLAM, your firm can categorize asset classes based on similar management attributes. For example, assets might be classified together if they are financial, fit into a particular space within the IT structure, or if a specific group of users tends to rely on them.

TLAM helps you look across your IT portfolio, enabling you to assess each asset's lifecycle. TLAM can be used to evaluate asset-management strategies and best practices at each stage of life, focusing on better managing total lifecycle cost.

For example, when looking at actual properties, an organization might conclude after completing a total-lifecycle analysis that constructing a data center is more attractive than leasing one. Or a Chief Information Officer (CIO) working for a state agency might figure that, based on maintenance cost projections, procuring a new system is preferable to repairing and maintaining existing servers.

FIGURE 15.2 (Total) asset lifecycle management model.

Using TLAM guidance, the structure is wrapped around IT asset management. A series of crucial IT management phases, which span the technology lifecycle, are implemented systematically:

- The *strategy* phase focuses on understanding an asset's role and value to the organization.
- The *planning* phase looks at how best to integrate assets into business plans.
- The *design* phase occurs when assessing product performance or design.
- The *construct/procure* process compares asset purchase and capacity requirements.
- The *commission* phase includes all processes to prepare and confirm that the asset operates as designed.
- The *operate* phase includes all processes required to keep an asset operational.
- The *maintain* phase encompasses any process required to sustain or reuse an asset.
- The *decommission* phase ensures that asset disposal complies with all security, legal, and contractual requirements.

ITIL also looks at managing the entire asset lifecycle and has been evolving to raise the level of applicability in support of a changing world and leveraging the best practices of system and service management. Its evolution has taken it from the system-based processes to the new scheme that is services based for managing and supporting technology assets.

Each of these volumes focuses at different levels on the overall target of improved service (maintenance) excellence. ITIL will align well with integration within the business to COBIT and TLAM to deliver the best tools to fulfill the organization's needs to manage its IT assets and services to their optimal value. It reinforces the alignment with the goals of maintenance excellence in supporting IT assets: strategic, tactical, and continuous improvement across the organization within the information technology realm of the organization.

15.7   SERVICE-LEVEL MANAGEMENT: THE OPPORTUNITY TO RECOVER COSTS

Service level management consists of planning, coordinating, drafting, agreeing, monitoring, and reporting on *service-level agreements (SLA)*. It also consists of periodic reviews of *service achievements* to ensure that the identified services are maintained and improved in a cost-justifiable manner. The SLAs provide the foundation for managing the relationship between the service provider and the business (customer).

Through these reviews, the SLA can be used to identify where improvements have resulted in improved cost controls and overall business efficiencies (savings) that can be used to recover the costs of implementing these technological solutions. Further, the *SLA* can be used as a basis for *charging* and assists in demonstrating what value is being received for the money.

One of the most effective ways to manage services is to manage the assets that derive or enable those services. Leveraging proven practices and maximizing support capabilities will allow for the further enhancement and controls needed to address the IT services being provided, meet the drivers of service-level management, and support the organization's overall mission.

Where TLAM and ITIL capabilities overlap, they can dramatically improve an enterprise's ability to meet its overall mission. TLAM provides a methodology for monitoring and streamlining IT assets and capabilities throughout each asset's lifetime. At the same time, ITIL offers insights on how to best implement and manage these assets in support and delivery of IT services. As a result, organizations can weave TLAM and ITIL strengths together to generate an overarching view of their capabilities in ITAM.

ITAM helps your organization better combine its financial and mission objectives. The benefits include:

- More efficient data-sharing across the enterprise
- Enhanced asset visibility and security
- Improved planning around technology refresh and upgrade needs
- More efficient technology repair and deployment

- Lowering total lifecycle costs
- A better view of audit compliance
- More uptime to support the mission
- Doing more with less

TLAM and ITIL have many areas of overlap. By merging the two frameworks, your organization gets the best from both. Use the TLAM model to address internal IT asset issues. And look to ITIL guidance to see how your IT organization works and to identify where and how IT services can be better managed.

For example, your IT systems department might be most concerned about the purchase, use, and maintenance of a particular PC, while customers or internal end users, such as those in the tax department answering customer questions, care most about the PC's service can provide. By merging TLAM and ITIL guidance and capabilities, you can better unite user and business interests – ensuring that hardware, software, and service requirements are more effectively met.

The benefits of merging ITIL and TLAM are the most useful – and powerful – for enterprise change management. Using the ITAM methodology, you can better plan and manage change, using the best industry standardized practices from beginning to end.

For example, for an organization to anticipate or document the movement of its IT assets, it must first be able to track the available technology tools. However, most organizations today rely on spreadsheet data or have to make numerous ad-hoc phone calls to find this information. Because of the inherent inefficiencies in this approach, competing initiatives to capture the same information often result. Using TLAM and ITIL in tandem can help coordinate asset and change management efforts and drive efficiencies in near-term projects and long-term processes and technologies.

TLAM's asset repository, a TLAM Create/Procure process component, overlaps with ITIL's configuration management database (CMDB), which administers IT infrastructure change. As a result, TLAM's asset repository data – information about an organization's IT tools, where they are, and what they are used for – form the cornerstone of ITIL's change management database. Without these records, the configuration management database is incomplete and ineffective in supporting essential ITIL processes. Additionally, asset repository data can support critical ITIL processes, including Configuration Management and Release Management.

15.8 SUMMARY

The IT Service Management lifecycle consists of two unique pieces: the assets, which differ little from other enterprise assets, and the processes, similar to the methods we use for other maintenance solutions. TLAM is fundamental in ensuring the maintenance of IT assets and delivering sustainable solutions.

The next level of understanding in driving maintenance excellence is allowing maintenance to focus on service to the organization and how the physical assets of that organization support delivering service to the organization. Information technology is already a services-based part of the business; whether serving a single customer or multiple, the rules of operation remain the same. In maintenance excellence,

IT provides good asset management and service support, optimizing the assets that deliver good IT services.

PERMISSIONS

* *Asset Management Excellence: Optimizing Equipment Life-Cycle Decisions*, 2nd Edition by Editor(s), John D. Campbell, Andrew K. S. Jardine, Joel McGlynn Copyright (2011) by Imprint. Reproduced by permission of Taylor & Francis Group.

BIBLIOGRAPHY

AXELOS, https://www.axelos.com
COBIT, https://www.isaca.org/resources/cobit
IT Service Management Forum (itSMF), https://www.itsmf.ca/page-18070

16 Information Technology Asset Management*

Brian Helstrom
Previous contributions from Ron Green and
Brian Helstrom

Chapter Summary

This chapter focuses on what Information Technology (IT) assets are and why we would want to manage them. We will spend some time here to understand why we include IT within the realm of maintenance excellence and why IT asset management has become a part of the enterprise asset used to deliver productivity to the organization.

Managing and optimizing IT assets is critical to delivering cost-effective support to the business. A multidisciplinary approach to managing these assets and processes allows IT organizations to minimize costs while maximizing return on assets and achieving targeted service levels of support to the business.

All IT assets have a total lifecycle similar to the other assets within an organization. They move through the same cycle of events from planning, procuring, and implementing or commissioning to maintenance and upkeep, disposition, and replacement. The difference is that IT assets generally expend their lifecycle over a much shorter period and are undergoing more radical changes in support of newer technology and the changes in how we use technology. But like other plant assets, an IT asset needs continuous service and maintenance to operate effectively and at peak performance.

16.1 INTRODUCTION

While we are on the topic of enterprise assets, it is essential to consider one of the expansive areas in every business today: IT assets. The challenge here is less about capturing these assets into inventory and more about managing them and understanding the level at which to manage them. Within the realm of IT, your company needs to recognize that there are many views of an IT asset and its range in scope. But in the following few pages, we will elaborate on what an IT asset is and why we need to consider these in our overall asset management strategy.

DOI: 10.1201/9781032679600-20

Let us review what comprises a typical enterprise asset. An enterprise asset is something in the organization that we need to manage because it is critical to the viability of our business. With it, our company will succeed in performance and capability to deliver to our customers. The enterprise asset has both a value in terms of financial consideration and where it lies within the company's ability to produce goods and services for the marketplace.

In today's world, this clearly describes how IT assets belong within the overall enterprise view. Most organizations today cannot survive or deliver to their customers without the help of technology. Information technology has become embedded in overseeing and supporting an organization's ability to provide and offer its output to its customers.

To achieve a sustainable IT Asset Management (ITAM) program, an organization must leverage and apply four guiding principles:

- Reduce the total number of ITAM Tracking Methods.
- Simplify and standardize processes across the organization.
- Consolidate Control and Accountability for IT assets.
- Establish a "single source of truth" for ITAM data.

The governance component will be fundamental for a successful transition to ITAM and the long-term health of an Enterprise Asset Management (EAM) program. The Asset Management Maturity Model tasks and capabilities illustrated in this chapter and the recommendations for the way forward heavily depend on a solid governance structure within the enterprise.

16.2   WHAT IS ITAM?

Information Technology Asset Management (ITAM) is precisely what it implies: the management of the assets that comprise the workings of the business's underlying technology delivery capabilities. These would include the software, the hardware, the network, and the data, each with its unique level of importance to the business. But ITAM is more than just managing all the computer equipment for the information technology group. It is also about managing the integrated technology that supports the business, from the networks and equipment that operate the machinery in the plant to the portable devices used to update inventory and logistics in the warehouses. IT is no longer a back office or management tool; it has become fundamental in an organization's operations. As computers and IT become more integrated into every part of the business, it now impacts the tradespeople on the shop floor and the production workers doing their job in managing the delivery of products or services to the customer; ITAM has become mainstream.

16.3   WHY INCLUDE ITAM IN MAINTENANCE EXCELLENCE?

For assets to have longevity in the organization, we must ensure they operate at their peak performance. It is essential to maintain the equipment and undertake various maintenance tasks. What makes maintenance on an IT asset different is that we rarely use wrenches to do the maintenance. But we still need to perform maintenance. Further, IT assets get more complex as they have many intertwining non-tangible pieces. Therefore, understanding the requirement for maintenance on IT assets and that IT assets are core to the excellence of business performance, why would an organization want to exclude ITAM as part of the drive to maintenance excellence? A standard of excellence in maintenance throughout the organization must also apply to the IT assets. Maintenance excellence in IT is as important as maintenance excellence in the manufacturing environment. A number of the processes we use to manage equipment maintenance are the same as in managing IT, but we use different labels or terms.

IT assets are widespread throughout most organizations and are already expanding within the shop floor of the production plant. Through supervisory control and data acquisition (SCADA) and other statistical control systems that use microprocessors and are running on the networks, the plant floor has become more capable of sharing its operation statistics with the management information systems (MIS). These systems in operation are now integrating with other significant components of the once-isolated interactions of MIS to provide detailed information to the plant and production engineers to offer more control of the delivery processes of an organization. Tracking from start to finish along the production process includes not just the production side of things but all the other factors, such as environmental control, power consumption, and more.

As the shop floor continues to look more like IT and IT starts to embrace the plant's operations, we begin to respect the need to include such IT assets as an integral part of the productivity of plant operations. By its very nature, this implies the need to have IT assets within the overall maintenance excellence program.

Today's IT assets rarely act in isolation but comprise complex integrated hardware, software, networks, and data components. As such, this complexity of integration becomes significantly limiting when any of these parts start to perform poorly or fail. Proper maintenance can detect and prevent such failures and keep your systems from running poorly. Like the components in your car, the parts in IT must all work together to keep the company moving. We have all experienced frustration when a car component starts to perform poorly and fails because we thought we could avoid doing some maintenance. We all have heard the mantra of "pay me now or pay me later" from the mechanic when it comes to maintaining your car. As a good maintenance strategy on your car helps prevent failures and lower costs, the same is valid for maintaining our IT assets.

IT asset maintenance has fundamentally become part of the overall operational plant maintenance. Many organizations realize synergistic advantages in combining the management of operational assets with those of IT assets, especially the underlying IT infrastructure that supports managing and controlling many operational assets. To improve manageability, many operational assets are taking on many IT attributes, such as microprocessors, operating systems, and Internet Protocol (IP) addresses. Plant assets such as environmental controls, power meters, and instrumentation are increasingly being networked and managed with IT software. Statistical processing controllers have been integrated into the shop floor for many years and are now being networked to provide continuous feedback to the business. Since these plant assets behave similarly to the IT assets, there is an opportunity to leverage IT business processes such as software distribution and patch management onto the shop floor to provide better overall asset management. This is the reasoning that can be used to converge on the managing of IT asset management within the maintenance excellence framework.

16.4 WHAT IS THE VALUE OF ITAM?

IT assets have become integral to delivering production output to many plants for managing orders and shipments to the functional delivery of running the machines through automation. IT assets have spread throughout the organization, and managing these assets and their sub-component parts effectively can make the difference between a successful organization and one that will become obsolete or fail to survive.

IT Assets are now, more than ever, an integrated component of the organization. With the need to find efficiencies in delivering productivity in a global economy, IT and its implementation in the productivity chain have become inseparable.

ITAM brings value to the organization by providing a means to manage these embedded and integrated technology systems within the organization and facilitating having a single set of strategies within a business to manage all IT assets.

16.5 PROCESS IS THE KEY TO SUSTAINED PERFORMANCE

Forming and managing IT assets has encompassed much effort and time throughout the decades. The construction of a whole library of processes within a guiding framework for managing IT assets resulted. This framework of processes became the Information Technology Infrastructure Library (ITIL) and has become the de facto standard for the processes involved in managing IT assets and expanded to the whole of IT management. The entire framework is not just for asset management, as it entails much more as it delves into the delivery of service management to the community that uses the IT assets within an organization and much more. Embedded in the principles and practices of ITIL is the use of strategic, tactical, and continuous improvement methods in managing IT assets and delivering services to organizations. ITIL has, in its evolution, adapted some of the concepts covered in this book at striving to the ideas of maintenance excellence.

16.6 UNDERSTANDING THE ROI OPPORTUNITY

The return on investment (ROI) of ITAM is similar to the ROI for other enterprise assets. The criticality of IT assets and the need to track and manage the maintenance of those assets can identify what gains occur with proper management. Assessing the value placed on needing these IT assets is essential in evaluating the ROI. If these assets are not working, what impacts the business? What are your current maintenance costs for managing these assets, and what improvement would you make in better managing these assets? Understanding the ROI of any ITAM implementation means understanding the value the IT assets have in the delivery and management of the business.

Another way of stating the ROI equation of IT is to understand the cost of the IT infrastructure that needs to be added. The impact would be far-reaching if an organization could not create products, pay employees, provide regulatory reports, or purchase/manage inventory. For this reason, the ROI equation must be considered across a broader context. Due to this extended impact, each solution must be evaluated against business continuity. From the outset, redundancy, backup, and manual processes should be part of all planning activities.

Poorly managed IT can result in much money wasted on software licensing and improper care and maintenance of computer equipment which could lead to early replacement or lost assets. On the other hand, the value of proper management of IT assets and resources can save a large company millions of dollars.

16.7 WHAT TYPES OF TOOLS SHOULD YOU CONSIDER?

Many tools used for managing enterprise assets are the same tools we can use for managing IT assets. Since the issues of managing IT assets become embedded with managing service to the various components and supporting the productive activities for business, it is critical to address this complete combination. The difference in the tools needed for assets management extends beyond just managing the physical assets but includes the entire configured system of components, including software, hardware, networks, services, and data.

In addition, leveraging many specialized tools can assist in managing IT assets. Starting with specialized tools that can go out and discover all the assets on a network can be extremely helpful. Generally, these same tools can manage the various configurations of the software and devices on the associated network. IT has been managing its environment for years, and leveraging these tools to address these new application devices appearing in plant operations makes sense.

Even at the writing of this text, we are seeing further convergence of tools that will discover and monitor the expanding IT infrastructure. Anticipating that there will be further convergence in the tool sets and software asset management capabilities will become more refined over time to offer more comprehensive capabilities to the organization.

Implementing ITIL as a tool within the organization provides many targeted processes that manage the services and components associated with the IT assets throughout the organization. Within ITIL are concepts around housing configuration management information in a database. Understanding what configuration extends beyond just the physical components but also goes into adding information about how the parameters and settings within that asset have been set up to operate to maximum effectiveness for the organization. Technology assets are sometimes different even though they may look and act the same; a computer on one desk may be very different in its settings than the same computer on another desk or the plant floor.

Standardizing how IT assets are deployed, managed, and configured is essential to an organization's success. This would include applying configurations that might restrict deviating the asset's purposes. For example, banning a computer from browsing outside the bounds of the organization (i.e., browsing the Internet) might need consideration to ensure the device (and the people operating it) stays focused on its purpose to avoid negative impacts on its ability to function as intended.

16.8   HOW DO YOU BEGIN AN ITAM PROGRAM?

Beginning an ITAM program can appear imposing, and leveraging the concept of continuous improvement can be one of the fundamental drivers for this. Understanding that you cannot solve all past ills in a single step will go a long way toward setting expectations for all involved. Establishing the roles and empowering a person or position to create/an enforceable policy is vital. These policies should work within the existing job frameworks as much as possible. Creating redundant, cumbersome processes will result in poor results.

Starting an ITAM program begins with identifying two significant pieces: the depth of detail required and the breadth of involvement.

- When we talk about depth, we refer to what equipment components get serialized, that is, at what level are the individual items tracked and managed. Some organizations may wish to keep track of lower levels than others; for example, some may want to serialize the mouse on every computer, whereas others will consider this just a part of the main asset of the computer. Even the specific computer may only be a part of the workstation, thereby not even getting serialized for the enterprise asset, making it easily swapped. This is one of the crucial decisions needed in starting an ITAM program.
- The breadth of involvement refers to how much the organization the program will include and what asset types to consider. Hoping that eventually, the breadth becomes the whole of the organization and its assets. Still, in the first phases of the rollout, it might be appropriate to limit the scope to specific assets in one location or department and stage the changes for managing IT assets over a period of time to make the transition to the whole organization more feasible.

Implementing an ITAM program must start from the top and have the full support and understanding of the benefits of such a program to drive this down through the organization to the implementation to ensure the success of such a program. Establishing an executive sponsor of the initiative is critical to implanting any new program in the organization, and beginning an ITAM program is no different.

The second challenge is establishing the team that will deliver and develop the boundaries (breadth) of what the group is to work from. Attacking the entire organization, depending on the size, is likely a daunting task, so a strategic approach to discovering and implementing would be essential to ensure success. It is like the old question: How do you eat an elephant? One bite at a time. The bites you take must be small enough to be digestible but big enough to make a visible difference.

It is essential to recognize that depending on the organization and the current organization's state, implementing ITAM could run from several months to several years. So, it is crucial to have a strategy that will support this long-term initiative and see the tangible results that such implementation achieves its intended benefits. A steady path toward maintenance excellence will not happen overnight, but the rewards of achieving it will benefit the bottom line of the organization's financials.

EAM and IT asset management are fundamentally the same. However, because IT assets have such varied definitions and costs, it is likely necessary to manage the deployment of ITAM differently and separately.

To facilitate a further understanding of an approach to beginning an ITAM program, the capabilities and tasks outlined below are a possible roadmap for the enterprise to launch an asset management program successfully.

An ideal starting point may be leveraging a process maturity model, as these are central to several prominent methodology structures heavily used in the IT industry. This concept is beneficial to show measurable progress in the various areas that make up the Total Lifecycle Asset Management (TLAM) model for ITAM.

"Begin with the goal in mind" is a central tenet of most modern quality management philosophies. If an organization has no clear goal or purpose, it becomes impossible to define measurable key performance indicators (KPI) and articulate the progress against the processes that lead to success. Leveraging a maturity model is one method for driving toward a quantifiable measure for reaching this goal that can be universally understood within the organization.

The maturity model defines "capable" in terms of the definition and communication of goals and "aware" in understanding the targeted direction. The steps that must be taken to achieve a "capable" mission definition are as follows:

- Disseminate the goals of asset management to key stakeholders;
- Disseminate the goals of asset management throughout the organization or when managing breadth, at least through the target area;
- Prioritize the mission elements;
- Select the technical goals; and
- Select the broad business process goals.

Whereas a mission statement can be painted with broad brushstrokes, with goals that are very much "end-state" in their orientation, objectives are more process-oriented. Objectives should begin to answer the question, "How do we achieve our goals?" In other words, objectives are the mile markers and road signs that tell us we are on the correct route to achieve the mission.

The following are the steps that must be taken to achieve a "capable" state for the definition of objectives for the implementation of a program:

- Gain consensus on specific business process objectives;
- Consider/identify the services and service levels that are needed to meet the business objectives;
- Establish measurable performance indicators; and
- Set the timeline for near-term objectives to be achieved.

Once the asset management mission and objectives are established, the actual heavy lifting of establishing the asset management processes begins. These processes will provide the structure to enable members of the organization to support the business objectives that result in the achievement of the mission.

Establishing the asset management framework will address the governance issues identified as a significant gap for the enterprise in establishing a successful asset management program. Since the only governance framework is currently localized and fragmented, creating an enterprise-wide asset management framework will need to proceed through the "aware" stage first to arrive at the "capable" stage of the maturity model.

The following are the steps that must be taken to achieve an "aware" state for the asset management framework:

- Define the asset management framework, including process roles and standard operating policies and procedures (SOPs);
- Define any necessary services and service level agreements (SLA) that are needed to meet the asset management objectives (services may be either internally or externally provided); and
- Disseminate the asset management framework to your stakeholders.

Having made the above steps to achieve an aware state, then the following are the steps that must be taken to achieve a "capable" state for the asset management framework:

- Use a feedback spiral to remove process failures from the Asset Management framework;
- Review and revise policies, procedures, and roles regularly to reflect changing business conditions;
- Disseminate the asset management framework and processes across the breadth of the organization; and
- Define conforming processes for each mission element and organization unit.

Once the management framework is established, the actual asset lifecycle management can begin. TLAM would apply nicely here and is a fundamental concept that should be included. As with the previous steps, you must progress through the "aware" state to reach the "capable" state on the maturity model.

Starting with the "aware" state, the following are the steps that must be taken to achieve this for managing the asset lifecycle:

- Identify asset tracking repositories;
- Consolidate asset tracking information; and
- Nominate a final asset management repository (one source for the "truth").

Progressing to a "capable" state for managing the asset lifecycle requires implementing processes to support every phase of the asset management lifecycle and implementing metrics to track costs, disposal, and recovery.

The verification and audit processes should be part of the established asset management framework. These processes utilize the asset data specified in the asset tracking repositories and help ensure that the data has high accuracy and reliability.

The verifying and auditing of asset information follows again within our maturity model with the following steps that must be taken to achieve an "aware" state:

- Update supplier contact and contract information annually;
- Track software license volume;
- Track procurement volume;
- Check software licenses as part of any IT asset change;
- Schedule and execute audits on significant assets; and
- Track recognized deficiencies.

Once the achievement of the "aware" state has again been reached, the following are the steps that must be taken to achieve a "capable" state for verifying and auditing asset information:

- Track the asset lifecycle, capturing asset class, asset source, and supplier;
- Track performance against SLAs;
- Track software licenses by usage and deployment;
- Track software licensing exceptions and compliance failure;
- Enforce license policies and communicate policy enforcement agency-wide;
- Establish agency-wide audit procedures that are codified and repeatable;
- Trigger automatic audits when unauthorized IT assets are connected to the network; and
- Work continuously to remove conflicts between AM, corporate mission, and legal requirements.

Suppose all stakeholders have a high degree of confidence in the accuracy and reliability of asset data (because of the processes that are in place). In that case, supporting the demand for reporting and analysis of capabilities becomes necessary.

The beginning is through the "aware" state to Analyze and Provide Asset Information. The following are the steps that must be taken to achieve this state of "aware" for analyzing and providing asset information:

- Document and implement existing automated reporting;
- Implement repeatable analysis functions; and
- Implement project-based reporting (history, storage, disposal, value, cost, licenses, assets by location, owned and leased assets, contract or license expiration, inventory variance).

Like all reasonable steps, it is expected that once you move through the "aware" state, you will start to take the following steps to achieve the "capable" state for analyzing and providing asset information:

- Implement automated methods for most reporting;
- Expand analysis automation; and
- Implement standardized asset reporting (history, storage, disposal, value, cost, licenses, assets by location, owned and leased assets, contract or license expiration, inventory variance).

Once standardized asset management reporting and analysis tools are developed, the processes of continuous improvement by management in monitoring performance metrics can occur.

To achieve an "aware" state for evaluating asset management performance, the organization must consolidate and document existing measurements and performance indicators to track asset management efforts.

The steps to achieve the "capable" state for evaluating asset management performance are to determine measurements and performance indicators that match decision support requirements and then report and disseminate measures and performance indicators to key stakeholders.

16.9   ORGANIZATIONAL CLARITY AND CAPABILITY

Direction and Control Roles define how the organization establishes accountability for asset management. For example, reaching the "aware" state requires the departments to have an assigned owner of departmental assets, while achieving a "capable" state requires the Departmental Asset Owner to be appointed to manage the core IT assets.

Execution Roles are used to define responsibility for asset management within the organization. The steps needed to reach an "aware" state require the following:

- Asset Management roles and responsibilities have been discussed but may not be fully agreed to by all parties;
- Connections between actual asset management roles and mission and role descriptions may be vague; and
- Measurements exist, but there may be few meaningful measurements.

Whereas the steps to achieve a "capable" state require the following:

- All parties agree to asset management execution roles and responsibilities but may not yet be documented;
- Responsibilities generally tie to mission and role descriptions; and
- General qualitative measurements exist.

The area of skills and desired behaviors references the skills and behaviors needed for an organization to perform at the various levels of asset management maturity. For example, the "aware" state requires that the asset management skills are good but relate to only some of the data requirements and that the skills are primarily technical-based.

However, a "capable" state requires the following to be included:

- Determining technical and financial skills needed by asset management resources;
- Determining the technical and financial skills needed by asset management resources to be effective in the function of asset management;
- Documenting technical and financial skills of asset management resources;

- Documenting the technical and financial skills needed by asset management resources to be effective in the function of asset management;
- Comparing technical and financial skills needed with technical and financial skills that the asset management resources have, identify any gaps, and develop training to address any gaps; and
- Implementing technical and financial skill training for asset management resources.

16.10 MEASURING AND IMPROVING THE PROCESS

16.10.1 MEASUREMENT

Metrics provide the means to determine the performance of our processes, set meaningful improvement goals, and measure whether improvements to the process made a difference. Discussions of product or service expectations with the process customer help focus process performance measures on those characteristics the customer values. Their understanding of the causes that impact process performance provides insight that aids in setting preventive measures and defining procedure requirements. Since metrics drive performance by improving the focus on that which is measured, care should be given to avoid sub-optimizing the system by applying flow or efficiency measures over capacity processes.

The organization should carefully monitor the metric data values, especially in the first few reporting periods after the metric has been established. It is essential to validate the data gathered to ensure an accurate process assessment, and it is also necessary to validate the consistency of measurement collection methods. Before stabilizing the process, the organization should determine each metric's current "baseline" performance. The organization can then see the result of stabilizing actions by comparing the baseline performance to the stabilized version.

The following steps must be taken to achieve an "aware" state in the measurements arena:

- Directorate and mission-based asset management performance are formally tracked;
- Begin the project to identify key asset management measurements;
- Meet with the process customer and establish the end-item product or service measures. Negotiate the minimum acceptable performance levels for each. *Minimum acceptable levels are not to be confused with goals; and*
- Define and document each measure by clearly describing what is being measured and the calculation formula used.

The following steps must be taken to achieve a "capable" state in the measurements arena:

- The measurements feed the decision support loop for process mission and objectives;
- The measurement crosses directorate and process boundaries and addresses efficiency and performance;

- A repeatable and scheduled process that gathers and aggregates project-level measurements;
- Measurements include business and operational data; and
- Communicate with the "creators" and "collectors" of the measurement data to confirm the definitions and agree to a data collection method that ensures consistent use of the explanations over time.

16.10.2 UNDERSTANDING CUSTOMER SATISFACTION WITH THE PROCESS

The actual tasks/actions taken to complete a process must be documented so they can be internalized across the organization. They must be repeatable to the point that anyone, given the proper skill and experience, can complete the process correctly and report customer satisfaction on a repeatable basis. Documenting process steps is part of the puzzle often left out when defining processes. Many organizations develop process maps but never establish efforts to understand customer satisfaction. The result is a process that fails its goal and must revise steps to accurately reflect the actions/tasks that must occur throughout the process to reach that goal.

The following steps must be taken to achieve an "aware" state in understanding customer satisfaction within the process:

- Gather historical customer satisfaction data;
- Gather and document baseline customer satisfaction information;
- Identify all contractual, regulatory, and program requirements that apply to the process and customer satisfaction; and
- Identify any undocumented customer needs and expectations.

To achieve a "capable" state in understanding customer satisfaction with the process, the organization must implement repeatable customer satisfaction reporting and include the non-financial performance costs.

16.10.3 IMPROVING THE PROCESS

An improvement plan should be documented, accessible, and communicated to all involved parties to ensure progress continues even if resources are reallocated. It is necessary to provide a balanced set of metrics by reviewing customer measures, preventive process measures, and product control point measures. The goal is to create a balance to identify when you are improving one at the expense of another while focusing on the business strategy. Further, it is necessary to determine how each cause element will be managed. Controlling the right causes will enable organizations to prevent or predict errors before they occur. Items such as tool setting or employee skill level could be identified as a requirement in a procedure. This could result in other causes needing to be measured, inspected, or audited.

Process stability and improvement are process characteristics calculated from performance data using easy-to-learn statistics. These are not subjective assessments of a process. Stability in the process does not exhibit unusual data values or patterns that indicate the presence of special causes. Once the process is stable, analyzing the variation of process data helps identify what to control to meet the requirements of the process. The process performance, once stable, is compared to needs and aids in deciding whether to invest in improvements.

Once annual metrics have been set, multi-year and comparative thresholds for processes can be considered. The organization can treat each potential area of improvement as a problem/issue to be carefully analyzed and ultimately improved. This can help to align with the rollout across other areas of the organization for managing the breadth of delivery.

To achieve an "aware" state in improving the process, the organization must ensure the process is addressed when incidents make process failures evident and that process improvement begins when technical or financial issues are apparent.

The following steps must be taken to achieve a "capable" state in improving the process:

- Complete routine process improvements;
- Establish measures for the asset management processes;
- Implement a repeatable process for trend, control, and Pareto reporting;
- Personnel performance is included as a discrete analysis and is included in process improvement;
- Determine the process average for more than three reporting periods. This provides a representative summary of past performance that can be compared to the new performance after improvements are made. The resulting improvement in performance can be easily calculated using this baseline;
- Develop an improvement plan;
- Initiate the actions in the improvement plan to make the selected process changes;
- Communicate process changes to all stakeholders; and
- Document lessons learned.

16.11   DELIVERY OF EXCELLENCE

Having described the work breakdown structure of tasks that are a necessary part of the building to "capable" in the Asset Management Maturity Model, the following are the next steps that are needed to address the significant "gaps" that were previously identified and move to a higher level on the maturity model.

1) **Governance**

The steps to forming a good governance model consist of the following:
- Forming a governance body with decision-making authority that can develop, prioritize and shape asset management efforts to support agency-wide objectives;
- Gaining consensus from key stakeholders on the structure and scope of the governance body;
- Forming an agency-spanning committee for asset management efforts;
- Defining the channels for communicating customer needs and policy acceptance to the committee and garnering process and status information from the committee; and

- Charging the governance body to develop, implement, and control asset management vision, strategy, and processes.

2) **Organization Asset Management Vision**

The governance body needs to:
- Define the asset management vision;
- Define the asset management mission;
- Establish the asset management objectives; and
- Gain stakeholder consensus on the asset management objectives.

3) **Organization Asset Management Strategy**

The organization's asset management strategy will consist of the following:
- Defining the asset management framework, including asset management policies, business process roles, and responsibilities;
- Defining any necessary services and SLAs that are needed to meet the asset management objectives (services may be either internally or externally provided);
- Disseminating the asset management framework to stakeholders;
- Defining high-level KPIs and process metrics to allow measurement and monitoring of asset management performance.

4) **Organization Asset Management Processes**

The organization's asset management processes will:
- Establish a central repository for asset management data that becomes the "single source for the truth." (This will likely mean a selection of a single integrated technology tool);
- Define necessary SOPs and work instructions to support the agreed-upon policies, roles, and technology tools;
- Use system engineering processes to consolidate process redundancies caused by systems with duplicate or conflicting roles;
- Define data standards and naming conventions;
- Map existing systems to data types, asset classes, access types, and locations;
- Perform initial baseline physical inventory;
- Populate the central data repository from "cleansed" asset data based on the physical inventory (this can include verified/reconciled network discovery);
- Implement business process controls for asset management processes to ensure complete and usable information; and
- Implement the KPIs and process metrics defined in the asset management strategy to include the reporting process and management roles/responsibilities for monitoring and corrective action.

5) **Rationalization of Projects and Systems**

The individuals and divisions within the organization must recognize the challenges within asset management. Divisions must initiate multiple asset management programs and systems to address the asset management challenges. This is necessary to manage the breadth of implementation across the organization. Unfortunately, these efforts' uncoordinated and non-integrated nature is agitating rather than resolving the organizational asset management challenge. After addressing Governance, Vision, Strategy, and Processes, the organization should be able to rationalize existing asset management projects and systems.

Extensive collections of systems, such as asset management within the enterprise, are co-created by teams from various disciplines and departments. System development is an emergent process. This process's lack of central control may result in incompatible systems and processes that must be simplified and connected to support the organization's objectives. Rationalization examines current efforts against a developed course of action and determines which efforts support the course of action, which efforts detract from the course of action, and what additional steps need to be included to make the necessary changes to gain the desired capabilities. From the rationalization analysis, an executable plan is developed to remove systems and projects that do not support the organization, accelerate efforts that do support the organization, and expand existing efforts or initiate new efforts to fill gaps.

16.12 SUMMARY

As can be seen, IT assets are not just found in big data centers or offices but have infiltrated the core of manufacturing and plant floor operations. Through statistical process controllers, SCADA networks, and handheld devices for controlling inventory, IT assets are fundamental to the success of all plants and businesses. Properly managing and maintaining these assets can be core to the business's success. An effective ITAM solution that aligns with the TLAM principles will fulfill this need and support the needs of the enterprise.

ITAM is only a subset of the overall integration of Enterprise Asset Management and the TLAM for the enterprise with a specialized focus on a new technology that is explosive in its impacts on the organization. ITAM and TLAM must be part of every company's efforts in managing their assets and achieving maintenance excellence.

Through this rendition, you understand why information technology is part of the five asset classes that any organization needs to consider when managing assets and driving toward maintenance excellence. IT has become and will continue to expand its role in the successful organizations of the future. The importance of managing the maintenance of those IT assets through the lifecycle of the asset is the foundational philosophy behind this book and TLAM for ITAM.

PERMISSIONS

17 Achieving Asset Management Excellence*

Don M. Barry
Previous contributions from Don M. Barry and John D. Campbell

An asset-intensive organization must have a framework, standards, and principles by which it intends to apply asset management to achieve its objectives. They should plan to set out an asset management policy to achieve their goals.

Some organizations will need to develop a strategic asset management plan (SAMP). The suggested components of this document are outlined later in the chapter.

The preceding sections described: the evolution from reactive to proactive maintenance, managing equipment reliability to reduce failure frequency, optimizing equipment performance by streamlining maintenance for total life-cycle economics, and aligning asset management to drive operational excellence. In this chapter, we look at the specifics of implementing the concepts and methods of these processes.

We take you through a three-step approach to put a maintenance management improvement program into place and achieve results:

- **Step 1: Discover** – learn where you are in an asset management maturity profile, establish your vision and strategy based on research and benchmarking, and know your priorities, the size of the gap, and how much of it you want to implement now and what programs should be staged later
- **Step 2: Develop** – build the conceptual framework and detailed design, set your action plan for the prioritized initiatives and schedule to implement the design, obligate the financial resources, and commit the managers and skilled staff to execute the plan
- **Step 3: Deploy** – document and delegate who is accountable and responsible, fix milestones, set performance measures and reporting, select pilot areas, and establish detailed specifications and policy for procurement, installation, and training

The following chapter reviews the whys and hows of managing successful change in the work environment.

A friend who travels extensively in his work in the mining industry describes his most anxious moment abroad. "I was just back in San Paulo from the Amazon basin, driving and not paying attention, when it dawned on me that I hadn't a clue where I

DOI: 10.1201/9781032679600-21

FIGURE 17.1 Asset Management Excellence pyramid elements.

was. It was dark, I don't speak Portuguese, my rental was the best car in my line of sight, and I was sure I was getting eye-balled by some of the locals." Luckily, he returned to familiar territory, but not without much stress and wasted effort. Like this intrepid traveler, you must keep your wits about reaching your destination successfully. So, to achieve maintenance excellence, you must begin by first checking where you are.

The cornerstone of this described approach is the maintenance excellence pyramid. The most important principle embodied in this approach is that successfully implementing the pyramid's upper layers depends on a solid foundation at the lower levels. The original pyramid Asset Management elements were based on the book *Uptime, Strategies for Excellence in Maintenance Management* by Mr. John Campbell and are further enhanced in this book.

The process for achieving asset management (maintenance) excellence is depicted in Figure 17.2.

17.1 THE THREE STEPS

Before implementing reliability improvement and maintenance optimization, you must set up an organization or team mandated to effect change. First, you will need an executive sponsor to fund the resources and a champion to spearhead the program. You will need a steering group to set and modify direction. Members are typically representatives from the affected areas, such as maintenance, operations, materials, information technology, human resources, and engineering. A facilitator is invaluable, particularly one who has been through this process and understands the shortcuts and pitfalls. Last but certainly not least, you need a team of dedicated workhorses to execute the committed initiatives.

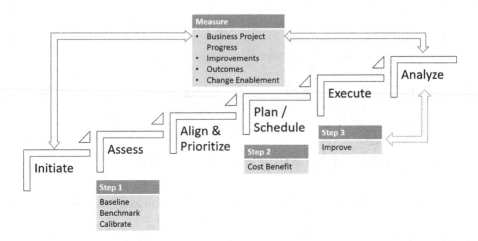

FIGURE 17.2 Suggested process for implementing asset management excellence.

17.2 STEP 1: DISCOVER

17.2.1 INITIATE

Initiating is defining an early hypothesis for what you are trying to improve. It is getting off the mark. It establishes enterprise and business unit scope, stakeholder scope, and location scope. It introduces design principles from which the initiative must work or achieve. Establishing the final project structure, staking out a work area, conference room, or office, and setting the broad "draft" project charter is also part of this mobilizing phase. You would also determine who may be the affected stakeholders and who should be part of the initiative. You then develop the "first cut" work plan, which will undoubtedly be modified after the next step, "Assess."

17.2.2 ASSESS

This step answers the "Where am I currently?" and follows a strict methodology to ensure completeness and objectivity.

17.2.2.1 Self-diagnostic

Based on a standard questionnaire, the plant, facility, fleet, or operation employees give feedback on asset management proficiency – identifying areas of strength and weakness at a reasonably high level. One is shown in Appendix A.22, designed as an initial improvement assessment of:

- Current maintenance strategy and acceptance level within the operations
- Maintenance organization structure
- Human Resources and employee empowerment
- Use of maintenance tactics – PM, PdM, CBM, RTF, TBM, etc.
- Use of reliability engineering and reliability-based approaches to equipment performance monitoring and improvement

- Use of performance monitoring, measures, and benchmarking
- Use of information technology and management systems focusing on integrating existing systems or any new systems needed to support best practices (i.e., EAM/APM, data analytics, machine learning, artificial intelligence, augmented reality, digital twinning document management, project planning, etc.)
- Use and effectiveness of planning and scheduling, and shutdown management
- Procurement and materials management that support maintenance operations
- Use of process analysis and redesign to optimize organizational effectiveness

The questionnaire can be developed with different emphases, depending on the area of focus. For example, for maintenance optimization, the research could be conducted on what the best practice companies are doing with expert systems, modeling techniques, equipment and component replacement decision-making, and life-cycle economics. Questions are posed to provoke thought and discussion and reflect on how leading practices are understood and adopted. This self-assessment exercise builds ownership of the need to change, improve, and close current and best practices gaps. The questionnaire acts as a change enablement tool by including a significant number and a cross-section of enterprise stakeholders. The questionnaire replies are summarized, graphed, analyzed, and augment information from the next activity.

17.2.2.2 Data Collection and Analysis

While the questionnaire is being completed, gather operating performance data at the plant for review. The data needed include:

- Published maintenance strategy, philosophy, goals, objectives, value, or other statements (design principles) that must be adhered to
- Organization charts and staffing levels for each division and its maintenance organization
- Maintenance budgets for the last year (showing actual costs compared with budgeted costs, noting any extraordinary items) and for the current year
- Current maintenance-specific policies, practices, and procedures (including collective agreement, if applicable)
- Sample maintenance reports currently in use (weekly, monthly, etc.)
- Current process or workflow diagrams or charts
- Descriptions or contracts concerning out-sourced or shared services
- Descriptions of decision support tools
- Summaries of typical spreadsheets, databases, and maintenance information management systems
- Descriptions of models and special tools
- Position/job descriptions used for current maintenance positions, including planning, engineering, and other technical and administrative positions and line functions.

Although the above information is usually sufficient, you sometimes discover additional needs once work starts at the plant. Compare against the background data and reconcile any issues that come to light.

17.2.2.3   Site Visits and Interviews

If you're unfamiliar with the plant's layout or operation, you will want to take a tour and learn safety procedures. A thorough tour that follows the production flow through the plant is best.

To facilitate the on-site interview process, present the self-diagnostic results to management and key personnel in a "kick-off" meeting. This will introduce best practice – the model presented as a perfect score on the self-diagnostic. Next, introduce what you are doing; describe generic industrial best practices and what you will do on-site. Then, conduct interviews of various plant personnel, using the self-assessment and documentation collected earlier as question guides.

The composition of the interviews can be driven by the areas of weakness and strength initially revealed through the self-assessment and any other input from the previously collected data. The interviews can delve into specific problem areas and their causes. In particular, you may identify organizational, systemic, or human factors at the root of any problems or areas of high performance.

The interviewees will generally be:

- Plant manager, human resources/industrial relations manager
- Operations/production managers
- Information Technology managers and systems administrators
- Purchasing manager, stores manager/supervisors
- Maintenance manager; superintendent, maintenance/plant engineers, planners, supervisors
- Several members of the maintenance workforce (at least two from each significant trade or area group)
- Representatives of any collective bargaining unit or employee association.

Conduct interviews in a private office or the plant while the interviewee walks through their job and workplace. You may also want to observe how a planner, mechanic, or technician, for instance, spends the workday. This can reveal many systemic and people issues and opportunities impacting best practices.

Interviews are best done privately. If the interviewer promises to collect all relevant comments and keep their source anonymous, the insights from these interviews are typically more candid and accurate. Examples of some of the interview questions that could be used are:

- Date, location, interview time, interviewee, business unit, etc.?
- What is the company's biggest issue?
- What have some of the more recent asset management activities been?
- How well where they implemented?
- What is the reason for this change initiative?
- Any issues that may have come to light from the pre-filled questionnaire?

17.2.2.4   Maintenance Process Mapping

A group session is one of the best ways to identify major processes and activities. The processes should be the actual ones practiced at the plants, which may not coincide with the process maps developed for an ISO 9000 certification. Treat existing charts or maps as "designed" drawings, not necessarily "as-is" ones. These are mapped graphically to illustrate how work, inventory, and other maintenance practices are managed and performed. Through the mapping process, draw out criticisms of the various steps to reveal weaknesses. This will help you understand what is happening now and can be used for the "as-is" drawings. Or, it can provide a baseline for process redesign, along with best practices determined from benchmarking or expert advice.

You will gain significant insight into the current degree of system integration and areas where it may help through process mapping. Redesigning processes will be part of any implementation work that follows the diagnostic.

The key processes to examine in your interviews and mapping include the following:

- PM development and refinement
- Procurement (stores, non-stores, services)
- Demand/corrective maintenance
- Emergency maintenance
- Maintenance prevention
- Work order management
- Planning and scheduling of shutdowns
- Parts inventory management (receiving, stocking, issuing, distribution, review of inventory investment)
- Maintenance long term planning and budgeting
- Preventive/predictive maintenance planning, scheduling, and execution

Through the interviews, identify and map other processes unique to your operations.

It is important to note that some organizations may prefer to focus on their future process opportunities versus investing in the cost of documenting the current process shortfalls. Optimally the existing processes could be reviewed by using a checklist that confirms that the considered processes are working, how they are being executed (i.e., manual, system, automated), and the health of the process execution.

17.2.2.5   Report and Recommendations

At the end of the site visit, compile all the results into a report showing overall strengths and weaknesses, the specific performance measures that identify and verify performance, and the most significant gaps between current practices and the vision. The report should contain an "opportunity map" that plots each recommendation on a grid, showing relative benefit compared to the level of difficulty to achieve it (see Figure 17.3).

17.2.3   BENCHMARKING AND CALIBRATING

An organization needs to understand what relevant leading practices can be used to compare to their current asset management disciplines and infrastructure.

In many companies, benchmarking is more industrial tourism than an improvement strategy. Even the most earnest may answer benchmark questions based on how

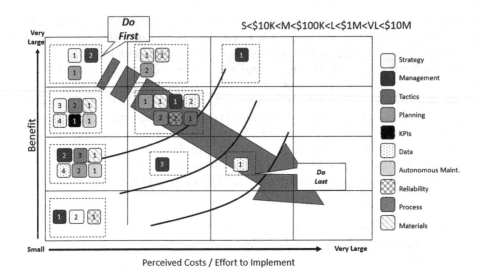

FIGURE 17.3   An example of an asset management initiative prioritization map.

they want to perceive themselves or how they want to be perceived. After completing a diagnostic assessment, select the critical factors for asset management success specific to your circumstances. It would be best if you focused precisely on gaining helpful information that can be implemented. Look within and outside your industry to discover who excels at those factors. Compare performance measures, their current process, and how they achieved excellence. However, recognize that finding organizations with excellent maintenance optimization can be challenging. For example, there may only be a few dozen who have successfully implemented condition-based maintenance optimization using stochastic techniques, completed risk and reliability programs, and have these automated within their EAM/APM software. Benchmarking can more often be a tool to help you understand the categories of ideas deployed by leading organizations outside your targeted process rather than the specific degree to which it is being accurately executed and measured. To this end, benchmarking can categorize what you wish to baseline measure in your operation as a starting point for measured improvements.

Focusing on "leading practices" may serve you as an external barometer than traditional benchmarking. With so many changes in asset and asset management technologies, picking the most effective asset management process or procedure at a given time would be like shooting at a moving target. Viewing how engineering, operations, and maintenance are working together to maintain functions, not assets, could help change the culture in your organization. An example of a process benchmark you could consider and adopt in your organization could be moving assets from manual inspections to onboard diagnostics and expert decision support systems that automatically initiate a call in the work management system. With prioritized assets, operational constraints, and external market and weather data, asset risk and reliability information could also be in the mix for an automated enterprise known for its operational excellence.

A training session with the stakeholders to calibrate discussions is essential. It helps with the solution development (**Step 2**) and could be leveraged to ensure the whole team has a standard understanding/definition of asset management leading practices.

17.3 STEP 2 – DEVELOP

17.3.1 PRIORITIZE

You will not be able to implement everything at once. Limited resources and varying benefits will whittle down your "shortlist." One technique to quickly see the highest value initiatives versus those with limited benefit and expensive price tags are shown in Figure 17.3.

The benefit is often measured in a log 10 savings scale, i.e., $10,000, $100,000, $1,000,000, or $10,000,000. The degree of difficulty could also be shown on a similar scale, an implementation timeline such as one year, two years, three years, or four years, or a factor related to the change required in the organization for the initiative.

This technique will identify the apparent "high benefit, low effort" opportunities (often called "low hanging fruit"). Many of these identified ideas are often executed for immediate savings and will help a project or significant initiative declare early identified savings.

17.3.2 STRATEGIZE, PLAN, AND SCHEDULE

The new operating model will be developed from the prioritized requirements that surface from the prioritization step. The proposed "new operating model" must be flexible and robust. Inputs into its strategy development could include:

- Information about related initiatives already underway;
- Current corporate strategy, mission/vision, values, desired behaviors, design principles, and design points;
- Existing relevant maintenance and asset management policy documentation;
- Relevant leadership or management training documents that would address the values and culture of "how things get done";
- The information available from existing systems to support the change; and
- Information about the regulatory environment and pressures that apply to the organization's operating environment.

Planning how you would implement the prioritized actions from the opportunity map may require considering whether there are any dependencies to accomplish some of the described "easy to implement and high benefit" improvements. At times, it may be determined that consolidating some of the early actions into a more substantial sequence of initiatives will make sense. For example, items such as "field training on numerous topics" or implementing a more rigorous integrated system/process may allow you to prioritize items and group them into program initiatives.

Grouping the identified initiatives into programs and then planning and scheduling these programs at a macro level will help the executive teams and staff communicate the sequence of the programs and why.

The organization's leadership team will need to understand and internalize the criticality of the initiatives, how and why they were grouped, and what "success" will look like when implemented. Often a diverse group of executives with varying agendas may not be on the same page when prioritizing an asset management set of initiatives. The notion of "design principles" and "design points" can help sort these discussions out.

- Design Principle:
  - A Solution Design standard that will not change in the enterprise or as an expected solution standard
- Design Points:
  - A Solution Design standard that could change as an expected solution standard to satisfy the business needs or operating context.

If they do not exist, a process owner and transformation leader (or initiative advocate) should be assigned to help with the deployment step.

Using software such as MS-Project will make the planning and scheduling process more rigorous. List the projects in priority, from Figure 17.3, and with reasonable timelines. No one will commend you for taking the fast track in the planning phase if the project fails for several months. Be sure to include people from other functions or processes related to maintenance if they can help ensure project success. Typically, you'll look to procurement, production planning, human resources, finance and accounting, general engineering, information technology, contractors, vendors, or service providers.

All program alignment will need a committed senior sponsor to promote the program initiatives actively. The displayed executive leadership will ensure the staff remains interested and committed.

17.3.3 Cost/Benefit

The priority assessment from Figure 17.3 can broadly group the recommendations. If there is a significant capital or employee time investment, calculate the cost/benefit.

TABLE 17.1

Roles and Responsibilities of Process Owners versus Transformation Leaders

Attributes of Process Owner	Attributes of the Transformation Leader
• Accountable and responsible for process actions and outcomes	• Empowered and has authority for related decisions
• Committed to the process life-cycle	• Willing, able (unique, relevant skill), credible
• Owns the risks and benefits of the transformation initiatives	• Management assigned and sponsored
	• Accountable and responsible for initiative actions and outcomes

Whereas costs are usually fairly straightforward – hardware, software, training, consulting, and time – estimating benefits can be more nebulous. For example, what are the benefits of implementing root cause failure analysis, a Reliability Centered Maintenance (RCM) program, or enhancing EAM/APM software? Chapter 1 described how to estimate the benefits of moving to more planned and preventive maintenance, which can be modified for specific projects targeting unplanned maintenance. The equipment and hardware manufacturer or software/methodology designer can often help based on their experience with similar applications. Benchmarking partners or websites focused on asset management are other sources of useful information.

17.3.4 DOCUMENTING THE ASSET MANAGEMENT PLAN

A SAMP format for formally documenting a strategic asset management plan can be leveraged from the published ISO55000 guidance. The areas suggested are:

- Enterprise organization scope and context
- Expectations of the assets, stakeholders, and the value expected from the critical assets
- The scope of the assets and asset management systems being included in the SAMP
- The planned asset management programs (or initiatives) committed by senior management that will contribute to the asset management plan
- The systems, financial, HR, HSE, and other resource contributions and impacts.

17.4 STEP 3 DEPLOY

17.4.1 EXECUTE

If possible, consider a pilot approach to improvement initiatives. This provides proof of concept and, more importantly, acts as an excellent marketing tool for a full rollout.

What does experience teach us about program management? Program management is managing a group of projects with a common theme. Some of its key elements are:

- Define the scope (one large enough to capture management's attention and get meaningful results)
- Follow a documented approach
- Delineate roles and responsibilities
- Put a lot of effort upfront in the discover and develop phases (the measure twice/cut once approach)
- Ensuring the program participants are engaged and motivated
- Assess the "do-ability" of the plan with all stakeholders
- Take no shortcuts or fast-tracking
- Estimate the risk beyond the budget
- Work to get the right champion who will be in it for the long run
- Keeping the executive sponsor engaged

17.4.2　Measure

The more detailed the implementation plan, the easier it is to measure progress. Include scope, detailed activities, responsibilities, resources, budgets, timelines for expenditures and activities, and milestones. Ensure maximum visibility of these measures by posting them where everyone involved can see them. Often forgotten are the review mechanisms: supervisor to subordinate; internal peer group or project team; management reviews; start of day huddle. They are all performance measures.

17.4.3　Analyze/Improve

After the initial round of piloting and implementation comes reality. Is the program delivering the expected results? How are the cost-benefit "actuals" measuring up? Despite best efforts and excellent planning and execution, adjustments still must be made to the original scope, work plan, staffing, or expected results. But if you have closely followed the Discover and Design stages and set up and managed performance measures, there will be few surprises at this stage.

17.4.4　Expand and Do It Again!

Several activities must be addressed when your program initiatives have been successfully implemented.

- Is there an opportunity to improve on what was implemented?
- What are the lessons learned from implementing each program?
- Have all the affected stakeholders achieved their expected objectives?
- Is additional change enablement needed? (If so, consider Chapter 18)
- Is there an opportunity to expand the successful change to other business areas?
- Has something changed (i.e., skill levels, tools, system support, etc.) that suggests upgrading this program?
- Can we rechallenge the business to drive new asset management programs and achieve operational excellence?

17.4.5　Managing the Change

Over and over, many organizations will implement process and technology change without considering the impact on the stakeholders. Unfortunately, this often means that the process user is not fully engaged when the change is executed, resulting in less-than-desired outcomes. Chapter 18 will review the many areas to consider when aligning all the essential stakeholders to facilitate the successful execution of change.

PERMISSIONS

BIBLIOGRAPHY

Campbell, J.D. *Uptime: Strategies for Excellence in Maintenance Management*, Portland, OR, 1995.

International Standard ISO 55002, Asset Management-Management systems—Guidelines for the application of ISO 55001, 2nd edition, 2018.11.

18 Enabling Change in Asset Management*

Don M. Barry

18.1 INTRODUCTION

Asset management and maintenance folks often define themselves as "problem solvers." Because they spend much of their time problem-solving: they become naturally cynical about being sold a new way to do things. Instead, experienced asset management and maintenance folks come to work looking to make a difference. They understand their role is essential and that they contribute to their organization's operational excellence goals and outcomes.

Change is a constant entity in any progressing business. So every day we come to work, we expect to progress some part of the business if we want to move the bar and make that difference. An old Chinese proverb suggests, *"Like rowing upstream, not to advance is to drop back."*

Change is needed as markets shift, technology advances, we learn from existing culture and process issues that need to be addressed, and people look for personal development. Assets are "everywhere" and deliver unique value when integrated into the emergence of new supporting technologies. The convergence of physical assets with operations systems drives the need to strive for operational excellence, and Information Technology (IT) applications require a new approach to managing infrastructure and services.

Physical asset management is not unique in its need for change enablement. The need for change management in asset management is acute with the many cultural dynamics of problem-solving cynics who may not want to embrace change enthusiastically. At the same time, they spend much of their time clearing the alligators from the swamp.

Change must be managed as global asset growth projections continue. The expected percent return on assets (ROA) remains with every asset investment. The expected percentage of ROA return will only be expected to grow. As a result, asset managers will need to be innovative to capitalize on the forecasted asset growth opportunity. If they do not, their replacements will likely be given the challenge.

Asset managers have constant pressures to improve their business area and contribute to operational excellence and profits to their company's bottom line. While they are working to manage their assets, associated risks, costs, and data, their working environment continues to swirl with external pressures, including:

1 Operations excellence demands
2 Regulatory requirements
3 Supply and demand

DOI: 10.1201/9781032679600-22

FIGURE 18.1 Pressures on the asset manager.*

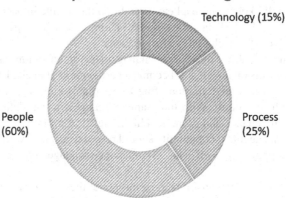

FIGURE 18.2 Percent focus mix for Asset Management Change enablement.*

4 Capital asset requirements
5 Restricted capital access
6 Limited revenue growth opportunities
7 Market/stakeholder pressures
8 Environmental demands

The enterprise investing in an asset management project will secure its long-term future and drive future improvements by doing the right thing (process effectiveness) and doing it well (process efficiency). To achieve their intended outcome of reduced costs, operational efficiency, and bottom-line benefits, they must pursue the people, process, and technology influencers with a tenaciousness and a passion for innovation. For example, they need to look to support:

- Their employees (**people**) with coaching, knowledge, support, and the tools to be effective and secure their future. Every asset-intensive enterprise is in the people business.
- Drive their business **process** agility by addressing those entrenched processes they have long known need to change and ensure they are standardized across our organization (driving simplified metrics and Key Performance Indicators - KPIs).
- Leverage industry-leading practice (**technology**) software, tools, and asset management disciplines/principles to increase asset performance and reliability. Ideally, this would be accomplished with minimal customization to align them for future software upgrades.

Asset management projects requiring change management support can originate in process change, people change, or technology change. Usually, if a change is happening in asset management, it will require understanding the impact of all three influences. Consensus suggests that even if the change is a technology initiative, the impact focus will likely require a heavyweight on the people impact side. For example, a typical EAM/APM, facilities, or IWMS project will suggest a 60% people, 25% process, and 15% technology change focus.

Changes in the culture of an asset management organization can evolve as the organization reflects on how it will become a new version of itself. There is an opportunity to inject a renewed definition of how the organization will contribute "value" as it moves into its future. With the planned changes, it can define its traditional approach to maintenance and promote a higher level of awareness and ownership of how the asset management community should respond going forward. John Moubray described this opportunity well in his "Maintenance Management – A New Paradigm" article.

With every asset management change initiative: there is an opportunity to up the value perception of maintenance and its' contribution to ROA. For example, it is correct that an asset management professional would suggest that maintenance is about preserving physical assets (i.e., a pump). However, the same professional would agree that it is more accurate to say that maintenance is about preserving the functions of assets (i.e., the 300 liters per minute of functionality expected from that

TABLE 18.1

Example of Asset Management Systems That Would Need Change Management Support*

Enterprise asset management (EAM)	It supports the optimal management of an organization's physical assets to maximize value. It covers the design, construction, commissioning, operations, maintenance, and decommissioning/replacement of plants, equipment, and facilities.
Asset performance management (APM)	It is used to improve the reliability and availability of physical assets. It includes integration, visualization, and analytics with data capture capabilities. In addition, APM includes condition monitoring, predictive forecasting, and reliability-centered maintenance (RCM).
Facility management	It is an interdisciplinary field primarily devoted to maintaining and caring for large buildings. Duties may include the care of plumbing, air conditioning, electric power, and lighting systems; cleaning: decorating: grounds keeping, and security. Computer programs can assist with some or all of these duties.
An integrated workplace management system (IWMS)	It is an enterprise platform that supports the planning, design, management, utilization, and disposal of an organization's location-based assets.

same pump). There is an opportunity in every asset management project to improve the cultural standard of asset management (and maintenance) when we declare that: Maintenance affects all aspects of business effectiveness. Maintenance also affects risk, safety, sustainability (energy efficiency), environmental integrity, product quality, and customer service, as well as plant availability and cost. This statement is a higher expectation from the asset management team than just an objective to support optimized plant availability and minimum cost.

The expected benefits from an asset management program change can be significant. While this can motivate many in the organization to want to support a change, the discussion on a tangible savings opportunity versus an intangible savings opportunity is helpful to declare. Doing this will help all the stakeholders manage the water

TABLE 18.2

Example of Maintenance Paradigm Influences*

Traditional Approach	New Approach
Maintenance is about preserving physical assets.	Maintenance is about preserving why the assets exist (Functions).
Routine maintenance is about preventing failures.	Routine maintenance is about minimizing the consequences of functional failures.
The maintenance function is to support plant availability at a minimum cost.	Maintenance also affects risk, safety, sustainability (energy efficiency), environmental integrity, product quality, and customer service, as well as plant availability and cost.
On its own, the maintenance department can develop a successful, lasting maintenance program.	Maintainers and users of the assets must work together to develop and achieve a successful, lasting maintenance program.

TABLE 18.3
Example of Tangible and Intangible Benefits from Asset Management Improvements

Tangible Benefits	Intangible Benefits
• Improved fixed asset utilization	• Improved sustainability (including HSE)
• Lower maintenance costs	• Improved financial controls
• Lower parts costs	• Improved customer experience
• Improved data analytics alignment	• Improved data integration and ML/AI analytics alignment
• Improved operational efficiencies	• Improved spending controls

cooler discussions about why the committed change needs to be well executed. An example of potential benefits from an asset management focus is listed in Table 18.3.

We understand from experience that many process-intensive change projects fail. Conventional wisdom suggests that 60% of failures are people-related issues. A study[1] was conducted against large successful EAM implementations with two open-ended questions asking:

- What did you do that made your project a success?
- What did you wish you had done more to become even more successful?

The surveys were sent to asset-intensive companies across various industries (i.e., government, oil and gas, forestry, communications, and packaged goods operations). The responses were interesting as the open-ended questions harvested some very similar answers. The suggestions that ranked 50% or higher from these survey results were consistently positioned as part of any future asset management (or process-intensive) project.

The essential elements of a successful change program were senior management leadership (buy-in), active and committed super-user communities, a well-organized and executed training program, and a robust communications plan. Performing a stakeholder analysis was also vital to understand the level of understanding and commitment evident in the hearts and minds of those affected. The mitigating actions could be identified and planned with the issues and risks identified.

A successful organizational change (people) program will have some specific disciplines included in the program. These program attributes will include[1] the following:

- A focus on the "What's in it for Me" by role when developing communications, training, and change readiness plans for each constituent.
- A focus on face-to-face engagement via our sponsor, leaders (change champions), and members of the change agents in an organization.
- Training the change agents to understand the process and technology impacts of the changes, how people move through the phases of change, and how to help their colleagues adopt.

TABLE 18.4

Survey Feedback from Successful Asset Management Systems Projects with a Change Focus*

What We Did to Achieve Change Success		Would Do More to Improve the Outcome	
69%	Senior management buy-in	69%	Senior management buy-in
69%	Super-user communities	69%	More robust, more frequent communications
69%	Scalable training and analysis	46%	Leveraging change management as an essential project component
69%	Strong communications plan		
62%	Train-the-trainer (T3)		
62%	Aligned communications with the program rollout		
54%	A stakeholder analysis was done		

- Leveraging the lessons learned from previous EAM/APM or other operational implementations.
- Leveraging experience and best practices in change to ensure that you successfully adopt and sustain the benefits of the planned changes.
- Leveraging job aids and online help for pre and post-go-live support.

Successful change programs had evidence of:

- Early organization change enablement drives business ownership and accountability at the appropriate levels of the business and ultimately contributes to sustaining the changes the project will introduce.
- Appropriate participation of stakeholders in identifying business benefits and impacts will facilitate the ability for people to identify benefits and impacts for themselves and plan accordingly.
- Early identification of change management risk factors allowing for the development of a robust organizational change management strategy.
- The "people" systems and structures correctly aligned with technology, process, and system changes, providing a better chance of realizing the intended benefits of the system implementation.
- Creating a change team and network with clear roles, responsibilities, and competencies enhances the effectiveness of the change effort.

Lengthy asset management projects will have many inherent risks due to the time-critical resources committed to the project/program. Therefore, the organizational change management approach should expect to address some challenges organizations will face over an extended change effort. This effort will likely extend beyond the transformation go-live milestone. Some of the significant risks in an asset

TABLE 18.5
Potential Risks a Change Focus Can Mitigate*

Risk Type	Potential Mitigation
Executive Risk	• Maintain a strong, visible, aligned, and engaged executive commitment. • Ensure leadership continuity. • Build the capacity for both change leadership and organizational leadership.
Organizational Risk	• Drive significant change to internal processes that will effectively transform the organization. • "Keep the lights on" while maintaining a focus on change. • Manage integration, alignment, and prioritization of all projects and initiatives. • Move from a crisis-management, reactive approach to a more deliberate work style.
Employee Risk	• Ensure mission-critical employee continuity during a period of potentially high attrition. • Engage managers in owning the difficult organizational decisions. • Overcome uninformed expectations that technology/software will solve everything. • Ensure employee buy-in and maintain organizational energy for the duration. • Ensure effective resource utilization planning. • Address constituent change fatigue and cynicism.
Project Team Risk	• Avoid consulting partner dependency. • Ensure capacity building of the technology/software team. • Maintain effective team dynamics throughout a long-term project. • Effective decision-making.

management project include; executive commitment and visibility; sustaining the organization during the focused change; employee risks; and the project team.

18.2   THE ROLE OF THE CHANGE LEAD

The change manager's role is to help the affected stakeholders support the project plan and expected outcomes from the investment in the planned change.

The organizational change role is not the same as a project manager. It is not managing the technical change. It is more than coordinating communication or ensuring the training is organized and delivered. The change lead will:

- Help increase the success rate of the transformation initiatives
- Prepare the impacted people to modify their behavior and adopt a change
- Work to minimize the risks and barriers to adoption
- Work to achieve the expected benefits through sustained adoption

Depending upon the size of the change and the impact being managed, the change leads' formal responsibilities may include all of the above as well as:

- Manage the change schedule at the strategic level with the project management office and leadership

- Manage the strategy for change repositories
- Establish and maintain change standards, policies, and procedures
- Ensure adherence to change management methods
- Manage and track change budgets across functional organizations
- Identify and mitigates change risks across functional organizations
- Escalate change issues and concerns to the executive program leadership
- Develop program-level status reports and provide feedback
- Track change benefits across functional organizations
- Develop onboarding and standards for change advocates
- Sustains change programs post-go-live and ensures planned activities occur
- Collects lessons learned

For larger projects: a formal project organization chart will be established. An example of this chart is shown in Figure 18.3. It is expected that the change lead will be in the office of the executive sponsor significantly more often than the project manager. The project owners, sponsors, and leadership need to be seen as leading the initiative and being able to communicate its priority for success. The change lead is often their conduit to the pulse of the folks affected by the planned change. These leaders recognize the business and personal risk if the planned transformation does not execute well, and they need to drive the expected benefits.

The change management lead (or office) supports change management initiatives (people, processes, and technology), builds capabilities and competencies, and expands

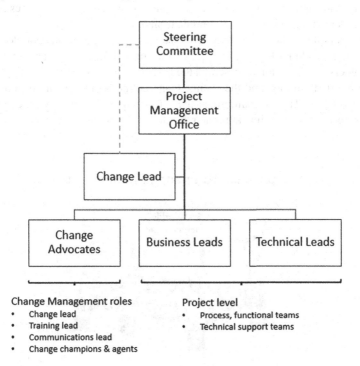

FIGURE 18.3 Example of a large project structure with a change lead.

the benefits of effective change management from a single initiative view to a broader portfolio view of change initiatives.

The change lead will be responsible to:

- Describe the change vision, objectives, and goals.
- Actively involve the leaders in owning the change (and coach them as appropriate).
- Assess change impacts and plan how the change will be managed.
- Engage and prepare the impacted employees to adopt the new way of working.
- Align the organization to enable and reinforce the desired behaviors.
- Monitor the adoption of the change to ensure the desired outcomes are realized and sustained.

Managing the enablement and the steps to support the change will differ from initiative to initiative. The change lead will work with the leadership and the stakeholder groups to establish the support programs (roadmap) to achieve the expected outcomes. The change role is complex and requires a tenacious approach, with the change lead expected to still be in their role long after the planned change has been implemented and the rest of the transformation team moves on to other things.

An example of a branded change approach is to add more than just training to the planned solution. The change approach could also include the need for communications and ensuring that the affected people are ready. But, of course, that change workload and support activity will be much more than just the four themes (solution, training, people readiness, and communications).

The example shown in Figure 18.4 suggests an agreed stakeholder support approach that includes four legs of the stool to support the asset management project stakeholders. The solution represented the EAM/APM solution, the training lead and team supported training, and the change team led the people readiness and communications efforts. This oversimplifies the effort required to achieve their expected goals but provides a baseline approach the stakeholders could embrace.

Approach to Change Enablement

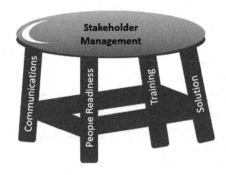

FIGURE 18.4   Example of change enablement branded strategic approach.

18.3 PLANNING FOR CHANGE IN AN ASSET MANAGEMENT TRANSFORMATION

At a high level, the change lead will plan for and then support the change execution. Typical change manager activities would span projects initiate, blueprint, design, build, deploy, and support phases with activities such as strategy, plan and design, go-live preparation, and sustain.

Figure 18.5 and Table 18.6 describe a representative overlay of these phases and a list of activities.

The phases of activities for planning and working on the change will not necessarily align with the project phases. For example, in planning for the change, it is suggested in Figure 18.3 that the phases could be the strategy, plan/design, go-live preparation, and sustain. The change plan versus the change execution activities will be very different activity sets. The detailed tasks are aligned in Table 18.6. The potential change execution tasks and their alignment with the project plan are shown later in this chapter.

The change management effort is often underestimated. Unfortunately, change management is also often underfunded and, as a result, often is the reason the project benefits are not realized. Figure 18.5 shows that the change strategy work should start when the project formally begins, if not before that time. For example, scope clarity will be required at the start, and training strategies will need to support the culture the organization may expect for a similar transformational project. Training is a topic that must have an agreed-to strategy and a plan. Then, the strategy and plan execution become execution activities in the change management project alignment (Table 18.6).

One of the many roles the change management office will need to coordinate is performing a change history and stakeholder analysis. Both are typically done face to face through interviews (where practical).

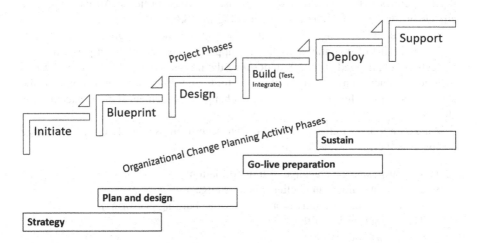

FIGURE 18.5 Project phases versus change planning phases.*

TABLE 18.6

List of Potential Organizational Change Activities Needed to Be Planned by Phase*

Organizational Change Planning by Phase	List of Potential Activities
Strategy	• Build clarity of the vision for the change and the leadership team's resolve to deliver • Define the vision and reason for the change • Define the training strategy • Informing the selection of change strategy based on an early diagnostic
Plan and design	• Plan the change management (stakeholder management, communication planning, training, team, etc.) • Designing the proper structure changes to behavior and Human Resources (HR) processes to fully integrate the new process and technology
Go-live preparation	• Manage the delivery of training, communications, etc. • Develop job profiles, security, updates to HR processes, etc.
Sustainment	• Work to confirm new processes are accepted ongoing habits • Report on benefits realized

The change history analysis asks questions about how change has been handled and received in the past (last 1-5 years). This provides insight into what the people being asked to change expect before any communications are started on this new initiative.

The change history categories of questions include: change vision, leadership, building commitment, sustaining change, configuring change, and managing change. When collected, the results can be displayed to help the executive team understand the challenges that should be expected based on past projects. An example of a change history summary chart is shown in Figure 18.6.

Implementing the change transformation project activities is a significant investment.

Often the list of potential activities to support change enablement gets cut when budgets are challenged. This can undermine the project's success and the end user's experience when transitioning to the new tools and processes. An example of the kinds of potential phases and tasks in the change management support execution is shown in Table 18.7.

An example list of change elements that have been used to describe the change attributes and effects could include:

• The purpose for the change or transformation
• The vision, values, and beliefs about change
• The overriding imperative for change (the "burning platform")
• The characteristics of the change:
  • who will be affected
  • why will they be affected
  • where will the change happen

Change History Observations

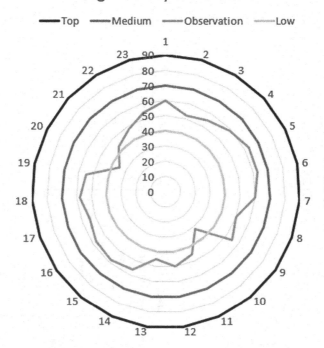

FIGURE 18.6 Example change history report.*

- what the change will look like
- when the change takes place
- identification of implementation stakeholders
- sponsor, agent, and target communities
- Discussing an approach for identifying and managing resistance to implementation

The stakeholder analysis will likely be done several times across the project timeline to get the pulse of the stakeholder issues and readiness for change. The stakeholder analysis requires understanding the business group and confirming their understanding of the expected change and impacts. The change lead will need to describe the expected strategic change and then, through questions, understand how this will impact the interviewee.

Over and over, many organizations will implement process and technology changes without considering the impact on the stakeholders. Unfortunately, this often means that the process user is not fully engaged when the change is executed, resulting in less-than-desired outcomes. An example of a stakeholder assessment is shown in Figure 18.8.

This stakeholder analysis was performed early in a project cycle and demonstrated that the maintenance and operations teams were not ready nor focused on the planned initiative (an EAM/APM implementation). The finance team was aware but only

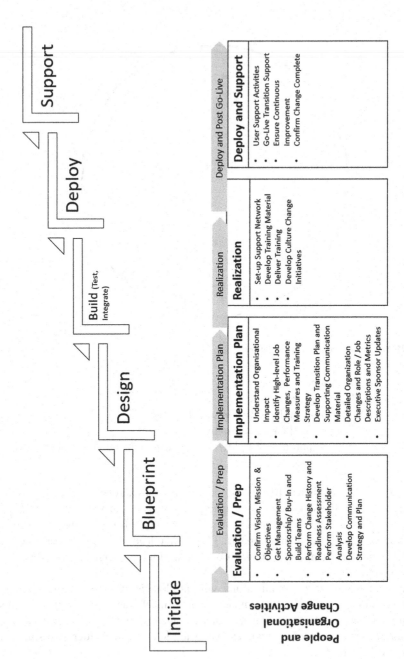

FIGURE 18.7 Example of change transformation activities in an EAM implementation.*

TABLE 18.7

Example of Change Activities by Phase*

Organizational Change Transformation Activities by Phase	List of Potential Activities
Evaluate and prepare	• Confirm how the change will be managed.
	• Identify potential organizational solutions to maximize change benefits.
	• Establish how the adoption of the change will be measured.
	• Establish what "success" will look like.
	• Confirm vision, mission, and objectives.
	• Get management sponsorship/ buy-in and build teams.
	• Perform change history and readiness assessment.
	• Perform stakeholder analysis.
	• Develop a communication strategy and plan.
	• Confirm the training plan.
Implementation plan	• Confirm that the leaders are committed to the change.
	• Confirm how the change will impact the people and the organization.
	• Confirm stakeholder readiness for the change.
	• Develop and deliver the key messages to be communicated and when.
	• Perform a stakeholder analysis.
	• Identify and mitigate change resistance.
	• Understand the organizational impact and adjust.
	• Identify high-level job changes, performance measures, and training strategies.
	• Adjust transition plan and supporting communication material.
	• Detailed organization changes, role/job descriptions, and metrics.
	• Executive sponsor updates.
Implementation realization	• Confirm how people impacted will be prepared with the required skills and knowledge to perform in the new environment.
	• Confirm that the organization is ready for the planned change.
	• Setup support network.
	• Develop training material, prototype the training, and revise.
	• Deliver training.
	• Develop culture change initiatives and deliver culture change activities.
Deploy and support	• Confirm that leaders are holding the impacted people accountable for performing in a new way.
	• Confirm how the impacted people will be incented to sustain the planned change.
	• Determine the actions needed to sustain the change and realize the expected value.
	• Evaluate the training program.
	• Ensure that the user support activities are in place.
	• Provide go-live transition support.
	• Sustain the training program.
	• Ensure continuous improvement.
	• Confirm expected benefits are experienced.
	• Confirm change complete.

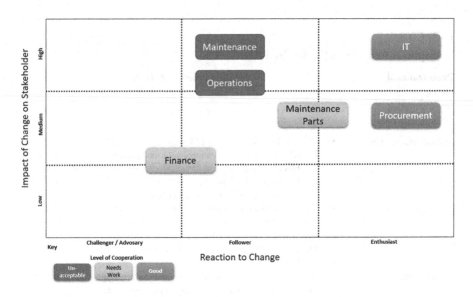

FIGURE 18.8   Example of a stakeholder analysis report for an EAM system implementation.*

worried about funding the project and seeing the benefits. The maintenance parts operation (which was outsourced) only worried about being blamed should the project not get executed well. This early identification of the status of the people impacted and understanding of their primary issues helped move the stakeholders that needed to become enthusiasts to the required level for the project's go-live execution. Ideally, we want at least the maintenance, operations, IT, procurement, and maintenance parts teams scoring in the upper right side of this matrix well before the planned go-live.

The stakeholder analysis chart effectively shows the status of the affected groups to the executive team. In addition, it will help to provoke discussion on potential mitigating actions throughout the transformation project.

This survey/interview set of steps can also help understand if there are common issues. It will identify folks with an issue where they may not be "able" or "willing" to bridge the planned change.

The collected data from the process can help identify individuals (or groups) that are:

- Enthusiasts (can and will)
- Followers/supporters (cannot and will)
- Challengers (can and will not)
- Adversaries (cannot and will not)

The recommended actions from these insights will help prescribe a course of action for each set of gaps.

The stakeholder (group) is identified as low and highly impacted by the planned change. Their commitment to engagement can be somewhere between challenger/ adversary to follower or enthusiast. Surveys and interviews can determine the individual or group's understanding, impact, and engagement level. This process can help identify the level of impact the change may have.

Some of the questions to consider for impact analysis can be:

- The level of exposure to the planned change (i.e., system, process, etc.)
- The gaps in needed skills, knowledge, tools, and technology
- The gaps in cultural change and needs from the current environment
- The degree that the planned change in policies, controls, KPIs, and systems will be altered to affect the day-to-day operation at various levels

A communications strategy is direction-setting and strategic. It defines the overall approach to managing communications from the program team to program recipients throughout the project. Further refinement and evaluation happen as the project adjusts and attempts to be agile.

The objectives of a communications strategy are to:

- Identify stakeholder communication needs
- Identify and provide consistent vital messages to all stakeholders
- Provide relevant and current information to reduce anxiety
- Circumvent the spread of miscommunication
- Build stakeholder ownership, commitment, and readiness to change
- Encourage trust and confidence in the organization during disruption and uncertainty.

The strategy may include campaigns to support the perceived gaps as the main project progresses. An example of separate communication campaigns could be to:

- Explain the need for the solution – to be sent during the project pre-project and Initiate phase
- Clarify the solution vision – to be sent during the project Blueprint, Design, and Build phases
- Inspire for commitment and action – to be sent during the project Deploy and Go-Live phases
- Sustain the momentum – to be sent during the project Post Deploy and Support phase

Communication objectives, principles, campaign themes, scope, design, and potential channels are all included in the communications strategy. A communications plan supports the various needs of the project's progression. Often campaigns are used to help ready the minds and skill levels of those affected. An example of a communications plan is shown in Table 18.8.

Care needs to be taken to provide the communication in a timely manner (i.e., just before it will be helpful to the end user).

Another consideration for end-user support is a transition management support structure. For example, Figure 18.9 shows four support steps the end user would recognize as available when the transformation goes "live." They include:

TABLE 18.8

Example of a Communications Plan to Support Change in an EAM Project*

Initiative	Audience	Medium	Frequency	Design Points (Example)
Program Newsletter	Operations, maintenance, finance, maintenance parts, procurement	E-mail, podcasts	Bi-Weekly	Branding of Project Project Updates
Poster Campaign	Operations, maintenance, finance, maintenance parts, procurement	Notice boards	Monthly	Branding of project
Extended Team Meetings	Operations, maintenance, finance, maintenance parts, procurement	Face to face	Monthly	Project design updates
Department Meetings	Operations, maintenance, finance, maintenance parts, procurement	Face to face	As scheduled	What this means to them
Executive Meetings	Executives and steering committee	Conf. call and face to face	Monthly	Project updates
Executive Letters	Operations, maintenance, finance, maintenance parts, procurement	E-mail, videos	Quarterly	Why is this important to our organization
Road Show	Operations, maintenance, finance, maintenance parts, procurement	By central location	Before the "new" process start	Branding readiness Q&A
Workshops	Operations, maintenance, finance, maintenance parts, procurement	Face to face, podcasts, videos	Before the "new" process start	Benefits to be process training
User acceptance testing (UAT)	Operations, maintenance, finance, maintenance parts, procurement	Face to face	Before the "new" process start	Validation of to-be processes
Training	Operations, maintenance, finance, maintenance parts, procurement	Face to face, T3	Before the "new" process start	As required, JIT T3
Helpdesk	Operations, maintenance, finance, maintenance parts, procurement	Phone. E-mail	Post transition support	Sustainment support

- A pre-transition training and checklist
- A copy of the training is kept available online to support the post-go-live timeframe (should anyone forget their training)
- A helpdesk and infrastructure sustainment support is in place should any issues arise post-go-live
- A list of the process owners should issue sharing or escalation be needed.

The transition management center (TMC) directs the end user to its local support hierarchy. Should the line management not resolve the issue, it can be escalated to the contingency plan leads, TMC communications team, support team, and ultimately to the executive steering committee and the boardroom.

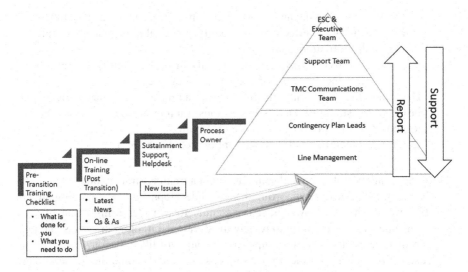

FIGURE 18.9 Example of an end-user go-live support path and a transition management center escalation path.*

18.4 TRAINING AND TRAINING SUPPORT

The training lead will be called upon to:

* Build the training strategy and plan and will advise and provide leadership to the development of the Training courseware and the training rollout and will assist in Train-the-Trainers (T3)
* Leverage the solution attributes (or, if the software is involved – leverage the IT integrator to provide SW), training assets, training documentation templates, and tools
* To tailor the training materials, related courseware, and data to align with cultural maturity and business requirements. In addition, they will often lead the training logistics and delivery so that the end user is taught by one of their own from their own culture.

Training in asset management is not change management. Training should not be an afterthought, even during the early stages of a project. Therefore, the "training lead" should be engaged to determine critical influences when implementing the training strategy. These include considering the following:

* Understanding the potential **training audience** – Identify and assess the communities of users that need to be trained
  * Casual users (i.e., occasionally users like managers or supervisors)
  * Core users (i.e., frequent users)
  * Super-users (i.e., high users, influencers, trouble-shooters, etc.)
* Understanding the required **training content**
  * For each audience, based on its needs and desired competencies

- Establish the training and performance objectives and define the strategy for measuring the effectiveness, quality, and quality of materials, training event, instructors, and learners
- Create a matrix of the type and quantity of information that should be delivered to the different roles and categories of end-users
- Determine the appropriate mediums and training technologies – identify instructional strategies and assess various training delivery mediums and technologies to support those training mediums
- Confirm the facilities and tools that are expected to be used

All of this could be collected and summarized in a training strategy document.

The plans and activities supporting the actual development of the training are not covered in this chapter. However, it is essential to consider using some of your expected super-user community to review the developed training materials and provide needed feedback. This same super-user group will likely become some of the target students for the train-the-trainer sessions and later become the trainers to the end-users. Staff typically are comforted to realize that a peer has seen the change and can teach and support the planned change.

Adults must be attracted to information before they will pay attention to it. Therefore, the training needs to tell them what they must gain from learning to attract adult learners and keep their attention. The training must also specify what they must memorize, understand, or do with the information presented. Adults are more motivated to learn when they know what is expected of them, how the new information can be applied on the job, or how it can improve their personal lives.

Training provides the knowledge, experience, and understanding needed for the affected users to learn and accept their new system, process, and environment. Students must use their new skills or knowledge from the training to be recognized as transferred. It must be demonstrated in the classroom before those students return to their workspaces. Training designers and developers must consider all learning preferences to assist this process.

The ADDIE Methodology is often suggested for adult training development and delivery.

In the ADDIE model, each step has an outcome that feeds into the subsequent stage.

The change teams work with the training lead and leverage the corporate culture to deliver:

- Comprehensive training plans with detailed role-based curricula
- Tools, materials, and job aids
- T3 (Train the Trainer) programs

ADDIE Methodology

FIGURE 18.10 ADDIE Stands for: Analysis > Design > Development > Implementation > Evaluation.

- Post-training support/plans
- Training evaluation to confirm capabilities and readiness
- Training resources (as required)

The American Society of Training and Development[2] suggests that adults have many different learning styles, including:

- **Print:** Student who learns from traditional text material. Unembellished written texts work well for this group.
- **Visual:** Student who learns from film, videos, pictures, and demonstrations. Artwork, color, and icons added to text will assist this group.
- **Aural/auditory:** Student who learns from lectures and audiotapes. Written materials supplemented with discussion and audiotapes enhance the ability to learn.
- **Interactive:** Student who learns from the discussion.
- **Tactile:** Student who wants hands-on experience and learns from modeling or sketching.
- **Kinesthetic:** Student who likes activities or physical games.
- **Olfactory:** Associates learning with smells and tastes.

A methodology that supports a best-practice adult learning approach incorporating several learning styles is shared below. This leading practice approach has five components (Prepare, Tell, Show, Let, and Help), as described in Table 18.9:

TABLE 18.9
Training Method List of Learning Styles*

Prepare me	• What can a student expect about their new roles and responsibilities? • What are the benefits of training?	This component deals with the change management and communication dialog that needs to occur before the training program.
Tell me	• What is the business process that I will be expected to know? • What is the big picture?	This component provides the conceptual information or sets the stage for what the student must do to perform their job.
Show me	• What do the transactions look like in the system? • Show me the systems I will be using to complete the business process.	This component deals with the instructors giving demonstrations of the transactions in the system. The instructor will show the system before the students complete their hands-on exercises.
Let me	• This component allows the student to practice the transaction they saw in the system.	The actual "hands-on" exercises help students retain what they have just seen.
Help me	• What do I do if I need help back at my desk? • What can I use to help me remember the new business process and transactions?	These components deal with providing help with work instructions and quick reference guides. These training components will be explained and used during the training class, facilitating retention of their use at their workplace.

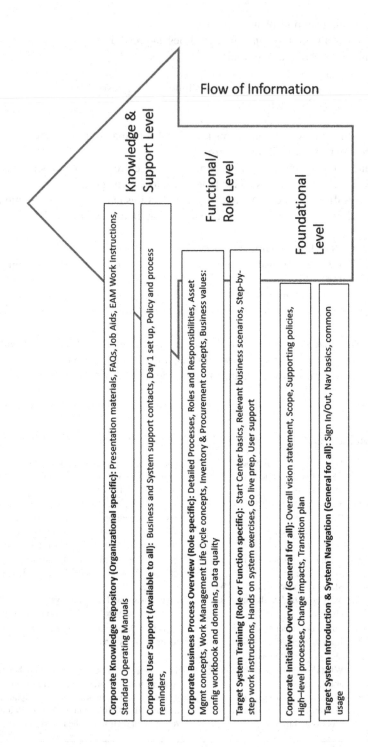

FIGURE 18.11   Example of an iterative knowledge transfer approach.*

Training can be set up to accomplish incremental levels of knowledge transfer with the information flow.

Foundational-level training would provide basic information about the corporate challenges, the reason for the change, expected impacts, and policies aligned with the change. As well a basic introduction to the expected changes can be provided.

A functional/role level training would provide more specific insights about the solution, including impacted processes, roles, responsibilities, relevant asset management concepts, and specific impacted tools insights. Targeted systems training (if in scope) would be included at this level.

Information transfer's knowledge and support level would include access to the presentation materials, FAQs, job aids, work instructions, standard operating manuals, and corporate user support materials (business and system support contacts, first-day setups, policy, and process reminders).

18.5 THE CHANGE CONTINUUM

Managing change can be a "never-ending quest." Recognizing your stakeholder groups and where they are in accepting and internalizing change is vital to ensuring that you continue to promote the process and culture change or take other appropriate risk actions until it produces the desired effect.

The change is incomplete until every stakeholder has worked through this change continuum and reached the point where the change is now a "habit." The change management resources are often removed from a project to manage the budget. The whole team often leaves the project (including the change lead) shortly after the project has passed the go-live milestone (Table 18.10).

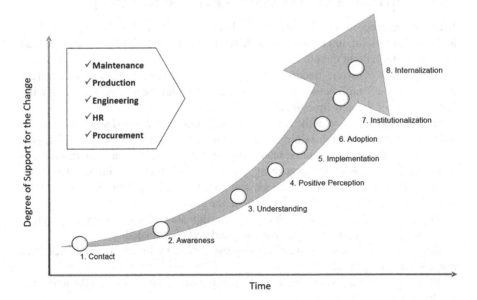

FIGURE 18.12 Change management continuum.*

TABLE 18.10

The Change Continuum Challenge to Making This a Habit[1]

Just Because:	It Does Not Mean:
You were contacted	You are aware of the change
You became aware	You understand the change
You understand	You feel positive about the change
You feel optimistic	You are ready to implement the change
You implemented (and everyone went home)	You are ready to adopt the change
You are adopting	You are adopting the standard expected
You are executing at the expected standard	You have made this change a habit!

Every implementation to improve the way we manage physical assets will be different. We are all individuals operating in unique company culture and implementing people-driven projects. In this chapter, we have tried to impart our knowledge of what works best in most cases for you to take and apply to your particular circumstances.

Although you will strive to use the most cost-effective methods and the best tools, these alone will not guarantee success. Success depends on a committed sponsor with sufficient resources, an enthusiastic champion from each affected area to lead the way, a project manager to stay the course, well-executed training when required, and a motivated team to take action and set the example for the rest of the organization.

18.6 CRITICAL SUCCESS FACTORS IN A CHANGE PROJECT

Some of the recommended elements to check are evident in your project to help achieve a successful outcome are listed here. These are essential for the change lead to confirm throughout the project.

Ensure that everyone understands the compelling need to change the current order, close the "gap" between what is done today and the vision, and clarify the justification for the new investment.

Build this vision of what the new order will be like so that it is shared, all accountable can buy into it, and there are understood long-term goals and scope of change.

Obtain visible and committed leadership so that: the implementation has a high-level executive sponsor or sponsoring group, the executive committee shares the same goals as the front-line managers, and an effective project office team and the financial resources are assigned to complete the job.

Promote broad-based stakeholder-wide participation toward a single program focus, with related activities effectively aligned and coordinated.

Get buy-in from those most affected by the change by linking performance with rewards and recognition.

Monitor performance and exercise leadership and control, especially when the implementation drifts off course.

Establish a strong project management discipline with consistent milestones; roles and responsibilities clearly defined and made visible; practical project goals; an enterprise-wide culture change being considered; and available skills to implement the change.

Communicate results at every step of the way and at regular intervals. Communication is the single most talked about pitfall by organizations that have stumbled in achieving the results hoped for at the outset of an improvement project. Provide targeted, effective communications so that: individual needs are met; consistency in the messages; effective two-way communications exist; successes are leveraged; enterprise-wide learning can occur.

PERMISSIONS

* This Chapter is adapted from Copywrite (Maintenance Parts Management Excellence Program Lecture Notes - 106 MPE – Impact of Change Management in Asset Management Training (2018) edited by (Barry). Reproduced by permission of Asset Acumen Consulting Inc.
1 Campbell, Jardine, McGlynn, *Asset Management Excellence: Optimizing Equipment Life-Cycle Decisions*, 2010, Boca Raton, CRC Press.

BIBLIOGRAPHY

1 John Moubray, Maintenance Management – A New Paradigm (now Aladon).
2 American Society of Training and Development, High-performance Training Manuals.

19 Sustainability and Asset Management*

Don M. Barry

19.1 INTRODUCTION

Sustainability has become a gate to global operations. It is a societal and business imperative. The world faces an environmental and social crisis, and consumers demand that companies take responsibility and action. Companies will work to compete and stay relevant in their path to sustainability.

Integral reporting of influences that govern sustainability in an asset-intensive organization includes trusted and auditable Environmental, Social, and Governance (ESG) Reporting. Investors increasingly apply their social values as part of their analysis process to identify material risks and growth opportunities. ESG data and reporting have increasingly become essential for investors seeking performance indicators. More public companies are positioning to attract investors by demonstrating their commitment to operational efficiency, decreasing resource dependency, and attracting new customers and employees.

ESG reporting primarily focuses on greenhouse gas emissions (GHG) but should also focus on all influences that create waste. A waste focus can be the traditional Reduce, Reuse, and Recycle. Asset-intensive organizations should also focus on asset-influenced waste and waste as a cost against operational excellence.

Resilient data strategies are needed to get a baseline and ongoing understanding of ESG for an enterprise.

Sustainability as an industry has become an exponentially growing opportunity for companies that support asset-intensive enterprises.

- Eight out of ten consumers believe procuring from good sustainable companies is essential.
- Seven out of ten companies see an excellent sustainability focus and reputation as a revenue enabler.
- Three-quarters of the "C" Suite executives believe that sustainability is essential to their business strategy.
- 100% year-over-year growth in supporting sustainability is projected for the foreseeable future.

With the momentum and focus on sustainability, the current challenge is integrating the right quality data to drive proper behavior and accurate reporting. Companies must tabulate the consolidated data while sorting through evolving global regulatory guidance, customer/employee/citizen acceptance, and changing standards.

DOI: 10.1201/9781032679600-23

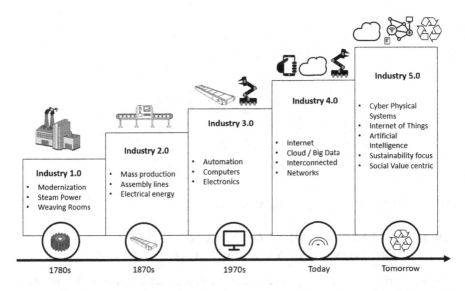

FIGURE 19.1 ESG (Environmental, Social, and Governance) influences in Industry 4.0 and 5.0.

19.2 ACCOUNTING FOR SUSTAINABILITY IN ASSET MANAGEMENT

Organizations have had environmental sustainability agendas as their responsibility over the past 50 years. Decades ago, the primary programs would have been around energy spend and "reduce, reuse, and recycle." But, of course, no organization supported the cost of waste as their value proposition.

Today the "environmental sustainability" agenda is a component of any "corporate sustainability" agenda. Corporations can:

- Capture opportunities to cut energy usage by radically improving how assets are managed
- Use data-driven insights to extend asset life and drive optimal replacement decisions
- Optimize maintenance strategies and work practices to drive down waste
- Manage and monitor their essential assets with programs that minimize asset functional failures that could create environmental incidents

Environmental accounting is a subset of financial reporting (accounting), its target being to incorporate both economic and environmental information.

Our environment is where we live. Our society is where we commune. Our economy is where we develop, work to flourish, and support commerce. The convergence of these three elements is where we look to confirm that we can sustain our business, society, and environment. Sustainable development becomes a goal for emerging developing nations and industrial ones. Environmental sustainability is a global responsibility and challenge.

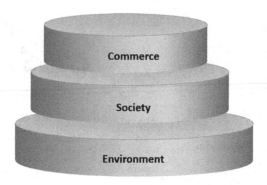

FIGURE 19.2   Environment versus commerce and society.

19.3   ASSET MANAGEMENT AND MAINTENANCE OPPORTUNITIES TO SUPPORT SUSTAINABILITY

Most recently, companies need to consider their place in the world and the minds of their shareholders for being resilient investments and excellent sustainable corporate citizens.

Asset management and maintenance play a significant role in a company's sustainability efforts by contributing to environmental, social, and economic aspects.

At a high level, the categories of opportunity to contribute to sustainability can include:

- Maximizing Energy Efficiency
- Optimizing Resources
- Optimizing Asset and Operational Reliability
- Recognizing the constant need to support Safety and Risk Mitigation
- Consider Asset Lifecycle Management
- Sense and respond to cost optimization opportunity
- Good ESG compliance and reporting

By prioritizing asset management and maintenance as an essential contributor to their sustainability strategy, companies can achieve operational efficiency, reduce environmental impact, ensure worker safety, optimize costs, and enhance their overall sustainability performance. A proactive and holistic approach to asset maintenance aligns with sustainability principles and contributes to long-term business success.

Here's how asset maintenance impacts a company's sustainability:

- Energy Efficiency: Properly maintained assets, such as machinery, equipment, and buildings, are more likely to operate efficiently. Regular maintenance, including cleaning, lubrication, and calibration, can optimize energy consumption, reduce waste, and minimize greenhouse gas emissions. By maximizing energy efficiency, companies can lower their carbon footprint and contribute to environmental sustainability.
- Resource Conservation: Asset maintenance ensures that equipment and materials are used optimally, minimizing resource consumption and waste

generation. Maintenance activities, such as repair, refurbishment, and asset optimization, can extend the lifespan of assets, reducing the need for replacement and conserving natural resources. This supports sustainable resource management and reduces the environmental impact associated with manufacturing and disposal.

- Asset and Operational Reliability: Well-maintained assets are more reliable, improving operational performance and reducing downtime. By implementing preventive maintenance practices and addressing issues promptly, companies can minimize disruptions, enhance productivity, and avoid costly breakdowns or emergency repairs. This reliability contributes to sustainable production and operational stability.

- Safety and Risk Mitigation: Regular asset maintenance helps identify and address safety risks associated with equipment or infrastructure. Companies can protect their employees, communities, and the environment from potential accidents or hazards by conducting inspections, replacing worn-out parts, and ensuring compliance with safety standards. A safe working environment supports social sustainability and enhances the company's reputation.

- Asset Lifecycle Management: Effective asset maintenance involves considering the entire lifecycle of assets, from procurement to disposal. Companies can optimize asset usage, reduce waste, and implement environmentally responsible practices by implementing strategies for maintenance planning, condition monitoring, and asset retirement. This holistic approach to asset management promotes sustainable resource allocation and reduces the environmental impact of asset-related activities.

- Cost Optimization: Leaders in asset management prioritize assets and related risks. Leading companies support reliability programs that prepare for unplanned repairs, minimizing downtime impacts, and extending the asset lifecycles. Reliability programs will identify and address issues before they impact the business which will in turn identify effective preventive maintenance programs, predictive maintenance techniques, optimize resource use, and reducing asset ownership costs. This economic sustainability contributes to long-term profitability and viability.

- ESG Compliance and Reporting: Asset maintenance practices often involve compliance with regulations, industry standards, and corporate social responsibility initiatives. Companies can meet legal requirements and enhance sustainability reporting by maintaining accurate records, conducting audits, and adhering to applicable regulations. Transparent reporting supports accountability and stakeholder trust.

19.4 GHG AND OUR LEVEL OF TOLERANCE

As an experiment with the frog kept in water that is slowly heating up, humanity is in a global environment that may be past our intellectual tolerance levels but have yet to recognize it.

Simple supply chains moving products from A to B are understood to produce volumes of Greenhouse gasses (GHGs). The shipping industry recognizes it is a significant contributor to climate change. These commercial organizations are experiencing pressure to reduce emissions across their supply chains through monitoring and management. Adopting policies removing inefficient vessels from a supply chain can lead to environmental and economic benefits.

Buildings represent the five asset classes' most prominent asset financial investment category globally and represent a significant GHG contribution opportunity. There's a recognition that real estate accounts for a significant portion of the world's energy usage – up to 40% according to the United Nations – and that property managers and operators have a significant role to play in contributing to a healthier, more resilient planet."

Production assets are not far behind regarding their perceived impact on GHG environmental impact.

If we were to put an environmental tolerance calculation to each organization's contribution to GHG and its aggregate to the global issue, perhaps the world would recognize the immediate and far-reaching negative cost impacts of assuming a run-to-failure tactic supporting environmental sustainability.

19.5   CLIMATE CHANGE IN THE PUBLIC SECTOR AND ASSET MANAGEMENT

As the climate changes, the world witnesses a redistribution of natural resources. (e.g., water, plants, animals). The emergence of a global climate change focus from multiple stakeholder groups has helped to bring sustainability to the minds of those who can affect it. When people understand how sustainability benefits their operation, it's easier to get engagement and action.

Everything from tourism, customer experience, and local production is impacted. More and more climate change, weather, and drought incidents are happening. Extreme heat days have grown four times or higher over the past two decades.

The public sector has seen the impact of climate change impacts, and their support needs to adapt. Unfortunately, the preparedness gap is growing larger. The number of heavy rain, snowfalls, landslides, forest fires, floods, and sinkholes is growing.

Evidence of climate change is all around us now. Temperature increases, floods, forest fires, sinkholes, and landslides are in the news seasonally. Understanding the probability and impact risks of its happening within the boundaries of a specific public infrastructure will help bring the issues to a higher priority to be addressed.

Asset Management has a significant role in challenging the risk and reliability of public infrastructure sustainability. It requires a concerted planning effort with all the stakeholders understanding the risks and impacts and prescribing mitigations. A Reliability Centered Maintenance (RCM) discipline can be applied here by adding climate change influences and events against the asset's operating context and functional requirements. In an RCM assessment, "climate change" and "sustainability issues" have become "reasonably likely failures." The mitigation strategies can become prescriptive by identifying the public assets with their operating context and associated risk potential and systemic causes from climate change.

Understanding the systemic causes will provide evidence of what could be done to avert or minimize the climate-related incidents impacting critical infrastructure.

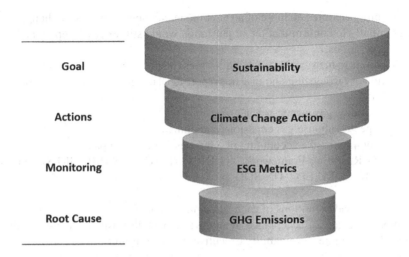

FIGURE 19.3 GHG versus sustainability goals.

Committing to proactively identifying infrastructure impacts from climate change is a significant step an organization can take to avert so-called natural weather impacts. This proactive action will enable the public infrastructure groups (i.e., municipalities) to plan and budget for preventive and mitigation actions.

Climate impacts must be part of every asset-intensive organization's Risk and Reliability program. The climate now is a must question when challenging the Failure Modes and Effects Analysis (FMEA) exercise and the causes of an asset's functional failure. Like RCM, this initiative should leverage a cross-functional stakeholder group that, as a team, will provide the insights needed to understand the direct and indirect impacts of climate change.

Climate change will not go away. It is here now and escalating. As a society, we need to collect insights from our constituents, staff, and planners. All must be heard to understand which issues should be addressed and when.

We must take stock of the critical assets and services required to support your committed infrastructure and services. Then, understand what the personal, business, and reputational impacts will be should a component of the public infrastructure or services become unavailable due to weather or other natural impacts. The lost hours or days could disrupt the current entitlements and impact future investment in your area.

The issue will require a cross-functional team effort. Being proactive in preparing for asset functional risk and reliability is no longer just an internal group exercise. For example, environmental and weather incidents were often dismissed as unlikely. That is no longer an option.

19.6 SUSTAINABILITY AND ESG

Sustainability is a broad term concerned with how we can meet our needs without compromising the ability of future generations to meet their needs. These days sustainability conjures up more of an environmental thought process, although it was likely initially to confirm that the organization would still exist to employ and serve future generations.

ESG refers to metrics that track our progress against sustainability goals and facilitate regulatory disclosure and investor reporting. The report ESG can include

- Electricity, gas, water, waste
- Scope greenhouse gas emissions and climate risk
- A comprehensive accounting of a business' societal impact
- ESG Reporting Frameworks – CDP, GRI, TCFD, SASB, GRESB, SEC, NABERS, EU EPC, UN SDG

See the roles and definitions of each acronym in Table 19.1.

To pass the gate of acceptance in an environmentally sustainable culture, and to thrive, the corporate sustainability executive (leadership) must:

- Define and align sustainability goals and strategies to their corporate strategies
- Establish an ESG data and systems architecture to track ongoing performance
- Embed sustainability decision-making in daily business operations

There is rising pressure from financial markets and consumer behavior on executives and their motivation to be seen proactively supporting the needs of the regulatory environment. As we observe a movement in Society 5.0, the business goals must consider co-creating a sustainability agenda and pathways supporting the global corporate social impact and business value trend. Supporting data must demonstrate the ability to manage environmental risk and display a system of record for ESG data. ESG should assist the enterprise in measuring, reporting, operationalizing, and achieving its sustainability roadmap.

Operational excellence will now include environmental sustainability in the asset value chain scenario.

Maintenance and operations groups will be expected to run sustainable supply chains. They will be integrated, automated, intelligent workflows for equitable, transparent, and carbon-regenerative value chains. Information Systems (IT) support will be "green," responsible computing to comply with the social impact agenda. Maintenance and operations leaders will enable intelligent assets, facilities, and infrastructure to enable clean energy, efficient waste management, and decarbonization agenda.

The reporting of corporate achievements against this new social agenda will become critical. Data and data collection is becoming paramount. There is now a need for a single version of the truth to legitimize corporate reporting. A single system of record for ESG data will establish a clear baseline. Without respected data, sustainability is not visible, actionable, or operational. Data is vital to delivering sustainability goals, and technology will evolve to help.

TABLE 19.1
Definitions and Roles of Entities Focusing on Sustainability

Terms	Roles and Definitions
Carbon Disclosure Project (CDP)	An international non-profit organization that helps municipalities and corporations declare their environmental metrics. It aims to make risk management and environmental reporting a specific business event, driving the disclosure, insight, and mitigating action toward a sustainable economy.
Global Reporting Initiative (GRI)	GRI Standards help organizations understand their impacts on the economy, environment, and society – including those on human rights. Using GRI increases accountability and enhances transparency in their contribution to sustainable development.
Task Force of Climate-related financial disclosures (TCRD)	TCFD recommendations focused on governance, strategy, risk management, metrics, and targets. It is a guidance framework that helps companies disclose climate-related financial risks to investors, lenders, and insurers.
Sustainability Accounting Standards Board (SASB)	The SASB Standards Board framework sets standards for sustainability accounting. SASB is part of the international financial reporting standards (IFRS) reporting requirements.
Global Real Estate Standards Benchmark (GRESB)	GRESB is an independent organization providing ESG performance data and accepted peer benchmarks for investors. Managers can leverage these standards with metrics to improve business intelligence, industry engagement, and decision-making. GRESB is an organization that produces internationally recognized benchmarks to track the ESG metric performance of commercial real estate and infrastructure.
Securities and Exchange Commission (SEC)	SEC works to positively impact America's economy, capital markets, and people's lives. Requires registrants to declare certain climate-related disclosures in their registration statements and periodic reports.
National Australian Built Environment Rating System (NABERS)	NABERS is a simple, reliable sustainability rating for the built environment. Like the efficiency star ratings on your fridge or washing machine, NABERS provides a rating from one to six stars for a building's efficiency across energy, water, waste, and indoor environments. They have a star rating system from 1-6, which is "starting" to "market-leading," respectively.
European Union European Policy Centre (EU EPC)	It is an independent, not-for-profit think tank that fosters European integration through analysis and debate, supporting and challenging European decision-makers at all levels. They make informed decisions based on sound evidence and analysis and provide a platform for engaging partners, stakeholders, and citizens in EU-focused policymaking.
UN SDG	UN SDG supports the goals and related thematic issues, including water, energy, climate, oceans, urbanization, transport, science and technology, the Global Sustainable Development Report (GSDR), partnerships, and Small Island Developing States. Home of the 17 development sustainability goals. They address global challenges, including poverty, inequality, climate, environmental degradation, prosperity, peace, and justice.

The data collection will become automated where ever practical. Putting a meter reading into a spreadsheet and emailing that spreadsheet to your sustainability team, you just spent time doing activities that technology should be more efficient at and, no doubt, burned efficient energy to accomplish the reporting. The data collection, data management, ongoing data health checks, and filling in missing data should be done by technology, where practical. Critical resources should spend their time on work that matters (i.e., efficiency projects, risk mitigation, and ensuring the building is as healthy, comfortable, and energy efficient). As businesses shift their priorities, their day-to-day activities shift.

Solutions will be available to:

- Measure and track ESG metrics such as greenhouse gas emissions and the contributions that may come from the impact of a related improvement project.
- Help organizations assess ESG related programs and projects in the appropriate language to help prioritize actions.
- Engage stakeholders by making your efforts to improve ESG metrics visible across the organization.
- Create confidence about the data used for sustainability target setting, with the operations and sustainability teams working from the same source of truth.
- Automate operational data input into a sustainability reporting platform, reducing time and effort.

Products now focused on this need can address the heavy lift associated with ESG metrics. They will ease the capture of sustainability-related data, convert it into helpful ESG metrics such as greenhouse gas emissions, and support reporting against internal and external reporting frameworks. In addition, they include the ability to analyze data to identify emissions savings opportunities.

For maintenance and operations folks with a sustainability mandate, they will be able to:

- Capture operational sustainability data for reporting purposes in a more automated process, reducing the cost and preparation time involved in manual data input
- Leverage the solid operational capabilities from their enterprise asset management EAM solution to help drive sustainability performance improvement for asset-intensive customers helping them to achieve challenging targets for emission reduction and ESG
- Improve compliance visibility with an integrated system that will reinforce their corporate reputation.

These outcomes are available now; however, the drive to embrace them is constrained by the business and political priorities to engage them. These capabilities will be embraced as global warming awareness and agendas grow.

19.7   EXAMPLE OF SUSTAINABILITY ACTIONS IN BUILDING ASSETS

Property managers need to know a bit about sustainability to be successful. They need to know what technologies or processes will make their buildings more efficient, and they must be able to sell sustainable strategies and investments to investors, owners, and other stakeholders.

Sustainability has become table stakes in property management. Buildings require a significant focus to help an organization drive environmental sustainability goals. Residential properties create tons of waste and consume significant amounts of energy. As a result, they can have a significant impact on local and global ecosystems. Now tenants of these buildings look to their hosts to own and drive the environmental mitigations by stepping up to environmental issues, reducing pollution, and cutting costs.

Building tenants and investors are increasingly intent on partnering with operators committed to greening their operations and taking the lead on climate issues. As a result, sustainability acceptance is growing for property management stakeholders – it can support return on investment, energy cash flow, and a positive impact on tenant relations.

Properties can consume massive amounts of energy and produce tons of waste. They have a significant impact on the ecosystem. Property owners are now looking to address the client's concerns about environmental issues, reduce pollution, and cut costs.

The pressure is on property managers to go further with sustainability and ESG. That means understanding what needs to be done and acquiring the insights and skills to take action.

Property managers need to understand how they relate to sustainability and how to get future ESG initiatives off the ground. It pays to put ESG strategies at the top of the property management agenda.

Some environmental sustainability initiatives declared by building managers are shown in Table 19.2.

19.8   CORPORATE IMPERATIVE CONSIDERATIONS IN MANAGING ENVIRONMENTAL SUSTAINABILITY

Leading corporations now need to be seen leading the environmental sustainability charge. They will do this by declaring their commitment and doing the actual work to transparently report (declare) and push lower the current and future impacts they influence.

ESG reporting will become the new Sarbanes Oxley imperative to report accurately with the expectation that audits will follow.

Some of the ESG reporting requirements may include:

- An established standard for sustainability reporting (i.e., a streamlined energy and carbon reporting capability)
- Demonstrated governance, enforcement, and assurance of ESG
- A climate-related disclosure that is included in a corporation's financial reporting

TABLE 19.2

Example of Some Actions That Could Be Taken in the Hospitality Building Asset Class to Support Environmental Sustainability

Initiative	What	Why
Environmental audit	Understand your energy consumption patterns	Identify issues and gaps, set workable goals, and choose the environmental practices that best work for their context
Use energy-efficient light bulbs	Use LED bulbs as a low-cost, high-impact practice	LED light bulbs use more than 30% less energy than traditional bulbs and can last up to 20 times longer
Consider a linen program	Hotels can run a towel and linen program-very common in Europe	You'll save energy, detergent, and water, allowing your guests to reuse their towels and linens. Let the clients know by placing water-saving information cards in the rooms
Use daytime cleaning	Implementing daytime cleaning among building tenants	Decreases energy spent on evening shifts
Eco-friendly cleaning appliances	Hospitality can invest in eco-friendly dishwashers and laundry systems Replacing old washing machines and dishwashers with energy and water-saving models. Prioritize low-consumption machines made of recyclable materials promise longer asset lifecycles	Many hospitality buildings and suite hotels have fully equipped kitchens and bathrooms. Residential buildings consume a lot of water
Support a paperless culture	Technology makes it easy for business owners to reduce their paper consumption. Centralized, cloud-based building management systems can manage all aspects of your properties in one central location	Audit trails are maintained with No printing required
Purchase green culture	Adopt a green purchasing policy by considering ESG metrics. Choose products and services that have a reduced impact on the environment	Avoids products containing toxic substances. Promotes products that can be recycled. Choose energy-efficient and Fairtrade products
Waste management and recycling culture	Buildings with tenants create significant waste. Developing a powerful recycling program (including composting) and waste management (reduce, reuse, recycle) program is a leading practice	Can engage tenants or guests to contribute by communicating the policies and making recycling bins available and in convenient locations throughout your property. Builds community brand and reputation
Train staff on green culture	Train staff to recognize green opportunities. Provide courses on green management initiatives and environmentally friendly practices. Involve them in prioritizing potential actions	Employee morale will be high when engaged and recognized that their organization is sustainable and they are contributing to this reputation

(Continued)

TABLE 19.2 (CONTINUED)
Example of Some Actions That Could Be Taken in the Hospitality
Building Asset Class to Support Environmental Sustainability

Initiative	What	Why
Install and monitor leak detection	Install leak detection devices, either Internet of Things (IoT) or alarm alerts. Manual water readings are often inaccurate and inconsistent in frequency	For example, a minor plumbing leak can cause significant refurbishment damage and cost thousands of dollars to fix. Often leaks take time to detect, and the damage is done (wasted water and capital damage)
Monitor energy consumption	Energy is the second highest operating expense for hospitality buildings (after employee costs). Use the tools to drive down these costs. Automating energy consumption usage is critical	People and systems drive behavior to do what you expect if you inspect (sense and respond)
Report your environmental sustainability achievements monthly	Sustainability goals can be filtered down to daily operations, providing context for performance targets	The performance of daily operations can then be filtered back up, informing organization-level decision-making and strategy

- Climate-related metrics that include greenhouse gas (GHG) emissions
- Supported frameworks such as Task Force for Climate-related Financial Disclosures

The ESG reporting system needs to help a global organization at least align with US SEC requirements, EU Corporate Sustainability Reporting Directive (CSDR) requirements, and UK Financial Conduct Authority requirements.

Systems will support the ESG reporting with data from Operations Management Systems, Enterprise Asset Management, and Integrated Workplace Management systems. System automation tools will respond to anomalies monitored with the data collected and aligned in a leading ESG reporting tool.

Environmental sustainability will become an additional operational and asset management resource pool skill set.

These resources will address the climate change initiatives to support sustainability. Asset-intensive corporations will need:

- People to intervene and mitigate the growing number of GHG sources or enhance the solutions (sinks) to store them. (Climate Change Mitigation)
- Influence the actual or expected climate impacts to reduce harm and increase beneficial opportunities (Climate Change Adaption)
- To understand the effects that may happen given the projected climate changes, if we do nothing (Climate Change Impacts)

- To be resilient. To accommodate, (resist, absorb,) and recover from the effects of climate-related hazards in an efficient (timely) manner at the community and systems levels (Climate Impact Resilience)

A real estate approach to mitigating climate change could use the suggested environmental sustainability goal categories and initiatives in Table 19.3.

TABLE 19.3

Example Environmental Sustainability Goal Categories and Initiatives

Environmental Sustainability Categories	Examples of Corporate Sustainability Initiatives
Energy and climate	• Sourcing their electricity from renewable sources • Measuring and driving their GHG emission outputs lower year/year • Drive to a society-accepted goal to become zero GHG emissions by a specific year (i.e., ten years out) • Incent your real estate team to create and execute projects that drive down tangible megawatt-hour usage Year/Year (including data center energy spend) • Replace energy-inefficient equipment with efficient equipment • Review and drive fleet intensity levels to lowered targets
Conservation and biodiversity	• Manage year/year water withdrawals, particularly in water-stressed regions • Manage paper usage and reduce where practical • Source paper from certified sustainable forests • Consider creating pollinator gardens as a volunteer employee-led program • Look for verified certifications from your construction and renovation project vendors
Management systems	• Maintain a single version of the truth in your ESG reporting enterprise-wide (globally) • Drive to an ISO14001 standard for Environmental Management Systems (EMS) • Drive to conform to ISO50001 standards for Energy Management Systems
Pollution prevention and waste management	• Diverting non-hazardous waste from landfill • Managing the waste-to-energy process ratios • Managing the end-of-life product waste going to landfill or incineration • Eliminate nonessential single-use plastics for retail products and cafeteria usage
Production supply chain (value chain)	• First-tier suppliers are monitored to maintain their environmental management system • Confirming that preferred suppliers in emission-intensive business sectors are incented to reduce their emissions • Participate in annual sustainability leadership symposiums

19.9   IN SUMMARY

Prioritizing sustainability initiatives also makes business sense. Assets that embrace energy-saving equipment and eco-forward practices are more resilient and cost-effective. Global ESG assets are expected to exceed $50 trillion by 2025, representing over a third of the world's total assets under management.

To progress on sustainability, you must understand how to engage stakeholders and communicate financial and social benefits. Gaining buy-in for sustainable initiatives can be a gate constraint for any asset class (see Chapter 2) but is particularly true for property management teams.

"Selling sustainability" is a skill that requires insight into how investments in cleaner, more efficient operations translate into long-term savings, brand recognition, and acceptance. Not a slam-dunk sell!

There are roles for asset management staff to consider against the environmental sustainability agenda.

Sustainability initiatives and commitments are becoming a compliance project gate to significant asset additions or replacements. Risk and reliability will now include sustainability as a potential asset failure. Overall Equipment Effectiveness (OEE) will still be a key metric, but now the question will be, "At what sustainability cost."

A high-level description of the sustainability roles that could pertain to asset management is shown in Table 19.4.

TABLE 19.4
Roles in Asset Management and Sustainability

Roles in Asset Management and Sustainability	Description
Sustainability Executive	Responsible for ESG, sustainability adoption, compliance matters, and reporting. An environmental sustainability focus ensures that relevant regulatory sustainability goals are met in prescribed timeframes to the delight of stakeholders.
Maintenance and Operations Leaders	Responsible for the entire asset portfolio, operational efficiencies, and minimizing Capital Expense (CAPEX) and Health, Safety, and Environmental (HSE) compliance matters related to assets. The maintenance and operations focus ensures that asset overall equipment effectiveness (OEE) is maximized while: <ul><li>Driving operational excellence;</li><li>Containing cost;</li><li>Minimizing energy use and waste;</li><li>Ensuring compliance with regulatory requirements, including ESG metrics; and</li><li>Meeting the business and environmental sustainability goals.</li></ul>

PERMISSIONS

*   This Chapter is adapted from copyright Sustainability Challenges in Asset Management Lecture Notes, 2023, edited by (Barry). Reproduced by permission of Asset Acumen Consulting Inc.

BIBLIOGRAPHY

Operationalizing Sustainability, Connecting the Systems to Support Your Efforts, IBM, 2022.

https://en.wikipedia.org/wiki/Carbon_Disclosure_Project

https://www.sasb.org/about/sasb-and-other-esg-frameworks/

https://www.epc.eu/en/

https://sdgs.un.org/

https://smartvatten.com/top-10-environmental-property-management-practices/

20 Technology Trends in Asset Management*

Don M. Barry
Previous contributions from Joel McGlynn and Don Fenhagen

20.1 THE ASSET AND PRIORITY FOCUS TIDES

As time and society evolve, the pendulum of what organizations want and need from their assets will also change. Societal expectations will undoubtedly be prioritized as the world moves from Society 4.0 to Society 5.0. Likewise, asset management needs will be influenced by the ever-present need to sustain shareholder investment in competitive markets. With many converging influencers, the asset focus of today will likely swing to a new priority depending on the change in seasons and environments (business and societal.) The traditional asset management focus needs may evolve, but many continue to be universally understood to be:

- Asset productivity (including return on assets – ROA)
- Regulatory compliance
- Production efficiencies
- Society brand reputation
- Waste (in all forms) minimized
- Safety and environmental impacts
- Workforce skills and attrition
- The ability to track costs and the skill to sense and respond to drive profits
- The constant focus on improved OEE

Regardless of the focus swing, technology will provide a pivotal enabling role. The need for data and the ability to leverage valuable data will separate the leaders from the laggards. The future will drive the continued convergence of Information Technology and Operations Technology. For the leaders, it will drive the ability to seamlessly integrate a leading organization's supply chain (value chain), operations focus, and asset constraints. Technology applied well at all levels of asset management will drive operational excellence. The entities that can do this well will be the future industry leaders.

20.2 INDUSTRY 4.0, 5.0, AND ASSET MANAGEMENT

Driving "Return on Asset" should be the primary corporate focus while maintaining safety, environmental, and personnel goals. Fundamentally, an organization purchases

DOI: 10.1201/9781032679600-24

443

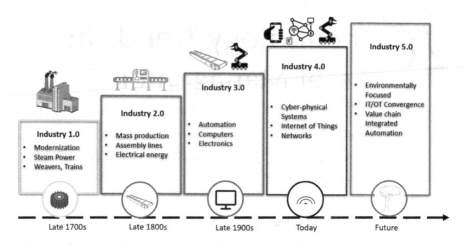

FIGURE 20.1   Industry 4.0 evolution to Industry 5.0 summary[1].

assets to create value. Maintenance is receiving a higher profile in their importance to their organization. Maintenance has many of the same responsibilities and pressures as were in place decades ago. However, the need to be competitive, drive operational excellence, customer experience, and ultimately stakeholder value adds to the pressure of this value chain.

Technology is leveraging the notion of Industry 4.0 to support what some call Asset Management 4.0. There is no doubt that leveraging technology is now the new focus element of maintenance management that cannot be ignored. Industry 4.0 acknowledges that manual production has given way to the notion of modernization with steam-powered trains, weaving rooms, and other production equipment in what is now re-labeled Industry 1.0. Industry 2.0 introduced electrical-powered equipment and assembly lines, and mass production. Industry 3.0 leveraged computers to support numeric controls, automation, and embedded electronics. In what is being called Industry 4.0, we see the leveraging of fully interconnected network production environments and cyber-physical systems taking advantage of the Internet of Things technology growth.

Industries are being challenged by the accelerated growth, improvements, and availability of technology to support their strategic missions. As a result, new technical improvements emerge in industrial, operational, consumer, and digital technologies.

This notion of Industry 4.0 driving Asset Management 4.0 will be disruptive. Industry 4.0 is growing and is seen as a powerful global force outside the control of any single organization. They will compel rapid change for customers, organizers, and providers. This emerging wave of technology will present a wave of pressing challenges for each organization to stay competitive. These influences must be acknowledged in the "near term." Each technology change can manifest as a disruptive force affecting the staff, organization, customer, enterprise, and market. Each change will influence the shape of the organization's success and culture.

Recently societal influences on Industry 4.0 are influencing the emergence of Industry 5.0, which will drive a more balanced intention between humans and production goals and outcomes.

TABLE 20.1

Emerging Technologies in the 21st Century

Industrial Technology	Operational Technology	Consumer Technology	Digital Technology
• Solar	• Smart grids	• Smart appliances	• Internet of Things
• Storage	• Embedded	• Electric vehicles	• Mobile
• Micro operations	micro-processing	• Applications (apps)	• Big Data and analytics
• Wind	• Automated demand/response	• Social networks	• Quantum computing
• Fuel cells	• Advanced network management	• Prosumer enablement	• Security
	• Situational awareness		• Machine learning/artificial Intelligence
			• Hybrid cloud
			• Automation, hyper-automation

Even in Industry 4.0, every company is a data company now. Imagine the challenges as each company moves to Society 5.0 and, with it, Industry 5.0, how the data needs will expand to include being more people and environment centric. At the same time, it drives its value chain with asset optimization.

Macro trends in data management and support include:

- Leveraging clouds and inter clouds to use hybrid data architectures for Business Intelligence (BI), machine learning (ML), and artificial intelligence (AI)
- Data lakes and warehouse convergence, decoupling data storage to support BI, ML, AI
- AI governance and risk compliance to manage AI lifecycles end to end

Data quality is an imperative, not an afterthought. Care must be taken to ensure that the data collected does not have an inherent bias that will skew the insights and decisions that may come from the data.

Industry 5.0 will drive more automation aligned with people and environmental influences. Automation tools will be expected to integrate with existing tools to extract better insights from your data to support OEE, Operational excellence goals, and production goals

ML/AI is the automation of human process analysis. ML/AI build process is typically to find data, prepare data, build and test insight and decision models, deploy (and pilot), and monitor (and adjust). Working on this process will establish trust in the data, models, and processes. ML/AI models must be able to:

- Use trusted data points
- Be easy to understand and qualify the outputs and decisions
- Handle dynamically changing model parameters
- Be monitored to challenge the accuracy of the model outcomes to minimize the model drift over time.

20.3 DIGITAL TECHNOLOGY TRENDS AND ELEMENTS

The Digital Technology elements will be the primary focus of this chapter.

Mobile technology is a prominent element of society's daily lives. Now, perhaps regrettably, even parent/child communication can find the mobile phone as either the communication enabling tool or the communication gap between the function of a family. Mobile technology can be a phone, photo transmitter, owner locator, personal data collector, barcode, or radio-frequency identification (RFID) reader. It clearly can be part of an asset management strategy as well.

Digital technology has significantly evolved over the past decade. More and more organizations are leveraging the Internet and the "Internet of Things" (IoT) to bring opportunities to leverage data in real-time dynamically. Other technologies that have emerged include situational awareness technologies (i.e., mobile tech), cloud computing and data storage, analytics, and the process convergence of Information Technology (IT) with Operational Technology (OT). In addition, ML and AI add to this mix of technical options as the asset management stakeholder can now consider automating the analysis (sense) and the reactive (response) activities.

20.3.1 THE INTERNET OF THINGS (IoT) ENABLES THE FOLLOWING

- Instrumented, Interconnected, and Intelligent Device/Asset Data (i.e., raw sensor and curated sensor data)
- Sensor and data explosion – spawns the data-driven industry of the future
- Consumption Data (i.e., Data to be consumed by a decision tree)
- Internet of people (i.e., People/worker support data such as from a wrist monitor or RFID tag)

20.3.2 SITUATIONAL AWARENESS TOOLS ENABLE

- Expanding to critical social customer interaction (i.e., mobile smartphone technology)
- Higher level of awareness (i.e., two-way communications and support)
- Automated awareness of breaching operational or asset health thresholds
- Health and safety of workforce and public (i.e., real-time updates and awareness)

20.3.3 HYBRID CLOUD AND DATA ANALYTICS TOOLS ENABLE

- Management and governance of data
- Improved accuracy and assessment of data (i.e., single source, data integrity analysis)
- An explosion of new data sources
- Leveraged data to enable disruptive innovation, creating the new industry order and further automation opportunity
- Cost-effective, collaborative, and secure data management
- Increasing the depth of analytics
- Instant, cheap, and agile systems required for industry disruption

- Agile enterprise for deployment and testing of new business models
- The leveraging of multiple platform options (i.e., IaaS, PaaS, SaaS)
- The leveraging of multiple cloud platforms (private and public clouds combined)

20.3.4 MACHINE LEARNING AND ARTIFICIAL INTELLIGENCE TOOLS ENABLE

- Computer/sensor assisted decision-making
- Broad-based data refinement and curating
- Embedding of ML/AI analytics for situational awareness
- Automated and integrated alarms and actions
- Industries to reconceptualize customer and operations dynamics

AI is already evident in many life experiences. AI assists in every area of our lives, whether reading emails, receiving driving directions, or getting music or movie recommendations. Many of us experience AI in day-to-day applications such as:

- Social media (including chatbots)
- Digital assistance (i.e., Siri)
- Web searching (i.e., Google search)
- Self-driving and parking vehicles (i.e., Tesla)
- Offline experiences (i.e., Google Maps)
- Visual Inspections (i.e., Facial recognition)
- Email communications (i.e., auto-answer, spam filtering)
- Online stores and services (i.e., product recommendations)

20.3.5 IT AND OT (OPERATIONAL TECHNOLOGY) CONVERGENCE TOOLS ENABLE

- Supporting IT and OT as they undergo structural transformation to align processes and data
- Interactions between IT and OT to be supported as they move away from previously well-defined interactions to flat and multi-variable interactions
- Integration and influence of operations, market, and weather data

20.3.6 AUTOMATION AND HYPER-AUTOMATION ENABLE

- The leveraged IT/OT converged systems/processes to integrate across multiple IT platforms
- Layers of automation to work together, managed by AI and IT control towers
- The journey to operational excellence and customer value chain
- The ultimate value competitive edge while managing the environmental, social, and governance ESG imperatives of an organization

Statistically, IT is needed to support over 80% of the asset management processes in a leading organization. Yet, typically only about 12% of maintenance staff take an overt interest in understanding the journey needed to get IT to prioritize and lead in the many IT solution options that could be applied.

Likewise, as an IT leader, investing in understanding the asset management needs of your business will help you align the value of your organization's assets against why your organization exists and its commitment to its stakeholder value and customer experience.

20.4   ASSET MANAGEMENT SPECIFIC SOFTWARE TOOLS

Regrettably, the maintenance organization is often left to drive their responsibilities without the committed help of their IT organization. The traditional role of IT in many asset-intensive organizations has been to facilitate the selection, project implementation, and support of the enterprise asset management (EAM) solution. It is time for the IT organization to step up and help drive its value proposition and a better asset value proposition supported by these emerging and available tools.

With the asset management stakeholders working through the current Industry 4.0 era, the enterprise should encourage IT to be a significant player in the corporate value proposition focusing on asset management. The process owner, the asset management leadership, still drives this.

Most large asset-intensive corporations leverage an EAM tool. For example, finance operations manage their capital investments manually (i.e., by spreadsheet) or with an Asset Investment Planning (AIP) tool. However, few entities have matured to leverage, or at least fully leverage, an Asset Performance Management (APM) solution.

EAM, APM, and AIP are related asset management solutions but separate systems and should not be confused about their purpose and value.

Definitions:

Enterprise Asset Management
- Software applications/tools that manage and facilitate asset registers, work order management transactions, maintenance parts inventory activity, and procurement functions.

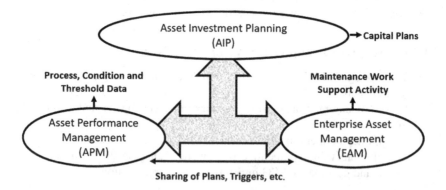

FIGURE 20.2   Marketed software solutions to manage asset management (EAM, APM, AIP).[1]

Asset Performance Management
- APM is a set of disciplines and supporting solutions to help prioritize assets, monitor asset health, and predict the need for related asset support activity
- Software applications/tools to track asset health and prioritize maintenance actions needed to optimize the availability and risk of operational assets essential to the operation of an enterprise. (i.e., Assets from industrial plants, equipment, and infrastructure)

Asset Investment Planning
- AIP is a set of disciplines and supporting solutions to help plan an asset's lifecycle
- Software applications/tools to forecast and optimize capital expenditures. (i.e., asset depreciation, repair versus replacement, rethinking asset strategies versus replacing assets).

Typically an asset-intensive enterprise would start with a significant investment in an EAM solution and add either an AIP or APM solution. Many of the leading organizations will have all three.

At a macro level, the IT and asset management organizations should look at solutions to manage the information that provides insights into "what" maintenance should be executed. Typically the EAM solution manages the asset data hierarchies, the maintenance transactions, the supporting maintenance parts activities, and the related metrics around this activity. In addition, the EAM solution manages the maintenance activity against a specific asset in a specific location. In other words, the EAM solution helps the enterprise execute maintenance "efficiently" but does not always hold the data to help the organization understand the most "effective" maintenance action or when.

Data will play a significant role in the future of asset management. So will the term "curate." Often, data is not set up correctly in any asset management system, which influences the ability and confidence of future decision-making.

The haste to implement a system without a good understanding of the data quality can compromise a significant investment. Therefore, care must be taken to ensure that the asset data quality and asset organization (asset hierarchies) are carefully managed when supporting such an initiative.

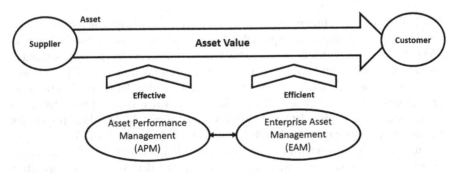

FIGURE 20.3   APM versus EAM functional roles to the Asset Value Proposition.[1]

Typically EAM solutions provide process functionality to help the enterprise manage the execution of its maintenance activities. The scope of a leading EAM solution would include process support for:

- Asset Management
- Work Management
- Parts Management
- Vendor Management and Procurement
- Maintenance execution metrics

20.5   IT AND ASSET MANAGEMENT'S BIGGER STRATEGIC PICTURE

There is always a gap between an organization's digital maturity and what needs to be considered "leading" in its industry or market. Most organizations have some level of an EAM solution in place but likely have not fully taken advantage of its functionality. For example, a typical list of functionality that a well-run maintenance operation would include is an implemented EAM solution (i.e., Maximo, SAP, Oracle, etc.) with: auto Planning and Scheduling, key asset prioritization, integrated maintenance parts management, maintenance key performance indicators (KPIs), some reliability activity, new asset data, limited data mining, some field support, and some operations alignment. However, to move toward a leading practice operation, they should want to include a solution with their installed EAM integrated with an APM and AIP solution set. Integrating these asset management software tools with supply chain planning and production software would enable an automated prescriptive response capability. A partial list of this functionality is as follows:

- Optimized operations scheduling with maintenance resource scheduling
- Total reliability rules leverage in asset activity prioritization
- Optimized operations and maintenance KPIs by Business Dimension
- IoT and AI data analytics with machine learning all IT leveraged
- Leading field support dynamic "asset constraints" factored into supply chain management (SCM) risk and reliability commitments
- More

At a high level, the solutions available to an enterprise to help manage its maintenance follow the progression displayed in Figure 20.4. If you are starting without an EAM solution, then much of what you track in your maintenance activity is manual. On the other hand, suppose you only have an EAM solution to focus on. In that case, you are set up only to enhance maintenance execution or the "efficiency" of the asset maintenance for which you are the custodian. Then, when you start to collect asset health, risk, and reliability data, you need a place to store it. EAM solutions typically are not set up to manage asset health or risk and reliability data well. For example, suppose you have performed a Reliability-centered Maintenance (RCM) team analysis. In that case, you should want a place to keep this data so that your operation can use this asset-specific data to help prescribe the appropriate alarm and activity

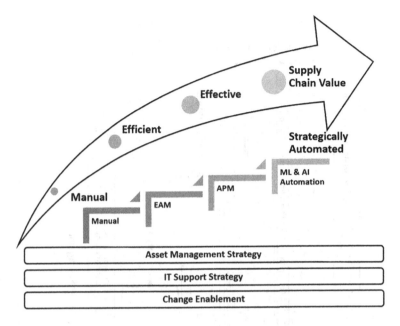

FIGURE 20.4 The asset management and IT continuum.[1]

response. Outcomes from an RCM analysis allow the organization to select which maintenance to perform. In other words, the analysis allows the organization to select the most "effective" maintenance needed for a specific asset in a specific operating context. An APM solution will support the data elements and analysis required to help an organization dynamically manage the asset risk and reliability data aligned to asset health and, if available, other external data such as operational inputs, market data, or weather influences.

Every critical asset is acquired to provide value. Strategically, the maintenance and operational teams can jointly apply ML and AI to analyze and influence the maintenance schedules and optimize the operational schedules where practical.

In this model shown above, the architecture of a combined EAM/APM solution should be developed or, at least, aligned to the organization's asset management strategy and enabled by their IT group. In addition, every change involving a cultural impact on the asset management stakeholders should be supported with a mindful eye on "change enablement."

The tactics and initiatives that could be included in a more detailed model are described in Figure 20.5. This model tracks the organization's maintenance maturity against their IT operation's willingness and commitment to commit and support asset management success in achieving leading practices. In addition, the model emphasizes the initiatives that promote maintenance efficiency versus effectiveness and how that can be integrated with operations.

The Art of the Possible model suggests that many manual maintenance operations use AIP (i.e., spreadsheets), local asset prioritization, and preventive maintenance activities. EAM solutions allow maintenance work to be more efficient than manual

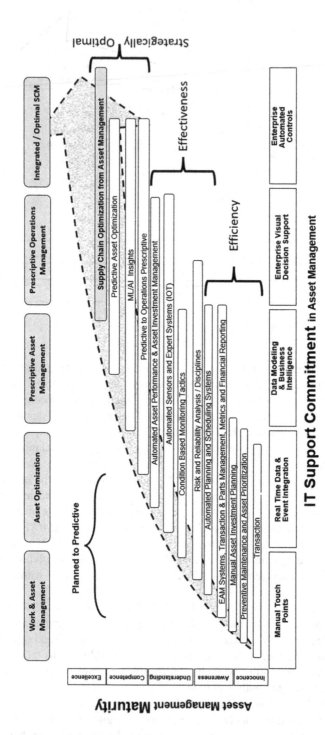

FIGURE 20.5　The Art of the Possible asset management IT continuum.[2]

operation. All asset transaction data is collected, and automating work order planning and scheduling help to effect higher planned versus unplanned ratios. If you have only installed an EAM system, you are, in effect, only four or five bullets from the bottom of this Art of the Possible continuum. While some EAM solutions can hold Condition-based monitoring thresholds, they typically do not hold the data that justifies the context of influences of the threshold, so changes in asset placement or use can create erroneous recommendations.

APM solutions focus on what maintenance to do. The scope of an APM solution will include whole asset lifecycle system optimization, reactive to proactive decision assist notifications and automated actions, and the ability to qualify asset data, external environments data, and operational data to generate actionable intelligence.

They can hold asset risk and reliability data to dynamically influence and prescribe alerts and actions. For example, if an RCM analysis has been completed, these systems can hold the mitigating tactic prescribed actions for a specific asset in its operating context. This model reinforces that while EAM solutions help maintain efficiency, APM solutions support asset management activity effectiveness.

Ultimately, assets (and people) create value for the enterprise. Supporting IT and OT convergence, including external data sets (i.e., market, operations constraints, or weather) with asset analytics allows the enterprise to automate its schedules and activities to optimize its focused production metrics. The continuum suggests that the organization can support why they strategically exist.

Today, many maintenance organizations focus on their "percent of planned maintenance" as a critical metric. It is a critical metric. However, think of the power of moving the maintenance culture to focus on the percentage of predicted maintenance. You still need to succeed at having a significant percent planned maintenance disciplines, but the degree to which you can successfully predict asset functional failure and address the failure before it happens, the higher the productivity and profit for the enterprise.

This Art of the Possible continuum focus promotes work and asset management, asset optimization, prescriptive asset management and operations, and an integrated (and optimal) supply chain management mindset.

20.6 ASSET RISK AND RELIABILITY DATA PURSUITS

Risk and Reliability pursuits are reviewed in an earlier chapter. However, this technology section expands on the data extracted from a well-executed RCM analysis. Too often, the feeling that an RCM analysis is too much effort: kills the start of such an initiative or influences a poor outcome. The total amount of data is not properly collected and held in a system that will allow it to be exploited later.

The prescriptive mitigations that come from an RCM analysis include the following five potential outcomes:

1. Predictive Maintenance (PdM) – condition monitoring
2. Preventive Maintenance (PM) – age or usage-based restoration or replacements
3. Failure Finding Tasks (FF) – periodic checks to see if usually "dormant" devices are still functional

4. No Scheduled Maintenance (NSM) – run the asset to failure if consequences (risks, costs, customer disruption) are more acceptable than being proactive (i.e., it costs less and has an acceptable reliability impact)
5. One-time changes – design, procedural, or training outputs that generally avoid the failures altogether or manage the consequences better than maintenance

The timing frequency of the above-selected mitigations is also captured from an RCM analysis.

Data that should have been logged from the analysis is often overlooked. This is the working data that came from the teams' review.

RCM analysis data elements collected can include:

- Asset data – location, risk references, operating context
- Asset functions – primary, secondary, standards
- Functional failures – failure types
- Failure Modes, Effects (FMEA) – frequency, severity, impact, HSE impacts, operator awareness, early warnings, likeliness, predictability, operational costs, maintenance costs, diagnose timelines, repair timelines, secondary damage
- Failure impacts – safety, environmental, hidden, operational, or non-operational
- Confirmation that the proactive prescribed solution is – technically feasible and worth doing
- Prescribed solution – aligned to asset location and risk prioritization
- Solution application – frequency, assignment, risk weighting, work order base, prediction thresholds
- Support requirements – training, tools, safety needs, parts/procurement

If you collect this data, ensure you have an APM solution to hold the data for your future needs. Where else can you get these insights if you are not doing an RCM analysis to collect this data? RCM can feel onerous, but it is worth the effort and sets the organization up to automate decisions with automation if you have the correct and adequately curated data.

20.7 APM FUNCTIONAL INSIGHTS

APM solutions are marketed to complement the data elements already in place in an organization's existing asset management solution. They help the organization hold the data and perform the analysis to manage "what" maintenance should be done and "when." As such, APM solutions help manage the "effectiveness" of the maintenance executed in the EAM solution.

An APMs role is to:

- Leverage existing asset data from asset management systems
- Collect and centralize asset risk and reliability strategy decision data
- Collect and centralize asset health data

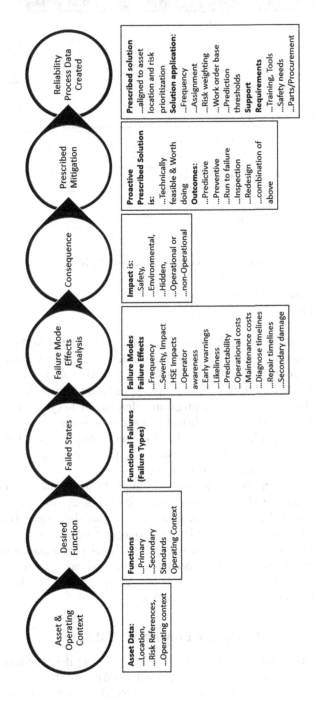

FIGURE 20.6 The potential data that can come from a Reliability-centered Maintenance (RCM) analysis.[2]

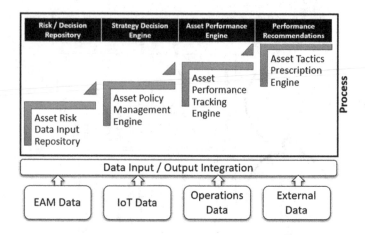

FIGURE 20.7   High-level model of APM functionality and process.[2]

- Track asset health and predict and prioritize asset failures
- Prescribe asset mitigation action execution
- Optimize asset maintenance planning
- Enhance asset performance strategies

A leading APM solution will: collect and **centralize asset-related health/risk** data. This can include holding the RCM analysis data such as Asset Risk Criticality settings, Safety/Environmental elements, HazOps, Fault Tree Analysis data, Risk/Reliability (RCM) data, Risk-Based Inspection, FMEA, and Mitigating Tactics.

The APM solution can create a **strategic decision engine** to cycle the Asset Risk data and manage the prescribed predictive maintenance actions. This can include Asset Criticality policy, Maintenance policy by asset, Safety Integrity Level (SIL), Asset Health Manager, Asset Repair/Replace policy, and Asset policy workflow. Asset health tracking can easily be managed here.

The APM solution can optimize asset maintenance planning with an **asset performance engine**. It does this by leveraging Asset Criticality analysis, Production Loss Accounting, Monitored and Condition data, SIL analysis, Inspection Documents, Corrosion Data, Test Result Data, and Digital Twinning Analysis.

The APM solution will support the **enterprise's asset performance strategies** and execution. It will provide recommendations for monitoring policy execution, Reliability Analysis, Root Cause Analysis (RCA), specific APM Metrics, Alert Generation, and creating Work Order Generation.

20.8   DATA ANALYTICS IN ASSET MANAGEMENT

The data scientist is a relatively new add as a stakeholder to the asset management success influencers group. If you have quality curated data, the data scientist can be invaluable. For example, you have completed the RCM analysis mentioned earlier in this chapter. In that case, the data analysis could bring you dramatically closer to leading practice regarding asset optimization, operational excellence, customer experience, and shareholder value. However, bringing in the data scientist too early may make them the data builder and tester, which is not a very gratifying role. Data scientists can spend 80% of their time finding, cleaning, and preparing data.

If handed asset data without context, the data scientist would often start by determining what data they have, what information they can glean from the available data, what insights could become evident from the information, and ultimately lobby for a decision that could be made from the insights.

The Reliability team would naturally start from an asset risk and reliability review and should have started their assessment from the opposite end of this process. Starting with what decision I need, what insights I need to make this decision, what information I need to get these insights, and what data is needed to get the information.

Many leading production organizations have invested in asset management and data analysis. A recent study suggested that leading operations have invested in data scientists. The same study suggested that these organizations want to grow their resource base. The study suggests that data scientists are in high demand and should be used in roles that help them recognize their value to the business.

Data scientists, by nature, are curious and enter their field to make a difference and impact. They are also driven by efficiency, which means they don't like to do things twice if they don't need to. Data scientists like to be doing investigative productive, valuable work. High turnover can be expected if they only spend time playing with data in a sandbox (never seeing their projects in production impacting actual data), spend time on data cleaning and prep instead of problem-solving or cutting-edge technologies, or do repetitive work.

Reducing this cost is a matter of proper tooling: providing the resources for staff to capitalize on past projects and reuse work.

One of the disciplines/tools that a data scientist could use is the cross-industry process (CRISP) for data mining approach. It is considered a robust and well-proven methodology and a model that does not try to capture all possible routes through the data mining process. The CRISP process steps are not unique to asset management

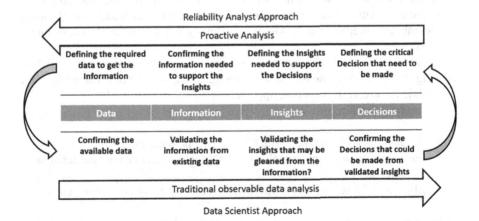

FIGURE 20.8   Data to decision cycle.[1]

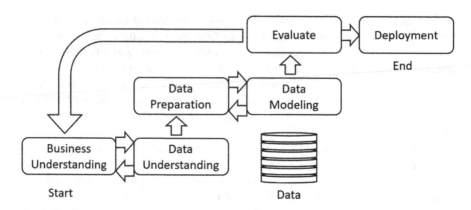

FIGURE 20.9 CRISP process model.

but can be easily applied. The process starts with understanding the business and the related data. From there, prepare data and models that align with the scenario you are looking to gain insight into and evaluate the outputs from the modeling. This process cycles until an understanding of the business, data, models, and outcomes are comfortably understood and lobbied across the decision-makers. Once accepted, the decision can be made, and deployment can happen.

20.9 DATA LANDSCAPES

Data is the foundational (lifeblood) of asset management and the IT and OT convergence evolution. Today, the creation of data from all available systems and tools leaves more than two-thirds not analyzed. Much of the data is collected as both structured and unstructured (e.g., documents, images, videos). Much of the data is kept in business unit silos, where the IT organization may be unable to access and qualify.

Often the perceived solution to managing critical data is to collect it into separate workable databases. However, this will perpetuate data sprawl with multiple locations and silos. Often data experts spend too much of their skills and time cleaning, integrating, and preparing data. Often data quality is a barrier in data integration projects.

An architecture approach is needed to facilitate access to trusted master data while limiting the resources to house the data and minimize data sprawl. The notion of a data fabric is the current promoted approach.

A data fabric architecture would:

- Access your data, in place, with built-in data protections
- Ensure governance and compliance that's centrally defined but executed in a distributed way
- Efficiently funnel your data wherever it needs to be – often automatically

Data fabric is an architectural approach to simplify data access in an organization and make it self-service available. The architecture should support multiple data environments, processes, utility, and geography, all while integrating end-to-end data-management capabilities. The architecture enables the automation of tasks like data discovery, governance, consumption, and decision-making, enabling enterprises to use data to maximize their value chain. In addition, it will help enterprises elevate the value of their data by providing the correct data at the right time, regardless of where it formally resides.

Some data principles to consider in a leading practice include:

- End-to-end data ownership belonging to the applicable line of business teams and departments
- Data tied to business goals and the value they generate
- Any data consumer (end-user), regardless of role, has easy access to the data they need
- Computation governance that enables self-service for easy end-user data consumption and collaboration

20.10   MACHINE LEARNING AND ARTIFICIAL INTELLIGENCE

Asset-intensive organizations are using AI to:

- Recognize the exceptions and navigate the unexpected
- Create user loyalty
- Innovate future ideas
- Automate effective business processes
- Help IT – innovate and modernize apps and infrastructure to gain insights
- Help operations – continuously improve automated processes for order of magnitude faster speed-to-insight and decision
- Complete the IT and OT convergence
- Drive and automate operational excellence

There is an emerging trend around Artificial Intelligence Operations (AIOps). AIOps is growing year over year in terms of market interest. However, it is still in its infancy. AIOps is not just a one silo function to be automated; it combines various aspects and automates it to provide value. Once embraced AIOps will drive integrated operational excellence converging asset management data, maintenance data, supply chain data, and market data with operational data to drive operational efficiencies and market advantage.

With ML/AL-proven algorithms driving the integrated value chain, operations will be able to boast the following:

- Faster (automated) decision-making with continual observation and response abilities
- Resource and process cycle-time optimization with continuous, automated resource allocation and optimization
- Proactive sense, response, and resolve capabilities with problem determination, remediation, and avoidance
- Automated actions that are prescribed and executed

The IT infrastructure to make ML/AI happen is available now. It is the ultimate goal of the continual IT and OT convergence that is happening in leading asset-intensive organizations.

Data is always the great unknown in working through a critical asset analysis. You can monetize data available in your plant and huge complexes for any process and power generation industry.

Using AI and ML will expand experience and expectations. ML and AI make sense when your enterprise can efficiently leverage data science techniques to contribute to asset optimization, operational excellence, and customer experience progression. It can make your asset management activities efficient and effective (leveraging EAM and APM, respectively) and integrate critical data elements from the operations process to help prioritize and align asset management. Asset support actions to why the enterprise has these assets in the first place.

Machine learning and artificial intelligence rely upon well-curated data, curated processes, and algorithm analysis. These analysis methods include Linear Regression, Logistic Regression, Decision Tree, Random Forest, Gradient, and Neural Networks. Each method has its place and pros and cons.

A linear model approach uses a simple formula to discover the best alignment through data points. This statistical approach has been used for decades and is considered because of its simplicity. Typically algorithm-based, this method can be used when the critical variable of the equation is understood and considered credible.

A decision tree-based model approach is like a series of conditions met (as in an analysis program). This approach is considered easy to follow. The variable may lead the final output to branch based on the conditions assessed through the logical progression.

A random forest approach takes advantage of many decision trees. Decision trees can result in a general wisdom outcome that may be complex to interpret.

Neural networks leverage interconnected neurons that exchange messages with each other. Machine Learning methods often adopt this approach. Neutral networks can be very complex but can handle highly complex tasks. No other algorithm method comes close to image recognition.

"Deep learning" suggests multiple layers of neural networks put one after another.

Model approaches can be unsupervised or clustered. Unsupervised learning means you are not trying to predict a variable; instead, you want to discover the hidden patterns within your data that will lead you to identify groups or clusters within that data. For example, marketing often uses clustering to group users according to multiple characteristics, such as location, purchasing behavior, age, and gender. The clustering approach could help look at asset class issues, procurement patterns, asset lost time, and parts use patterns versus operator, technician, or season.

Leveraging machine learning, AI, and data analysis will accelerate the return on investment. This is particularly true when applied to operational excellence, where the outcomes drive business profits. The data analysts and leadership team need to be focused on ensuring that the effort applied is relevant to the prioritized business needs. Openly communicating the analysis approach and intent, confirming when the model is proven and successful, and declaring the model more broadly applicable across the company are critical in achieving enterprise-wide sponsorship of the initiatives.

20.11 ASSET MANAGEMENT VISUAL INSPECTION AND AUDIO MONITORING TOOLS

Leveraging AI delivered with mobile or edge technologies means making it available where and when users need it. With the help of modern mobile phones and edge computing capabilities, visual inspection tools can provide immediate technician feedback from the user's hand. Visual inspection tools, once trained, enable the user to facilitate visual interpretation with high accuracy.

The operators of production assets are constantly looking for ways to improve overall equipment effectiveness (OEE). The ability to detect errors and immediately notify a responding technician or system creates the ability to respond to early detections of potential failure and creates high asset availability, higher operational excellence, and increased brand reputation.

The road to leveraging visual inspection tools includes properly aligned visual inspection tools (or audio monitoring tools) to collect target data, train the interpreting (AI) models, and deploy the working models into edge technology (e.g., mobility solutions). For example, recognizing a potential early indicator of failure in a rotating component (i.e., a car wheel) could require some AI model training. The AI training and detection process would be:

- Gathering and importing your datasets
- Labeling the data
- Reviewing and training the data
- Choosing the detection method for the operating context

In the case of visual inspection, images would be used. These images and datasets are used to train and refine custom AI models (i.e., what does good look like versus various images of degradation.) Machine learning can initiate automated training on the test data and create accurate models for inspection with the models understood.

Detecting and correcting self-learning machine algorithms can accelerate the confidence and productivity derived from AI and visualization automation. Production quality, throughout, and asset availability (the elements of OEE) are potential benefits of successfully applying an automated audio or visual inspection setup.

20.12 BLOCKCHAIN CONSIDERATIONS IN AM

Blockchain's application in asset management is particularly well suited to the services and parts procurement roles. A blockchain is a digital ledger of parallel transactions distributed across the entire network of computer systems on the blockchain. The blockchain records information in a system, make it difficult or impossible to hack, change, or fraudulently cheat the system.

Blockchain derives from the process involving a block data series (chain). The information contained in a block depends on the type of blockchain. For example, A parts purchasing block could contain information about the Sender, Receiver, and the number of parts to be acquired.

20.13   GIS (SPATIAL)

Geographic information systems (GIS) already play a significant role in many asset management environments across utilities, energy, government, transportation, telecommunications, and many other asset-intensive industries, by providing the capability to gather and summarize data about the diverse geographic locations and movements of strategic assets.

Organizations with linear assets can "spatially enable" a wide range of enterprise applications, including asset and service management solutions. Spatially enabled applications can support complex data analysis based on geographic location, such as representing data on maps in various spatial or geographic contexts and determining proximity, adjacency, and other location-based relationships among objects. One of the critical things occurring now in the asset management world is the convergence of the GIS business units and the asset management business units within organizations. As asset management systems integrate with GIS to enable the asset management teams spatially, GIS is becoming a part of asset management. GIS users are now asset management system users and vice versa. The footprint of GIS tools and technology is growing across industries as all companies now want to know where their physical assets are while also noting the asset status and health and risk ratings.

A powerful geospatial solution can be created by combining GIS with asset and service management business processes in a modern, service-oriented architecture (SOA). This enables decision-makers across the enterprise to make better-informed decisions, helping organizations increase productivity and efficiency while improving customer service.

20.14   DIGITAL TWINNING

In the world of asset management, the notion of twinning a monitorable process can come with multiple hypotheses. Twinning can include:

- Supply Chain optimization process constraint tracking and forecasting
- Operational optimization process twinning
- An asset's operational data points to predict critical operational degradation (i.e., OEE metrics, specific risk and reliability failure mode, and effects tracking)
- Maintenance parts inventory simulations
- Asset engineering and design lifecycle twins
- Energy spend analysis, supporting, tracking, and forecasting a facility

There is an unlimited number of processes that can be modeled in a twinning process or simulation

Digital twinning is not a new concept. Process simulations have been around for over 50 years, leveraging process simulation software and armed with quality data from an accepted model source. Spreadsheets have taken up less accurate process modeling but can help leadership account for past activities and roughly forecast near-future outcomes. Today, digital twinning is again emerging to simulate outcomes from modified variables assuming constant data models. Early lessons from performing successful digital twinning initiatives can be:

- Leveraging agile project management and remote teams
- Starting with a Proof of Concept (PoC), identifying gaps, and finishing with closing gaps and the rollout
- Developing a roadmap to manage the bridging of the gaps
- Understanding the digital and organizational maturity for operational readiness
- Understanding that a significant effort is needed to establish/create the foundational elements
- Ensuring that the end-user involvement is committed and active to drive a successful outcome
- Mobile solutions and wearables can be crucial to new working methods and the new normal
- Promoting the culture of "digital transformation," particularly when teams must work remotely from each other

20.15   OTHER DATA ELEMENTS LIKE WEATHER

Weather can influence airport or construction operations, when a farmer fertilizes their crops, influence a highly personalized just-in-time food or retail offer, or control how an energy company mobilizes its crews to prepare for a power outage.

For businesses, weather is more than just an environmental issue – it's also a factor that can have a powerful economic impact and potentially makes or break an organization's bottom line.

- Production impacts (including mining, construction, farming, etc.)
- Transportation (trains, planes, logistics)
- Facilities (heating, cooling, etc.)
- Infrastructure (roads, pipelines, overhead wires, etc.)
- IT assets impacted by the weather impact from the supporting asset classes (power, static, humidity)

Since the ultimate goal is to keep the assets running and support operational excellence, integrating weather data points into the automation tools can help an organization drive millions of dollars in production and sales.

PERMISSIONS

* Chapter adapted from pages 229–249, *Asset Management Excellence: Optimizing Equipment Life-Cycle Decisions*, 2nd Edition by Editor(s), John D. Campbell, Andrew K. S. Jardine, Joel McGlynn Copyright (2011) by Imprint. Reproduced by permission of Taylor & Francis Group.

1 Adapted from copyright *Physical Asset Management Lecture Notes* (Part 1 and Part 2), UofT (2019), edited by (Barry). Reproduced by permission of Asset Acumen Consulting Inc.

2 Adapted from copyright *IT and IoT in Asset Management Training* (2019) edited by (Barry). Reproduced by permission of Asset Acumen Consulting Inc.

Appendix A

References, Facts, Figures, and Formulas*

A.1 MEAN TIME TO FAILURE

The expected life, or the expected time during which a component will perform successfully, is defined as

$$E(t) = \int_0^\infty tf(t)\,dt$$

$E(t)$ is also known as the mean time to failure (or MTTF).
  Integrate

$$\int_0^\infty R(t)\,dt$$

by parts using

$$\int u\,dv = uv - \int v\,du$$

and letting $u = R(t)$ and $dv = dt$:

$$\frac{du}{dt} = \frac{dR(t)}{dt}$$

but by examining Figure 10.2, the probability density function, we see that

$$R(t) = 1 - F(t)$$

and

$$\frac{dF(t)}{dt} = f(t)$$

therefore,

$$\frac{du}{dt} = \frac{dR(t)}{dt} = -\frac{dF(t)}{dt} = -f(t)$$

Then, $du = -f(t)dt$
Substituting into

$$\int u\,dv = uv - \int v\,du$$

$$\int_0^\infty R(t)\,dt = \left[tR(t) \right]_0^\infty + \int_0^\infty tf(t)\,dt$$

but $\lim_{t\to\infty} tR(t) = 0$ because the part will fail eventually. That is, the reliability at infinity is 0.
   Therefore, the term

$$\left[tR(t) \right]_0^\infty$$

in the above equation is 0. Hence,

$$\int_0^\infty R(t)\,dt = \int_0^\infty tf(t)\,dt = E(t) = \text{MTTF}$$

A.2 MEDIAN RANKS

When only a few failure observations are available (say ≤20, use is made of the median rank tables:

	1	2	3	4	5	6	7	8	9	10	11	12
1	50	29.289	20.630	15.910	12.945	10.910	9.428	8.300	7.412	6.697	6.107	5.613
2		70.711	50.000	38.573	31.381	26.445	22.849	20.113	17.962	16.226	14.796	13.598
3			79.370	61.427	50.000	42.141	36.412	32.052	28.624	25.857	23.578	21.669
4				84.090	68.619	57.859	50.000	44.015	39.308	35.510	32.390	29.758
5					87.055	73.555	63.588	55.984	50.000	45.169	41.189	37.853
6						89.090	77.151	67.948	60.691	54.811	50.000	45.951

Example: Bearing failures times (in months): 2, 3, 3.5, 4, 6. From median rank tables:

7	90.572	79.887	71.376	64.490	58.811	54.049
8		91.700	82.018	74.142	67.620	62.147
9			92.587	83.774	76.421	70.242
10				93.303	85.204	78.331
11					93.893	86.402
12						94.387

From a Weibull analysis, μ = 3.75 months and σ = 1.5 months.

Median Rank (%)		
1st failure time	13	2 months
2nd failure time	31.5	3 months
3rd failure time	50	3.5 months
4th failure time	68.8	4 months
5th failure time	87.1	6 months

Benard's formula is a convenient and reasonable estimate for the median ranks.

$$\frac{\text{Cumulative Probability Estimator}}{\text{Benard's Formula}} = \frac{i - 0.3}{N + 0.4}$$

A.3 CENSORED DATA

Hours	Event	Order	Modified Order	Median Rank
67	F	1	1	0.13
120	S	2		
130	F	3	2.5	0.41
220	F	4	3.75	0.64
290	F	5	5.625	0.99

A.3.1 PROCEDURE

$$I = \frac{(n+1) \cdot (\text{previous order number})}{1 + (\text{number of items following suspended set})}$$

where I = increment.

The first order number remains unchanged. For the second, applying the equation for the increment I, we obtain

$$I = \frac{(5+1)\cdot(1)}{1+(3)} = 1.5$$

Adding 1.5 to the previous order number 1 gives the order number of 2.5 to the second failure:

$$\text{Cumulative Probability Estimator} \atop \text{Benard's Formula} \quad = \frac{i-0.3}{N+0.4}$$

$$I = \frac{(5+1)\cdot(2.5)}{1+(3)} = 3.75$$

$$I = \frac{(5+1)\cdot(3.75)}{1+(3)} = 5.625$$

Applying Benard's Formula to estimate median ranks
    For the first failure:

$$\text{median rank} = \frac{1-0.3}{5+0.4} = 0.13$$

Similarly, for the second, third, and fourth failures, respectively, we have (2.5−0.3)/5.4 = 0.41 (3.75−0.3)/5.4 = 0.64, and (5.625−0.3)/5.4 = 0.99.

A.4   THE 3-PARAMETER WEIBULL FUNCTION

Failure Number	Time of Failure	Median Ranks $N = 20$
i	t_i	$F(t_i)$
1	550	3.406
2	720	8.251
3	880	13.147
4	1,020	18.055
5	1,180	22.967
6	1,330	27.880
7	1,490	32.975
8	1,610	37.710
9	1,750	42.626
10	1,920	47.542

Failure Number	Time of Failure	Median Ranks N = 20
i	tᵢ	F(tᵢ)
11	2,150	52.458
12	2,325	57.374
13–20	Censored data	

The curvature suggests that the location parameter is greater than 0.

Failure Number	Time of Failure	Median Ranks N = 20
i	tᵢ	F(tᵢ)
1	0	3.406
2	170	8.251
3	330	13.147
4	470	18.055
5	630	22.967
6	780	27.880
7	940	32.975
8	1,060	37.710
9	1,200	42.626
10	1,370	47.542
11	1,600	52.458
12	1,775	57.374
13–20	Censored data	

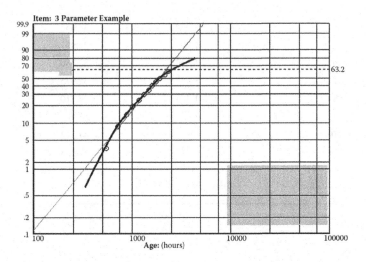

FIGURE A.1   Parameter example.

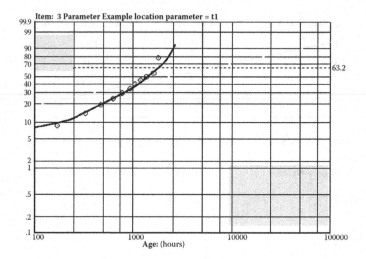

FIGURE A.2   Location parameter = t_1.

Now we get a curved line, proving that the location parameter has a value between 0 and 550. $\gamma = 375$ yields a straight line, as shown below.

Failure Number	Time of Failure	Median Ranks $N = 20$
i	t_i	$F(t_i)$
1	375	3.406
2	495	8.251
3	705	13.147
4	845	18.055
5	1,005	22.967
6	1,155	27.880
7	1,315	32.975
8	1,435	37.710
9	1,575	42.626
10	1,745	47.542
11	1,975	52.458
12	2,150	57.374
13–20	Censored data	

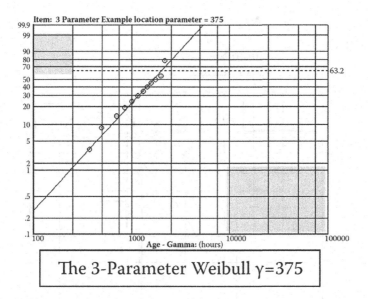

The 3-Parameter Weibull γ=375

FIGURE A.3   3-Parameter Weibel *y* = 375.

A.5   CONFIDENCE INTERVALS

From the Weibull plot, we can say that at time $t = 7$, the cumulative distribution function will have a value between 30 and 73% with 90% confidence. That is after 7 hours in 90% of the tests, between 30 and 73% of the batteries will have stopped working.

If we want a confidence interval of 90% on the Reliability $R(t)$ at the time $t = 7$, we take the complement of the limits on the confidence interval for $F(t)$:

$$(100 - 73, 100 - 30)$$

TABLE A.1
Confidence Interval Data

Failure Number	Median Ranks	5% Ranks	95% Ranks	t_i
1	5.613	0.426	22.092	1.25
2	13.598	3.046	33.868	2.40
3	21.669	7.187	43.811	3.20
4	29.758	12.285	52.733	4.50
5	37.853	18.102	60.914	5.00
6	45.941	24.530	68.476	6.50
7	54.049	31.524	75.470	7.00
8	62.147	39.086	81.898	8.25
9–12	Still operating			

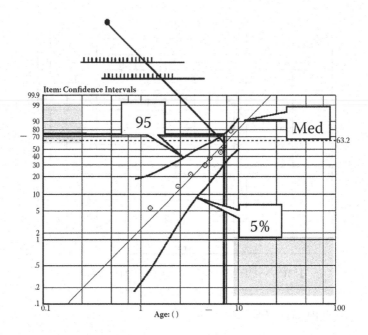

FIGURE A.4   Confidence interval data

So the 90% confidence interval for the reliability at time $wt = 7$ hours is between 27 and 70%. Or we can say that we are 95% sure that the reliability after 7 hours will not be less than 27%.

					5% Ranks						
	1	2	3	4	5	6	7	8	9	10	11	12
1	5.00	2.53	1.70	1.27	1.02	0.85	0.71	0.64	0.57	0.51	0.47	0.43
2		22.36	13.054	9.76	7.64	6.28	5.34	4.62	4.10	3.68	3.33	3.05
3			36.84	24.86	18.92	15.31	12.88	11.11	9.78	8.73	7.88	7.19
4				47.24	34.26	27.13	22.53	19.29	16.88	15.00	13.51	12.29
5					54.93	41.82	34.13	28.92	25.14	22.24	19.96	18.10
6						60.70	47.91	40.03	34.49	30.35	27.12	24.53
7							65.18	52.9	45.04	39.34	34.98	31.52
8								68.77	57.09	49.31	43.56	39.09
9									71.69	60.58	52.99	47.27
10										74.11	63.56	56.19
11											76.16	66.13
12												77.91

					95% Ranks						
	1	2	3	4	5	6	7	8	9	10	11	12
1	95	77.64	63.16	52.71	45.07	39.30	34.82	31.23	28.31	25.89	23.84	22.09
2		97.47	86.46	75.14	65.74	58.18	52.07	47.07	42.91	39.42	36.44	33.87
3			98.31	90.24	81.08	72.87	65.87	59.97	54.96	50.69	47.01	43.81
4				98.73	92.36	84.68	77.47	71.08	65.51	60.66	56.44	52.73
5					98.98	93.72	87.12	80.71	74.86	69.65	65.02	60.90
6						99.15	94.66	88.89	83.13	77.76	72.88	68.48
7							99.27	95.36	90.23	85.00	80.04	75.47
8								99.36	95.90	91.27	86.49	81.90
9									99.43	96.32	92.12	87.22
10										99.49	96.67	92.81
11											99.54	96.95
12												99.57

A.6 ESTIMATING (ALSO CALLED FITTING) THE DISTRIBUTION

A.6.1 MAXIMUM LIKELIHOOD ESTIMATE (MLE) METHOD

$$\frac{\sum_{i=1}^{n} x_i^{\hat{\beta}} \ln x_i}{\sum_{i=1}^{n} x_i^{\hat{\beta}}} - \frac{1}{n} \sum_{i=1}^{n} \ln x - \frac{1}{\hat{\beta}} = 0 \qquad \hat{\eta} = \frac{\sum_{i=1}^{n} x_i^{\hat{\beta}} \frac{1}{\hat{\beta}}}{n}$$

A.6.2 LEAST SQUARES ESTIMATE METHOD

$$\hat{\beta} = \frac{\sum_{i=1}^{n} x_i^2 - \frac{\left(\sum_{i=1}^{n} X_i\right)^2}{n}}{\sum_{i=1}^{n} x_i y_i - \frac{\sum_{i=1}^{n} x_i \sum_{i=1}^{n} y_i}{n}} \qquad \hat{\eta} = e^{\hat{A}}$$

where $y_i = \ln(t_i)$ and $x_i = \ln(\ln(1\text{-Median Rank of } y_i))$,

$$\hat{A} = \frac{\sum y_i}{n} - \frac{1}{\beta} \frac{\sum x_i}{n}$$

The MLE method needs an iterative solution for the Beta estimate. Statisticians prefer maximum likelihood estimates to all other methods because MLE has excellent statistical properties. They recommend MLE as the primary method. In contrast, most engineers recommend the method of least squares estimate. Both methods should be used because each has advantages and disadvantages in different situations. MLE is more precise. On the other hand, for small samples, it will be more biased than rank regression estimates (from Reference 2 in Chapter 10).

A.7 KOLMOGOROV–SMIRNOV GOODNESS-OF-FIT TEST

A.7.1 STEPS

1. Determine the distribution to which you want to fit the data. Then determine the parameters of the chosen distribution.
2. Determine the significance level of the test (α usually at 1.5% or 10%). It is the probability of rejecting the hypothesis that the data follows the chosen distribution assuming the hypothesis is true.
3. Determine $F(t_i)$ using the parameters assumed in Step 1. $F(t_i)$ is the value of the theoretical distribution for failure number i.
4. From the failure data, compute $\hat{F}(t_i)$ using the median ranks if appropriate.
5. Determine the maximum value of

$$\left\{ \begin{array}{l} \left| F(t_i) - \hat{F}(t_i) \right| \\ \left| F(t_i) - \hat{F}(t_{i-1}) \right| \end{array} \right\} = d$$

If $d > d_\alpha$, we reject the hypothesis that the data can be adjusted to the distribution chosen in Step 1. (d_α is obtained from the K-S statistic table.)

We have tested five items to failure. The failure times are $t_i = 1, 5, 6, 8, 10$ hours. We assume the data follow a normal distribution and will check this assumption with a K-S Goodness-of-Fit test.

A.7.2 SOLUTION

The solution is as follows:

Estimate the parameters of the chosen distribution: Estimate of $\mu = \Sigma t_i / n = 6$ and the estimate of $\sigma^2 = \Sigma(t_i - t)^2 / (n - 1) = s^2$

t_i	$F(t_i)$	$\left| \hat{F}(t_i) \right|$	$\left| F(t_i) - \hat{F}(t_i) \right|$	$\left| F(t_i) - \hat{F}(t_{i-1}) \right|$	d
1	0.070	0.129	0.059		0.059
5	0.390	0.314	0.076	0.261 d_{max}	0.261
6	0.500	0.500	0.0	0.186	0.186
8	0.720	0.686	0.034	0.220	0.220
10	0.880	0.871	0.009	0.194	0.194

$F(t_i)$ values are obtained from the normal distribution table.

Sample Size n	K-S Level of Significance ($d\alpha$)				
	0.20	0.15	0.10	0.05	0.01
1	0.900	0.925	0.950	0.975	0.995
2	0.684	0.726	0.776	0.842	0.929
3	0.565	0.597	0.642	0.708	0.828
4	0.494	0.525	0.564	0.624	0.783
5	0.446	0.474	0.510	0.565	0.669
6	0.410	0.436	0.470	0.521	0.618
7	0.381	0.405	0.438	0.486	0.577
8	0.358	0.381	0.411	0.457	0.543
9	0.339	0.360	0.388	0.432	0.514
10	0.322	0.342	0.368	0.410	0.490
11	0.307	0.326	0.352	0.391	0.468
12	0.285	0.313	0.338	0.375	0.450
13	0.284	0.302	0.325	0.361	0.433
14	0.274	0.292	0.314	0.349	0.418
15	0.266	0.283	0.304	0.338	0.404
16	0.258	0.274	0.295	0.328	0.392
17	0.250	0.266	0.286	0.318	0.381
18	0.244	0.259	0.278	0.309	0.371
19	0.237	0.252	0.272	0.301	0.363
20	0.231	0.246	0.264	0.294	0.356
25	0.21	0.22	0.24	0.27	0.32
30	0.19	0.20	0.22	0.24	0.29
35	0.18	0.19	0.21	0.023	0.27
Over 35	$1.07/\sqrt{n}$	$1.14/\sqrt{n}$	$1.22/\sqrt{n}$	$1.36/\sqrt{n}$	$1.63/\sqrt{n}$

Here we are engaging in Hypothesis Testing. A significance level is applied by some authority or standard governing the situation: either 0.01, 0.05, 0.10, 0.15, or 0.20.

We apply our K-S statistic, 0.261, to the row for a sample size of $n = 5$. Assuming we wish to conform to a significance level of 0.20, we note that 0.261 is not greater than 0.446. That means that if we reject the model, there would be a high probability (20%) that we are wrong (in rejecting a good model). Hence we say that the model is accepted based on a 20% significance level. Frequently, the less stringent 5% significance level is applied.

A.8   PRESENT VALUE

A.8.1   PRESENT VALUE FORMULAE

Consider the following to introduce the present value criterion (or present discounted criterion). If a sum of money, say $1,000, is deposited in a bank where the compound interest rate on such deposits is 10% per annum, payable annually, then after 1 year, there will be $1,100 in the account. If this $1,100 is left in the account for a further year, there will then be $1,210 in the account.

In symbol notation, we are saying that if $L is invested and the relevant interest rate is $i\%$ per annum, payable annually, then after n years, the sum $S resulting from the initial investment is

$$s = \$L\left(1 + \frac{i}{100}\right)^{n} \tag{A.1}$$

Thus, if $L = \$1,000$, $i = 10\%$, and $n = 2$ years, then

$$\$S = 1,000\left(1 + 0.1\right)^{2} = \$1,210$$

The present-day value of a sum of money to be spent or received in the future is obtained by doing the reverse calculation of that above. Namely, if $S is to be spent or received n years in the future, and $i\%$ is the relevant interest rate, then the present value of $S is

$$PV = \$S\left(\frac{1}{1 + \dfrac{i}{100}}\right)^{n} \tag{A.2}$$

where

$$\left(\frac{1}{1 + \dfrac{i}{100}}\right) = r$$

is the discount factor.

Thus the present-day value of $1,210 to be received 2 years from now is

$$PV = \$1,210\left(\frac{1}{1 + 0.1}\right)^{2} = \$1,000$$

That is, $1,000 today is "equivalent" to $1,210 2 years from now when $i = 10\%$.

It has been assumed that the interest rate is paid once per year. Interest rates may, in fact, be paid weekly, monthly, quarterly, semi-annually, etc., and when this is the case, Equations A.1 and A.2 must be modified.

In practice, with replacement problems, it is usual to assume that interest rates are payable once per year, so Equation A.2 is used in the present value calculations.

It is usual to assume that the interest rate i is given as a decimal and not in percentage terms. Equation A.1 is then written as

$$PV = \$S\left(\frac{1}{1 + i}\right)^{n} \tag{A.3}$$

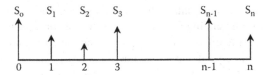

FIGURE A.5 Present value model.

An illustration of the sort of problems where the present value criterion is used is the following. If a series of payments $S_0, S_1, S_2, \ldots, S_n$, illustrated in Figure A.5, are to be made annually over n years, then the present value of such a series is

$$PV = S_0 + S_1\left(\frac{1}{1+i}\right)^1 + S_2\left(\frac{1}{1+i}\right)^2 + \ldots + S_n\left(\frac{1}{1+i}\right)^n \qquad (A.4)$$

If the payments S_j, where $j = 0, 1, 2, \ldots, n$, are equal, then the series of payments is termed an annuity, and Equation (A.4) becomes

$$PV = S + S\left(\frac{1}{1+i}\right) + S\left(\frac{1}{1+i}\right)^2 + \ldots + S\left(\frac{1}{1+i}\right)^n \qquad (A.5)$$

which is a geometric progression, and the sum of $n + 1$ terms of a geometric progression gives

$$PV = S\left[\frac{1 - \left(\frac{1}{1+i}\right)^{n+1}}{1 - \left(\frac{1}{1+i}\right)}\right] = S\left(\frac{1 - r^{n+1}}{1 - r}\right) \qquad (A.6)$$

If the series of payments of Equation (A.5) is assumed to continue over an infinite period (i.e., $n \to \infty$, then from the sum to infinity of a geometric progression, we obtain

$$PV = \frac{S}{1-r} \qquad (A.7)$$

We have assumed that i remains constant over time in all of the above formulae. If this is not a reasonable assumption, then Equation (A.4) should be modified slightly; for example, we might let i take the values $i_1, i_2, \ldots, i_n$ different periods.

A.8.2   EXAMPLE: ONE-SHOT DECISION

To illustrate the application of the present value criterion to decide on the best of a set of alternative investment opportunities, we consider the following problem.

A subcontractor obtains a contract to maintain specialized equipment for 3 years, with no possibility of extending this period. The contractor must purchase a special-purpose machine tool to cope with the work. Given the costs and salvage values of the table for three equally effective machine tools (A, B, C), which one should the contractor purchase? We assume that the appropriate discount factor is 0.9 and that operating costs are paid at the end of the year in which they are incurred.

Machine Tool	Purchase Price $	Installation Cost $	Operating Cost $			Salvage Value
			Year 1	Year 2	Year 3	
A	50,000	1,000	1,000	1,000	1,000	30,000
B	30,000	1,000	2,000	3,000	4,000	15,000
C	60,000	1,000	500	800	1,000	35,000

For machine tool A:

$$\text{Present value} = 50,000 + 1,000 + 1,000(0.9) + $$
$$1,000(0.9)^2 + 1,000(0.9)^3 - 30,000(0.9)^3$$
$$= \$31,570$$

For machine tool B:

$$\text{Present value} = 30,000 + 1,000 + 2,000(0.9) + $$
$$3,000(0.9)^2 + 4,000(0.9)^3 - 15,000(0.9)^3$$
$$= \$27,210$$

For machine tool C:

$$\text{Present value} = 60,000 + 1,000 + 500(0.9) + $$
$$800(0.9)^2 + 1,000(0.9)^3 - 35,000(0.9)^3$$
$$= \$37,310$$

Thus, equipment B should be purchased because it gives the minimum present value of the costs.

A.8.3   FURTHER COMMENTS

In the above machine tool purchasing example, note that the same decision on the tool to purchase would not have been reached if no account had been taken of the time value of money. Note also that many figures used in such an analysis will estimate future costs or returns. Where there is uncertainty about any such estimates or where the present value calculation indicates several equally acceptable alternatives (because their present values are more or less the same), then a sensitivity analysis of some of the estimates may provide information to enable making an "obvious" decision. If this is not the case, we may attribute other factors, such as "knowledge of the supplier," "spares availability," etc., to assist us in coming to a decision. Of course, when estimating future costs and returns, an account of possible increases in materials costs, wages, etc. (i.e., inflationary effects) should be taken.

When dealing with capital investment decisions, a criterion other than present value is sometimes used. For a discussion of such criteria (e.g., "pay-back period" and "internal rate of return"), the reader is referred to the engineering economic literature.

A.9 COST OF CAPITAL REQUIRED FOR ECONOMIC LIFE CALCULATIONS

Assume that the cost associated with borrowing money is an interest rate charge of about 20% per annum (p.a.). The reason for this high-interest rate is due, in part, to inflation. When economic calculations are made, it is acceptable either to work in terms of "nominal" dollars (i.e., dollars having the value of the year in which they are spent [or received]) or in "real" dollars (i.e., dollars having present-day value). Provided the correct cost of capital is used in the analysis, the exact total discounted cost is obtained whether nominal or real dollars are used. (This does require that inflation proceeds at a constant annual rate.)

To illustrate the method for obtaining the cost of capital to use if all calculations are done in present-day (i.e., real) dollars, consider the following:

- *Assuming no inflation*: I put $100 in the bank today, leave it for one year, and the bank will pay me 5% interest. Thus, at the end of 1 year, I can still buy $100 worth of goods plus an item costing $5.00. Thus, my return for doing without my $100 for 1 year is to buy an item costing $5 (and still have goods costing $100 if I wish).
- *Assume inflation at 10% p.a.*: I put $100 in the bank today. To obtain a "real" return of 5% by foregoing the use of this $100 today requires that I can still buy $100 worth of goods—which in 1 year would cost $110 since inflation occurs at 10%, plus the item initially costing $5.00—which 1 year later would cost $5.00 + 10% of $5.00, which is $5.50. Thus, at the end of 1 year, I need to have $100 + $10 + $5 + $0.50 + $115.50. Thus, the interest required on my $100 investment is

$$\$10 + \$5 + \$0.50$$
$$= \theta + i + i\theta$$
$$= 10\% + 5\% + 0.5\%$$
$$= 15.5\%$$

where θ = inflation rate, i = interest rate alternatively—if today the interest rate for discounting is 15.5%, and inflation is occurring at 10%, then the "real" interest rate is

$$i = \left(\frac{1+0.155}{1+0.10} \right) - 1$$

$$= \left(\frac{1.155}{1.10} \right) - 1$$

$$= 1.05 - 1$$

$$= 0.05 \left(\text{or} \, 5\% \right)$$

Formally, the appropriate cost of capital when working in present-day dollars is

$$i = \frac{1 + \text{Cost of capital} \left(\text{as a decimal and with inflation} \right)}{1 + \text{Inflation rate} \left(\text{as a decimal} \right)}$$

Example: Illustrate using "real" and "nominal" dollars. Assume the following:

- We are in the year 2024
- Cost of a truck in 2024 = $75,000
- Maintenance cost for a new truck in 2024 = $5,000
- Maintenance cost for a 1-year-old truck in 2024 = $10,000
- Maintenance cost for a 2-year-old truck in 2024 = $15,000

Assuming that the real cost of capital is equal to 15%, then the total discounted cost of the above series of cash flow is

$$\$75,000 + \$5,000 + \$10,000 \left(\frac{1}{1+0.15} \right)^1 + 15,000 \left(\frac{1}{1+0.15} \right)^2$$

$$= \$75,000 + \$5,000 + \$10,000 \left(0.8696 \right) + \$15,000 \left(0.8696 \right)^2$$

$$= \$75,000 + \$5,000 + \$8,696 + \$11,342$$

$$= \$100,038$$

Assuming inflation now occurs at an average rate of 10% per annum, then the cost of capital would be

$$0.15 + 0.10 + 0.10 \left(0.15 \right) = 0.265$$

The cost data would now be
    Cost of a truck in 2024 = $75,000
    Maintenance cost of a new truck for 1 year in 2024 = $5,000
    Maintenance cost of a 1-year-old truck for 1 year in 2025 = $10,000 (1 + 0.1) = $11,000
    Maintenance cost of a 2-year-old truck for 1 year in 2026 = $15,000 (1 + .1)^2 = $18,150

Using a cost of capital of 26.5%, then the total discounted cost would now be

$$\$75,000 + \$5,000 + \$11,000\left(\frac{1}{1+0.265}\right)^1 + 18,150\left(\frac{1}{1+0.265}\right)^2$$

$$= \$75,000 + \$5,000 + \$11,000(0.7905) + \$18,150(0.6249)$$

$$= \$75,000 + \$5,000 + \$8,696 + \$11,343$$

$$= \$100,038$$

which is identical to that obtained when present-day dollars were used along with a "noninflationary" cost of capital.

A.9.1   EQUIVALENT ANNUAL COST

The economic life model gives a dollar cost that is a consequence of an infinite chain of replacements or modifications (see following) for the first N cycles. It is useful to convert the total discounted costs associated with the economic life to an equivalent annual cost (EAC) to ease the interpretation of this total discounted cost.

In the above calculations, the total discounted cost over the 3 years was \$100,038 to an EAC. The formula for calculating the capital recovery factor (CRF) is

$$CRF = \frac{i(1+i)^n}{(1+i)^n - 1}$$

Rather than use the above formula, most books dealing with financial analysis include CRF tables for various interest rates. In such tables, one would find for $i = 15\%$ and $n = 3$ years, that

$$CRF = 0.4380$$
$$EAC = \$100,038(0.4380) = \$43,816.644 \text{ p.a.}$$

This means that regular payments of \$43,816.644 p.a. would result in a total discounted cost of \$100,038 if the capital cost is 15%. To check this:

$$T.D.C. = \$43,816.644\left(\frac{1}{1+0.15}\right)^1 + 43,816.644\left(\frac{1}{1+0.15}\right)^2$$

$$+ \$43,816.644\left(\frac{1}{1+0.15}\right)^3$$

$$= \$38,101.43 + \$33,131.678 + \$28,810.152$$

$$= \$100,043\,(\text{given rounding errors, this is equal to } \$100,038)$$

A.10 OPTIMAL NUMBER OF WORKSHOP MACHINES TO MEET A FLUCTUATING WORKLOAD

A.10.1 STATEMENT OF PROBLEM

From time to time, jobs requiring workshop machines (e.g., lathes) are sent from various production facilities within an organization to the maintenance workshop. Depending on the workload of the workshop, these jobs will be returned to production after some time has elapsed. The problem is determining the optimal number of machines that minimizes the system's total cost. This cost has two components: the cost of the workshop facilities and the cost of downtime incurred due to jobs waiting in the workshop queue and then being repaired.

A.10.2 CONSTRUCTION OF MODEL

1. The arrival rate of jobs to the workshop requiring work on a lathe is Poisson distributed with arrival rate λ.
2. The service time a job requires on a lathe is negative exponentially distributed with mean $1/\mu$.
3. The downtime cost per unit time for a job waiting in the system (i.e., being served or in the queue) is C_d.
4. The cost of operation per unit time for one lathe (either operating or idle) is C_i.
5. The objective is to determine the optimal number of lathes n to minimize the system's total cost per unit time $C(n)$.

$$C(n) = \text{Cost per unit time of the lathes} + \text{Downtime cost per unit time due}$$
$$\times \text{ to jobs being in the system}$$

$$\text{Cost per unit time of the lathes} = \text{Number of lathes} \times \text{Cost per}$$
$$\text{unit time per lathe} = nC_l$$

$$\text{Downtime cost per unit time of jobs being in the system}$$
$$= \text{Average wait in the system per job}$$
$$\times \text{ Arrival rate of jobs in the system}$$
$$\times \text{ per unit of time} \times \text{Downtime cost per}$$
$$\times \text{ unit time/job}$$
$$= W_s \lambda C_d$$

where W_s = mean wait of a job in the system. Hence

$$C(n) = nC_l + W_s \lambda C_d$$

This is a model of the problem relating the number of machines n to the total cost $C(n)$.

A.11 OPTIMAL SIZE OF A MAINTENANCE WORKFORCE TO MEET A FLUCTUATING WORKLOAD, TAKING INTO ACCOUNT SUBCONTRACTING OPPORTUNITIES

A.11.1 STATEMENT OF PROBLEM

The workload for the maintenance crew is specified at the beginning of a period, say a week. By the end of the week, all the workload must be completed. The workforce size is fixed; thus, a fixed number of person-hours are available per week. If demand at the beginning of the week requires fewer person-hours than the fixed capacity, then no subcontracting occurs. However, if the demand exceeds the capacity, the excess workload is subcontracted and returned from the subcontractor by the end of the week.

Two types of costs are incurred:

1. Fixed costs depending on the size of the workforce
2. Variable cost depending on the mix of internal/external workload

Because the fixed cost increases through increasing the workforce size, subcontracting is less likely necessary. However, there may frequently be occasions when fixed costs will be incurred, yet demand may be low, that is, considerable under-utilization of the workforce. The problem is determining the workforce's optimal size to meet a fluctuating demand to minimize the expected total cost per unit of time.

A.11.2 CONSTRUCTION OF MODEL

1. The demand per unit of time is distributed according to a probability density function $f(r)$, where r is the number of jobs.
2. The average number of jobs processed per man per unit of time is m.
3. The total capacity of the workforce per unit of time is mn, where n is the number of men in the workforce.
4. The average cost of processing one job by the workforce is C_w.
5. The average cost of processing one job by the subcontractor is C_s.
6. The fixed cost per man per unit of time is C_f.

The basic conflicts of this problem are illustrated in Figure A.6 from which it is seen that the expected total cost per unit time $C(n)$ is

$$C(n) = \text{Fixed cost per unit time} + \text{Variable internal processing per} \\ \times \text{unit time} + \text{Variable subcontracting processing cost} \times \text{per unit time}$$

$$\text{Fixed cost per unit time} = \text{Size of workforce} \times \text{Fixed cost per man} = nC_f$$

$$\text{Variable internal processing cost per unit time} = \\ \text{Average number of jobs processed internally} \\ \text{per unit time} \times \text{Cost per job}$$

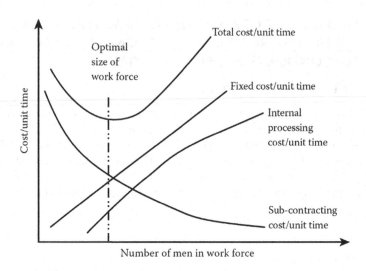

FIGURE A.6 Optimal size of work force.

Now, the number of jobs processed internally per unit of time will be

(1) Equal to the capacity when demand is greater than capacity
(2) Equal to demand when demand is less than, or equal to, capacity

$$\text{Probability of}\,(1) = \int_{nm}^{\infty} f(r)\,dr$$

$$\text{Probability of}\,(2) = \int_{-\infty}^{nm} f(r)\,dr = 1 - \int_{nm}^{\infty} f(r)\,dr$$

When (2) occurs, the average demand will be

$$\frac{\int_{0}^{nm} rf(r)}{\int_{0}^{nm} f(r)}$$

Therefore, the *variable internal processing cost per unit of time* is

$$\left(nm\int_{nm}^{\infty} f(r)\,dr + \frac{\int_{0}^{nm} rf(r)}{\int_{0}^{nm} r(r)}\int_{0}^{nm} f(r) \right) C_w$$

Variable subcontracting processing cost per unit time = Average number of jobs processed externally per unit time × Cost per job

Now, the number of jobs processed externally will be

(1) Zero when the demand is less than the workforce capacity
(2) Equal to the difference between demand and capacity when demand is greater than capacity

$$\text{Probability of } (1) = \int_{0}^{nm} f(r)\,dr$$

$$\text{Probability of } (2) = \int_{nm}^{\infty} f(r)\,dr = 1 - \int_{0}^{nm} f(r)\,dr$$

When (2) occurs, the average number of jobs subcontracted is

$$\int_{nm}^{\infty} (r-nm)f(r)\,dr \,\Big|\, \int_{nm}^{\infty} f(r)\,dr$$

Therefore, the *variable subcontracting processing cost per unit of time* is

$$\left(0 \times \int_{0}^{nm} f(r)\,dr + \frac{\int_{nm}^{\infty}(r-nm)f(r)\,dr}{\int_{nm}^{\infty} f(r)\,dr} \int_{nm}^{\infty} f(r)\,dr \right) C_s$$

and

$$c(n) = nC_f + \left(nm \int_{nm}^{\infty} f(r)\,dr + \int_{0}^{nm} rf(r) \right) C_w$$

$$+ \left(\int_{nm}^{\infty} (r-nm)f(r)\,dr \right) C_s$$

This model of the problem relates workforce size n to the total cost per unit of time $C(n)$.

A.12 OPTIMAL INTERVAL BETWEEN PREVENTIVE REPLACEMENTS OF EQUIPMENT SUBJECT TO BREAKDOWN

A.12.1 STATEMENT OF PROBLEM

Equipment is subject to sudden failure, and when a failure occurs, the equipment must be replaced. Because failure is unexpected, it is not unreasonable to assume that a failure replacement is more costly than a preventive replacement. For example, a preventive replacement is planned, arrangements are made to perform it without unnecessary delays, or a failure may cause damage to other equipment. Preventive replacements can be scheduled to occur at specified intervals to reduce the number of failures. However, a balance is required between the amount spent on the preventive replacements and their resulting benefits, that is, reduced failure replacements.

In this Appendix, it is assumed, not unreasonably, that we are dealing with a long period of time over which the equipment is to be operated, and the intervals between the preventive replacements are relatively short. When this is the case, we need to consider only one cycle of operations and develop a model for the cycle. If the interval between the preventive replacements were "long," it would be necessary to use the discounting approach, and the series of cycles would have to be included in the model.

The replacement policy is one where preventive replacements occur at fixed intervals of time, and failure replacements occur when necessary, and we want to determine the optimal interval between the preventive replacements to minimize the total expected cost of replacing the equipment per unit of time.

A.12.2 CONSTRUCTION MODEL

1. C_p is the cost of a preventive replacement.
2. C_f is the cost of a failure replacement.
3. $f(t)$ is the probability density function of the equipment's failure times.
4. The replacement policy is to perform preventive replacements at constant intervals of length t_p, irrespective of the age of the equipment, and failure replacements occur as many times as required in intervals $(0, t_p)$. The policy is illustrated in the figure below.
5. The objective is to determine the optimal interval between preventive replacements to minimize the total expected replacement cost per unit of time.

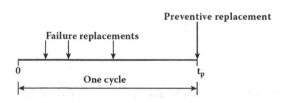

FIGURE A.7   Optimal interval work model.

The total expected cost per unit of time for preventive replacement at time t_p, denoted $C(t_p)$, is

$$C(t_p) = \frac{\text{Total expected cost in interval}(0, t_p)}{\text{Length of interval}}$$

Total expected cost in the interval $(0, tp)$ = Cost of a preventive replacement
+ Expected cost of failure replacements = $C_p + C_f H(t_p)$

where $H(t_p)$ is the expected number of failures in the interval $(0, t_p)$.

Length of interval = t_p

Therefore,

$$C(t_p) = \frac{C_p + C_f H(t_p)}{t_p}$$

This model of the problem relates replacement interval t_p to total cost $C(t_p)$.

A.13 OPTIMAL PREVENTIVE REPLACEMENT AGE OF EQUIPMENT SUBJECT TO BREAKDOWN

A.13.1 STATEMENT OF PROBLEM

This problem is similar to Section A.12, except that instead of making preventive replacements at fixed intervals, thus incurring the possibility of performing a preventive replacement shortly after a failure replacement, the time the preventive replacement occurs depends on the age of the equipment. When failures occur, failure replacements are made.

Again, the problem is to balance the cost of the preventive replacements against their benefits, and we do this by determining the optimal preventive replacement age for the equipment to minimize the total expected cost of replacements per unit of time.

A.13.2 CONSTRUCTION OF MODEL

1. C_p is the cost of a preventive replacement.
2. C_f is the cost of a failure replacement.
3. $f(t)$ is the probability density function of the failure times of the equipment.
4. The replacement policy is to perform preventive replacements once the equipment has reached a specified age t_p, plus failure replacements when necessary. The policy is illustrated in Figure A.8.
5. The objective is to determine the optimal replacement age of the equipment to minimize the total expected replacement cost per unit of time.

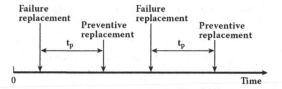

FIGURE A.8   Preventive replacement model.

In this problem, there are two possible cycles of operation: one is determined by the equipment reaching its planned replacement age t_p, and the other is determined by the equipment ceasing to operate due to a failure occurring before the planned replacement time. These two possible cycles are illustrated in Figure A.9.

The total expected replacement cost per unit time $C(t_p)$ is

$$C\left(t_p\right) = \frac{\text{Total expected replacement cost per cycle}}{\text{Expected cycle length}}$$

Total expected replacement cost per cycle = Cost of a preventive cycle
$\times$ Probability of a preventive cycle + Cost of a failure cycle
$\times$ Probability of a failure cycle $= C_p R\left(t_p\right) + C_f \left[1 - R\left(t_p\right)\right]$

Remember: If $f(t)$ is as illustrated in Figure A.10, then the probability of a preventive cycle equals the probability of failure occurring after time t_p; that is, it is equivalent to the shaded area, which is denoted $R(t_p)$.

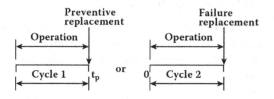

FIGURE A.9   Failure replacement model across two cycles.

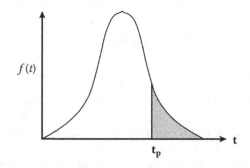

FIGURE A.10   Probability of failure curve after t_p.

The probability of a failure cycle is the probability of a failure occurring before time t_p, which is the unshaded area of Figure A.10. Because the area under the curve equals unity, the unshaded area is $[1 - R(t_p)]$.

Expected cycle length = Length of a preventive cycle ×
Probability of a preventive cycle +
Expected length of a failure cycle ×
Probability of a failure cycle

$$= t_p \times R(t_p) + (\text{Expected length of a failure cycle}) \times \left[1 - R(t_p)\right]$$

To determine the expected length of a failure cycle, consider Figure A.11. The MTTF of the complete distribution is

$$\int_{-\infty}^{\infty} tf(t)\,dt$$

which for the normal distribution equals the mode (peak) of the distribution. If a preventive replacement occurs at time t_p, then the MTTF is the mean of the shaded portion of figure above because the unshaded area is an impossible region for failures. The mean of the shaded area is

$$\int_{-\infty}^{t_p} tf(t)\,dt \,/\, \left[1 - R(t_p)\right]$$

denoted $M(t_p)$. Therefore,

$$\text{Expected cycle length} = t_p \times R(t_p) + M(t_p) \times \left[1 - R(t_p)\right]$$

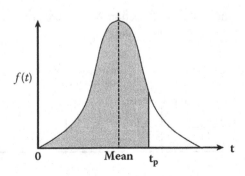

FIGURE A.11   Probability of failure before t_p.

$$C(t_p) = \frac{C_p \times R(t_p) + C_f \times \left[1 - R(t_p)\right]}{t_p \times R(t_p) + M(t_p) \times \left[1 - R(t_p)\right]}$$

This is now a model of the problem relating replacement age t_p to the total expected replacement cost per unit of time.

A.14 OPTIMAL PREVENTIVE REPLACEMENT AGE OF EQUIPMENT SUBJECT TO BREAKDOWN, TAKING INTO ACCOUNT THE TIMES REQUIRED TO EFFECT FAILURE AND PREVENTIVE REPLACEMENTS

A.14.1 STATEMENT OF PROBLEM

The problem definition is identical to Section A.13 except that, instead of assuming that the failure and preventive replacements are made instantaneously, the time required to make these replacements is taken into account.

The optimal preventive replacement age of the equipment is again taken as that age which minimizes the total expected cost of replacements per unit of time.

A.14.2 CONSTRUCTION OF MODEL

1. C_p is the cost of a preventive replacement.
2. C_f is the cost of a failure replacement.
3. T_p is the time required to make a preventive replacement.
4. T_f is the time required to make a failure replacement.
5. $f(t)$ is the probability density function of the failure times of the equipment.
6. $M(t_p)$ is the MTTF when preventive replacement occurs at time t_p.
7. The replacement policy is to perform a preventive replacement once the equipment has reached a specified age t_p, plus failure replacements when necessary. This policy is illustrated in Figure A.12.
8. The objective is to determine the optimal preventive replacement age of the equipment to minimize the total expected replacement cost per unit of time.

As was the case for the problem of Section A.13, there are two possible cycles of operation, and they are illustrated in Figure A.13.

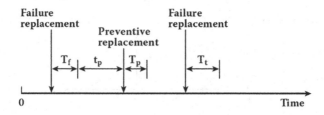

FIGURE A.12 Replacement policy model.

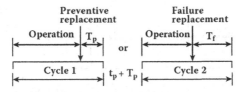

FIGURE A.13   Two operations cycle model.

The total expected replacement cost per unit of time, denoted $C(t_p)$, is

$$C(t_p) = \frac{\text{Total expected replacement cost per cycle}}{\text{Expected cycle length}}$$

$$\text{Total expected replacement cost per cycle} = C_p \times R(t_p) + C_f\left[1 - R(t_p)\right]$$

Expected cycle length = Length of a preventive cycle ×
Probability of a preventive cycle +
Expected length of a failure cycle ×
Probability of a failure cycle

$$= (t_p + T_p)R(t_p) + \left[M(t_p) + T_f\right]\left[1 - R(t_p)\right]$$

$$C(t_p) = \frac{C_p R(t_p) + C_f\left[1 - R(t_p)\right]}{(t_p + T_p)R(t_p) + \left[M(t_p) + T_f\right]\left[1 - R(t_p)\right]}$$

This model of the problem relates preventive replacement age t_p to the total expected replacement cost per unit of time.

A.15   OPTIMAL REPLACEMENT INTERVAL FOR CAPITAL EQUIPMENT: MINIMIZATION OF TOTAL COST

A.15.1   STATEMENT OF PROBLEM

This problem is similar to Section A.9 except that (1) the objective is to determine the replacement interval that minimizes the total cost of maintenance and replacement over a long period, and (2) the cost trend is considered discrete rather than continuous.

A.15.2   CONSTRUCTION OF MODEL

1. A is the acquisition cost of the capital equipment.
2. C_i is the cost of maintenance in the ith period from new, assumed to be paid at the end of the period, where $i = 1, 2, \ldots, n$.

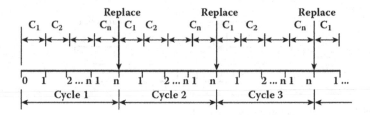

FIGURE A.14   Replacement policy illustrated.

3. S_i is the resale value of the equipment at the end of the nth period of operation.
4. r is the discount rate.
5. n is the age in periods of the equipment when replaced.
6. $C(n)$ is the total discounted cost of maintaining and replacing the equipment (with identical equipment) over a long period of time, with replacements occurring at intervals of n periods.
7. The objective is to determine the optimal interval between replacements to minimize total discounted costs, $C(n)$.

The replacement policy is illustrated in Figure A.14.

Consider the first cycle of operation. The total cost over the first cycle of operation, with equipment already installed, is

$$C_1(n) = C_1 r + C_2 r^2 + C_3 r^3 + \ldots + C_n r^n + A r^n - S_n r^n$$

$$= \sum_{i=1}^{n} C_i r^i + r^n (A - S_n)$$

For the second cycle, the total cost discounted to the start of the second cycle is

$$C_2(n) = \sum_{i=1}^{n} C_i r^i + r^n (A - S_n)$$

Similarly, the total costs of the third, fourth, etc. cycle, discounted back to the start of their cycle, can be obtained.

The total discounted costs, when discounting is taken to the start of the operation (i.e., at time 0), is

$$C(n) = C_1(n) + C_2(n) r^n + C_3(n) r^{2n} + \ldots + C_n(n) r^{(n-1)n} + \ldots$$

Because $C_1(n) = C_2(n) = C_3(n) = \ldots = C_n(n) = \ldots$, we have a geometric progression that gives, over an infinite period,

$$C(n) = \frac{C_1(n)}{1-r^n} = \frac{\sum_{i=1}^{n} C_i r^i + r^n (A - S_n)}{1-r^n}$$

This model of the problem relates replacement interval n to total costs.

A.16   THE ECONOMIC LIFE MODEL USED IN PERDEC

The economic life model used in PERDEC is

$$EAC(n) = \left[\frac{A + \sum_{i=0}^{n} C_i r^i - S_n r^n}{1-r^n} \right] * i$$

where
   A = acquisition cost
   C_i = O&M costs of a bus in its ith year of life, assuming payable at the start of
      the year, $i = 1, 2, \ldots n$
   r = discount factor
   S_n = resale value of a bus of age n years
   N = replacement age
   $C(n)$ = total discounted cost for a chain of replacements every n years

A.17   ECONOMIC LIFE OF PASSENGER BUSES

The purpose of this Appendix is to demonstrate how the standard economic life model for equipment replacement can be modified slightly to enable the economic life of buses to be determined, taking into account the declining utilization of a bus over its lifetime. Specifically, new buses are mostly utilized to meet base load demand, while older buses are used to meet peak demands, such as "rush hours."

The case study describes a fleet of 2,000 buses whose annual fleet demand is 80 million kilometers. The recommendations resulting from the study were implemented, and substantial savings were reported.

A.17.1   INTRODUCTION

The study occurred in Montreal, Canada, where Montreal's Transit Commission provides bus services to approximately 2 million people. To meet the bus schedules requires a fleet of 2,000 buses undertaking approximately 80,000,000 kilometers per year. Regarding North American bus operators, Montreal has the third largest fleet, falling behind New York and Chicago. The bulk of the buses used in Montreal are standard 42-seat General Motors (GMC) buses.

The study's objective was to analyze bus operations and maintenance costs to determine the economic life of a bus and to identify a steady-state replacement policy, one where a fixed proportion of the fleet would be replaced annually.

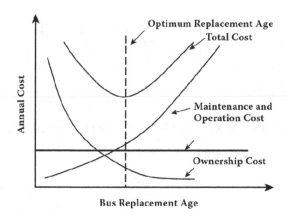

FIGURE A.15   Economic life model.

A.17.2   THE ECONOMIC LIFE MODEL

Figure A.15 illustrates the standard conflicts associated with capital equipment replacement problems.

The standard economic life model is

$$C(n) = \frac{A + C_{1+} \sum_{i=2}^{n} C_i r^{i-1} - S_n r^n}{1 - r^n}$$

where

A = acquisition cost

C_1 = operations and maintenance (O&M) cost of a bus in its first year of life

C_i = O&M cost of a bus in its ith year of life, assuming payable at the start of the year, then

$i = 2, 3, \ldots n$

r = discount factor

S_n = resale value of a bus of age n years

n = Replacement age

$C(n)$ = total discounted cost for a chain of replacements every n years

The total discounted cost can be converted to an EAC by the CRF, which in this case is i, the interest rate appropriate for discounting. *Note*: $r = 1/(1 + i)$.

A.17.3   DATA ACQUISITION

- *Acquisition cost*: In terms of 1980 dollars, the acquisition cost of a bus was $96,330.
- *Resale value*: The policy implemented by Montreal Transit was replacing a bus on a 20-year cycle, at which age the bus value was $1,000. There was

considerable uncertainty in the resale value of a bus, and for purposes of the study, two "extremes" were evaluated: one being termed a "high" trend in resale value and the other a "low" trend in resale value.

"High" Trend in Resale Value		"Low" Trend in Resale Value	
Replacement Age	Resale Value	Replacement Age	Resale Value
(Years)	($)	(Years)	($)
1	77,000	1	2,000
2	65,000	2	2,000
3	59,000	3	2,000
4	54,000	4	2,000
5	50,000	5	2,000
6	46,000	6	2,000
7	42,000	7	2,000
8	38,000	8	2,000
9	34,000	9	2,000
10	31,000	10	2,000
11	28,000	11	2,000
12	25,000	12	2,000
13	22,000	13	2,000
14	19,000	14	2,000
15	16,000	15	2,000
16	13,000	16	2,000
17	10,000	17	2,000
18	7,000	18	2,000
19	4,000	19	2,000
20	1,000	20	1,000

A.17.4   INTEREST RATE

The interest rate appropriate for discount was "uncertain." In the study, a range was used (from 0% to 20%) to check the sensitivity of economic life to variations in interest rates. The study's major conclusions were based on an "inflation fee" interest rate of 6%.

A.17.5   OPERATIONS AND MAINTENANCE COSTS

O&M costs are influenced by both the age of a bus and its cumulative utilization. O&M data were obtained for six cost categories: fuel, lubrication, tires, oil, parts, and labor. Analysis of the costs resulted in the following trend being identified:

$$c(k) = 0.302 + 0.723 \left(\frac{k}{10^6} \right)^2$$

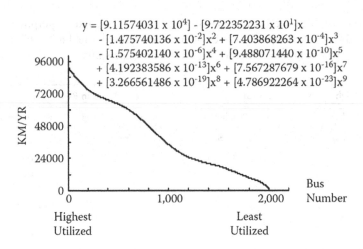

FIGURE A.16   Bus example trend line.

where

k = cumulative kilometers traveled by the bus since new

$c(k)$ = trend in O&M costs (in \$/km) for a bus of age k kilometers

A.17.6   BUS UTILIZATION

Figure A.16 shows the trend line that was fitted to the relationship between bus utilization (km/yr) and bus age (newest to oldest). This relationship is because new buses are highly utilized to meet base load requirements, with the older buses being used to meet peak demands.

In the analysis that was undertaken, it was assumed that the relationship identified in Figure A.16 would be independent of the replacement age of the bus. For example, using the present policy of replacing buses on a 20-year cycle, then, in a steady state, 2000/20 = 100 buses would be replaced annually. The total work done by the newest 100 buses in their first year of life would then be:

$$\int_{0}^{86,361} \left[0.302 + 0.723 \left(\frac{k}{10^6} \right)^2 \right] dk$$

Similarly, the O&M costs for the remaining 19 years can be calculated and inserted into the economic life model, $C(n)$, to enable C(20) to be calculated.

The entire process is then repeated for other possible replacement ages to enable the optimal n to be identified.

Figure A.17.3 shows the results when i ranges from 0% to 20%, and the "low" trend in resale value is used.

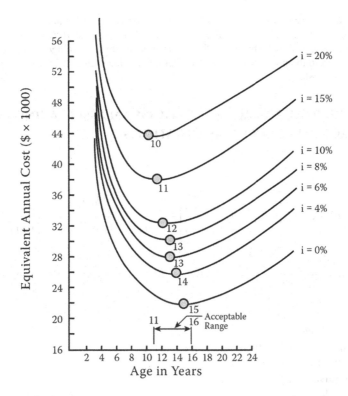

FIGURE A.17   Bus example result options.

A.17.7   CONCLUSIONS

Montreal Transit implemented a 3-year bus purchase policy based on a 16-year replacement age. The resultant savings were approximately $4 million per year. The class of problem discussed in this Appendix is typical of those found in many transport operations. For example:

- Haulage fleets undertaking both long-distance and local deliveries: When new vehicles are used on long-haul routes initially, and then as they age, they are relegated to local delivery work.
- Stores with their own fleets of delivery vehicles, with peak demands around Christmas time: The older vehicles in the fleet are retained to meet these predictable demands. Because of this unequal utilization, evaluating the vehicle's economic life by viewing the fleet as a whole rather than focusing on the individual vehicle is necessary.

BIBLIOGRAPHY

AGE/CON User's Manual, OMI, Specialized Software Division, Suite 704, 3455 Drummond Street, Montreal, Quebec, Canada H3G 2R6. (AGE/CON is a software package used for vehicle fleet replacement decisions. It includes the routines for handling problems of the class described in this Appendix.)

Simms, B.W., Lamarre, B.G., Jardine, A.K.S. and Boudreau, A., Optimal buy, operate and sell policies for fleets of vehicles, *European Journal of Operational Research*, 15(2), 183–195, February 1984. (This paper discusses a dynamic programming/linear programming approach to the problem that enabled alternative utilization policies to be incorporated into the model.)

A.18 OPTIMAL REPLACEMENT AGE OF AN ASSET TAKING INTO ACCOUNT TAX CONSIDERATIONS

The formula is

$$
\mathrm{NPV}(k) = \mathrm{A}\left[1 - \frac{td}{i+d}\left(\frac{1+i/2}{1+i}\right)\right] - \frac{it\,\mathrm{A}\left(1-\dfrac{d}{2}\right)(1-d)^{k-1}}{(i+d)(1+i)^k} - \frac{S_k(1-t)}{(1+i)^k}
$$
$$
+ \sum_{j=1}^{k} C_j \times \frac{(1-t)}{(1+i)^{j-1}}
$$

where
  i = Interest rate
  d = Capital cost allowance rate
  t = Corporation tax rate
  A = Acquisition cost
  S_k = Resale price
  C_j = Operations and maintenance cost in the jth year

$\mathrm{NPV}(k)$ = Net present value in the kth year

Remark: $A\left(1-\dfrac{d}{2}\right)(1-d)^{k-1}$ is the nondepreciated capital cost.

The EAC is then obtained by multiplying the NPV(k) by the CRF.

$$
\frac{i(1+i)^k}{(1+i)^k - 1}
$$

A.19 OPTIMAL REPLACEMENT POLICY FOR CAPITAL EQUIPMENT TAKING INTO ACCOUNT TECHNOLOGICAL IMPROVEMENT: INFINITE PLANNING HORIZON

A.19.1 STATEMENT OF PROBLEM

For this replacement problem, it is assumed that once the decision has been made to replace the current asset with the technologically improved equipment, this equipment will continue to be used, and a replacement policy (periodic) will be required for it. It will be assumed that replacements will continue to be made with technologically improved equipment. Again, we wish to determine the policy that minimizes total discounted maintenance and replacement costs.

A.19.2 CONSTRUCTION OF MODEL

1. n is the economic life of the technologically improved equipment.
2. $C_{p,t}$ is the maintenance cost of the present equipment in the ith period from now, payable at time i, where $i = 1, 2, ..., n$.
3. $S_{p,t}$ is the resale value of the present equipment at the end of the ith period from now, where $i = 0, 1, 2, ..., n$.
4. A is the acquisition cost of the technologically improved equipment.
5. $C_{i,j}$ is the maintenance cost of the technologically improved equipment in the jth period after its installation and payable at time j, where $j = 1, 2, ..., n$.
6. $S_{i,j}$ is the resale value of the technologically improved equipment at the end of its jth period of operation $(j = 0, 1, 2, ..., n; j = 0$ is included so that we can then define $S_{t,o} = A$. This then enables Ar^n in the model to be canceled if no change is made.

   Note that it is assumed that if a replacement is to be made at all, then it is with the technologically improved equipment. This is not unreasonable as it may be that the equipment currently in use is no longer on the market.
7. r is the discount factor.
8. The replacement policy is illustrated in Figure A.18.

The total discounted cost over a long period of time with replacement of the present equipment at the end of T period of operation, followed by replacements of the technologically improved equipment at intervals of n, is

$$C(T,n) = \text{Costs over interval}(0,T) + \text{Future costs}$$

$$\text{Costs over interval}(0,T) = \sum_{i=1}^{n} C_{p,t}r^i - S_{p,T}r^T + Ar^T$$

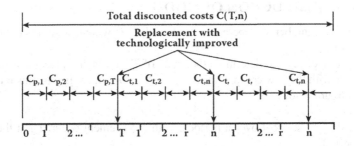

FIGURE A.18 Capital equipment replacement model.

Future costs, discounted to time T, can be obtained by the method described in Section A.16 where the equipment's economic life is calculated. We replace C_i by $C_{t,j}$ to obtain

$$C(n) = \frac{\sum_{i=1}^{n} C_{t,j} r^i + r^n (A - S_n)}{1 - r^n}$$

Therefore, $C(n)$ discounted to time zero is $C(n)r^T$ and

$$C(T,n) = \sum_{i=1}^{T} C_{p,t} r^i - S_{p,T} r^T + Ar^T + \left(\frac{\sum_{j=1}^{n} C_{t,j} r^i + r^n (A - S_n)}{1 - r^n} \right) r^T$$

This model of the problem relates changeover time to technologically improved equipment, T, and economic life of new equipment, n, to total discounted costs $C(T,n)$.

A.20   OPTIMAL REPLACEMENT POLICY FOR CAPITAL EQUIPMENT TAKING INTO ACCOUNT TECHNOLOGICAL IMPROVEMENT: FINITE PLANNING HORIZON

A.20.1   STATEMENT OF PROBLEM

When determining a replacement policy, there may be equipment on the market that is, in some way, a technological improvement on the equipment currently used. For example, maintenance and operating costs may be lower, throughput may be greater, quality of output may be better, etc. The problem discussed in this section is determining when to take advantage of the technologically improved equipment.

It will be assumed that there is a fixed period of time from now during which equipment will be required, and if replacement is with the new equipment, then this equipment will remain in use until the end of the fixed period. The objective will be to determine when to make the replacement if at all, to minimize total discounted maintenance and replacement costs.

A.20.2   CONSTRUCTION OF MODEL

1. n is the number of operating periods during which equipment will be required.
2. The objective is to determine the value of T, at which replacement should occur with the new equipment, $T = 0, 1, 2, \ldots, n$. The policy is illustrated in the figure below.

The total discounted cost over n periods, with replacement occurring at the end of the Tth period, is

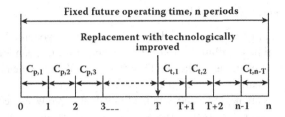

FIGURE A.19   Capital equipment replacement model policy.

$C(T)$ = Discounted maintenance costs for present equipment over period $(0,T)$
+ Discounted maintenance costs for technologically improved
   equipment over period (T,n)
+ Discounted acquisition cost of new equipment
− Discounted resale value of present equipment at the end of Tth period
− Discounted resale value of technologically improved equipment
   at the end of nth period

$$= \left(C_{p,1}r^1 + C_{p,2}r^2 + C_{p,3}r^3 + \ldots + C_{p,T}r^T \right)$$
$$+ \left(C_{t,1}r^{T+1} + C_{t,2}r^{T+2} + \ldots + C_{t,n-T}r^n \right) + Ar^T$$
$$- \left(S_{p,T}r^T + S_{t,n-T}r^n \right)$$

Therefore,

$$C(T) = \sum_{i=1}^{n} C_{p,t}r^i + \sum_{j=1}^{n-T} C_{t,j}r^{T+j} + Ar^T - \left(S_{p,T}r^T + S_{t,n-T}r^n \right)$$

This model of the problem relates replacement time T to the total discounted cost
$C(T)$.

A.21   OPTIMAL INSPECTION FREQUENCY: MINIMIZATION OF DOWNTIME

A.21.1   STATEMENT OF PROBLEM

The problem of this section assumes that equipment breaks down from time to time
and, to reduce the breakdowns, inspections, and consequent minor modifications can
be made. The problem is determining the inspection policy that minimizes the total
downtime per unit time incurred due to breakdowns and inspections.

A.21.2 CONSTRUCTION OF MODEL

1. Equipment failures occur according to the negative exponential distribution with mean time to failure (MTTF) = $1/\lambda$, where λ is the mean arrival rate of failures. (For example, if the MTTF = 0.5 years, then the mean number of failures per year = $1/0.5 = 2$, that is, $\lambda = 2$.).
2. Repair times are negative and exponentially distributed with mean time $1/\mu$.
3. The inspection policy is to perform n inspections per unit time. Inspection times are negative and exponentially distributed with mean time $1/i$.

The objective is to choose n to minimize total downtime per unit of time. The total downtime per unit will be a function of the inspection frequency n, denoted $D(n)$. Therefore,

$D(n) =$ Downtime incurred due to repairs per unit time
$+$ Downtime incurred due to inspection per unit time

$$= \frac{\lambda(n)}{\mu} + \frac{n}{i}$$

The above equation models the problem relating inspection frequency n to total downtime $D(n)$.

A.22 MAINTENANCE ASSESSMENT QUESTIONNAIRE

Department	Positions	How many?
Production	Manager	All
	Superintendents	All
	Supervisors	Minimum one for each area
	Operators	Minimum one for each area
Maintenance	Manager	All
	Superintendents	All
	Supervisors	Minimum one for each area
	Maintenance Planners/Scheduler	All
	Technicians (Inspection, vibration, etc.)	All
	Trades/Crafts	Minimum two for each trade
Other Departments	Managers & Superintendents	All

A.22.1 MAINTENANCE ASSESSMENT QUESTIONNAIRE INSTRUCTIONS

This questionnaire is divided into ten sections. Each section has several statements that relate to a specific maintenance topic. For each statement: assign a rating from 0 to 4 using the definition below for how well each statement describes your maintenance organization.

The ratings have the following definition:

	Score
Strongly agree	4
Mostly agree	3
Partially agree	2
Disagree	1
Do not understand	0
Not applicable	N/A

First, please tell us about yourself:

Name (optional): _____   Job Title: _____

E-mail Address
(optional): _____

Plant/Site: _____   Division: _____

Primary Responsibility	Department
____Executive	____Maintenance/Tech
____Management	____Operations/Production
____Supervision	____Engineering
____Trades/Hourly	____Purchasing/Warehousing
____Administrative	____IT Support
____Other	____Other

1. ASSET MANAGEMENT STRATEGY

Statement   **Score (4,3,2,1,0)**

1. The plant's or site's mission statement, vision, and values are communicated and understood by all.
2. All stakeholders communicate and understand the asset management and maintenance department's objectives and goals.
3. We effectively track and manage the company's assets through their entire lifecycle, from when assets are acquired to when they are retired.
4. We have established policies or guiding principles for maintenance.
5. Our approach to maintenance is proactive. We do our best to prevent breakdowns; we fix it immediately when something breaks.
6. The maintenance budget is related to expected performance, and indications are provided as to the likely outcome if work is planned to be deferred.
7. Capital equipment and maintenance strategy focuses on the total cost of ownership of the production assets, not just purchased or installed costs.
8. The policies for health and safety, quality, environment, and regulatory are effectively communicated to employees and contractors and effectively enforced.

Total (max. 32)

Notes: Strongly Agree (4), Mostly Agree (3), Partially Agree (2), Totally Disagree (1), Do Not Understand (0)

2. ORGANIZATION

Statement	Score 4,3,2,1,0

1. Maintenance staffing levels are adequate.

2. Maintenance personnel have the proper experience, training, and education.

3. Our employees understand what is expected of them.

4. The maintenance organization is primarily decentralized and organized by area or product line where appropriate, with centralized support functions as required.

5. First-line supervisors are responsible for 12 to 15 maintenance workers.

6. Overtime represents less than 5% of annual maintenance person-hours.

7. A formally established apprenticeship program addresses the asset management and maintenance department's future needs for qualified trades.

8. Part of the individual's pay is based on demonstrated skills, knowledge, results, and productivity.

9. Contractors augment plant staff during shutdowns or for specific projects or specialized jobs.

10. Every new employee and contractor is exposed to some orientation program covering all areas of health and safety, plant operation, maintenance, and management procedures.

Total (max. 40)

Notes: Strongly Agree (4), Mostly Agree (3), Partially Agree (2), Totally Disagree (1), Do Not Understand (0)

3. IT SYSTEMS SUPPORT

Statement	Score 4,3,2,1,0

1. A fully functional and integrated enterprise asset management system (EAM) is linked to the plant's financial and material management systems.

2. Our maintenance and materials management information is considered valuable and used regularly. The system is not just a "black hole" for information or a burden to use that produces no benefit.

3. Our asset management support systems (EAM, APM, AIP) are easy to use.

4. Most maintenance department staff, especially supervisors and trades, has been trained on it, can use it, and do use it.

5. Our planners/schedulers use the EAM system to plan jobs and select and reserve spare parts and materials.

6. Parts information is easily accessible and linked to equipment records. As a result, finding parts for specific equipment is easy to do, and the stock records are usually accurate.

7. Scheduling for major shutdowns is done using a project management system (linked to the EAM work orders) that determines critical paths and required levels of resources.

8. IT Technology (IoT, EAM, APM, AIP, Mobile, ML, AI, Process Automation) is actively pursued to help our organization compete in a market where external influences (i.e., Sustainability, Global Warming) expect higher standards.

Total (max. 32)

Notes: Strongly Agree (4), Mostly Agree (3), Partially Agree (2), Totally Disagree (1), Do Not Understand (0)

4. TACTICS

Statement	Score 4,3,2,1,0

1. We use a formal failure modes analysis program for determining the appropriate preventive maintenance (PM)s and predictive maintenance (PdM)s to perform.
2. The failure modes analysis program is used to fine-tune and continuously improve our PM and PdM performance.
3. Condition-based maintenance activities such as vibration analysis, oil sampling, non-destructive testing (NDT), and performance monitoring account for 15% of total maintenance person-hours.
4. 15% or more of total maintenance person-hours is devoted to performing time or usage-based PMs.
5. Compliance with the proactive maintenance program (PMs and PdMs) is high: 95% or more of the proactive work is completed as scheduled.
6. Less than 5% of total maintenance person-hours are devoted to emergencies.
7. Maintenance history is used to identify inadequacies in asset design, maintenance, and operating procedures.

Total (max. 28)

Notes: Strongly Agree (4), Mostly Agree (3), Partially Agree (2), Totally Disagree (1), Do Not Understand (0)

5. PARTS MANAGEMENT

Statement	Score 4,3,2,1,0

1. Parts and materials are readily available where and when needed. Stores catalog is up-to-date and readily available for use.
2. Stock-outs represent less than 3% of orders placed at the storeroom.
3. Parts and materials issued from stores are traceable to the equipment, tradesman, and time.
4. Distributed (satellite) stores are used throughout the plant for commonly used items (e.g., fasteners, fittings, common electrical parts).
5. Parts and materials are automatically restocked before the inventory on hand runs out without the maintenance crews' prompt.
6. A central tool crib is used for special tools.
7. Inventory management policy leverages order points, and quantities are based on the lead time, safety stock, economic order quantities, and asset risk dynamics and are automated.
8. Vendor performance reviews and analyses are conducted.

Total (max. 32)

Notes: Strongly Agree (4), Mostly Agree (3), Partially Agree (2), Totally Disagree (1), Do Not Understand (0)

6. MAINTENANCE WORK EXECUTION

Statement	Score 4,3,2,1,0

1. A plant equipment register lists all equipment in the plant that requires some form of maintenance or engineering support during its life.
2. Over 90% of maintenance work activities are covered by standard written work order, standing work order, PM work order, a PM checklist, or routine.

Statement	Score 4,3,2,1,0

3. Over 80% of planned maintenance work (preventive, predictive, and corrective) activities are scheduled in advance of execution with sufficient lead time to avoid the craftsperson doing part searches and arranging support resources.

4. Non-emergency work requests are screened, estimated, and planned (with tasks, materials, and tools identified) by a dedicated planner.

5. Realistic assessments of jobs are used to set standard times for repetitive tasks and help schedule resources.

6. Priorities are assigned to all work orders using pre-defined criteria. These criteria are not abused to circumvent the system.

7. Equipment criticality assessments have been done with ratings assigned to prioritize equipment-related work effectively. Work priorities are based on: their criticality to plant operations, safety, and probability of failure.

8. Work is scheduled for the upcoming period in consultation with production and is based on balancing work priorities set by production with the net capacity of each trade, considering PM work and emergency work.

9. Work backlog (ready-to-be scheduled activities) is measured and managed at 2 to 4 weeks (or other agreed-to-level).

10. Five-year or longer-term asset management plans forecast significant shutdowns and maintenance work and prepare the annual maintenance budget.

Total (max. 40)

Notes: Strongly Agree (4), Mostly Agree (3), Partially Agree (2), Totally Disagree (1), Do Not Understand (0)

7. METRICS

Statement	Score 4,3,2,1,0

1. Labor and material costs are accumulated and reported against crucial systems and equipment cost centers.

2. Downtime records, including causes, are kept on critical equipment and systems.

3. Measures are regularly reviewed to ensure they remain relevant to our goals and objectives.

4. Performance indicators are routinely measured and tracked to monitor results relative to the maintenance strategy and improvement process.

5. These performance indicators are developed from and support our maintenance and asset management strategy.

6. The performance measures result in positive and proactive behavior.

7. The performance measures used for my area or department are sufficiently limited in number to recall the measures and their importance with little or no assistance.

8. All maintenance staff have been trained in or taught the significance of the measures we use to a proficiency level where we can read the measures and trends and determine whether we are improving our overall performance.

Total (max. 32)

Notes: Strongly Agree (4), Mostly Agree (3), Partially Agree (2), Totally Disagree (1), Do Not Understand (0)

8. RISK AND RELIABILITY

Statement	Score 4,3,2,1,0
1. SAE JA1011-compliant Reliability-centered maintenance analysis or other reliability-centered maintenance (RCM) structured approach determines the failure management policies for critical equipment and systems and lower risk.	
2. Equipment history is maintained for all critical equipment showing failures, preventive replacement, proactive maintenance activities, causes of failures, and repair work completed.	
4. We successfully predict failures using condition-based maintenance techniques like vibration, thermal, and oil analysis.	
5. We experience very few failures that could have been predicted using condition monitoring or prevented using periodic restoration or replacement.	
6. We regularly check alarm and warning devices, emergency shutdown devices, trips, relief systems, voting logic systems, backups, standby equipment, and limiting devices to ensure they work correctly.	
7. Value-risk studies have been conducted as part of our RCM analysis to optimize maintenance program decisions.	
8. All equipment has been classified based on its importance to plant operations and safety. The classification is used to help to determine work order priorities, priority for RCM analysis, and direct engineering resources.	
9. We work on the most critical equipment problems first.	
10. Our reliability staff/engineer uses expert systems and decision support tools like Relcode, Perdec, AgeCon, Weibull, and Exakt are used by our reliability staff/ engineer to optimize our reliability performance.	
11. All maintenance work identified in our RCM analyses is planned in detail, even for the less likely failures that have not yet occurred.	

Total (max. 44)

Notes: Strongly Agree (4), Mostly Agree (3), Partially Agree (2), Totally Disagree (1), Do Not Understand (0)

9. TOTAL PRODUCTIVE MAINTENANCE

Statement	Score 4,3,2,1,0
1. We have a participative organization where decisions are made at the lowest effective level (no "rubber-stamping" required).	
2. Operators understand the equipment they run and perform minor maintenance activities like cleaning, lubricating, minor adjustments, inspections, and minor repairs (generally requiring simple tools).	
3. Multi-skilled tradespeople (e.g., electricians doing minor mechanical work, mechanics doing minor electrical work, etc.) are a feature of the organization.	
4. Call-outs are performed by an on-shift maintainer who decides what support is needed without a supervisor's guidance.	
5. Trades usually respond to call-outs after hours. As a result, operations can get needed support from maintenance trades quickly and with minimum effort.	
6. Self-directed work teams of operators, maintainers, and engineers perform much of the work.	
7. Leaders regularly discuss performance and costs with their work teams.	
8. Continuous improvement teams are in place and active.	

Statement	Score 4,3,2,1,0
9. Maintenance gets involved in new projects at the design stage and participates as required to ensure the assets' maintainability, reliability, and availability at a minimum Life Cycle Cost (LCC).	
10. Partnerships have been established with key suppliers and contractors. These partnerships facilitate the acquisition of parts and services directly by those requiring them.	

Total (max. 40)

Notes: Strongly Agree (4), Mostly Agree (3), Partially Agree (2), Totally Disagree (1), Do Not Understand (0)

10. RE-THINK ASSET MANAGEMENT

Statement	Score 4,3,2,1,0
1. We have a culture where "Change" is an expected constant as we strive to be competitive and align with Society 5.0 standards (i.e., ESG reporting).	
2. Key maintenance and maintenance, repair, and operations (MRO) materials management processes (e.g., planning, corrective maintenance, stock issues) have been identified, and "as-is" processes are mapped.	
3. The "as-is" process maps accurately reflect the typically used processes across our enterprise.	
4. Key maintenance processes are redesigned to reduce or eliminate non-value-added activities (e.g., duplicate data entry).	
5. Activities are challenged for effectiveness.	
6. The EAM/APM or other management systems automate workflow processes to reduce errors and administrative effort.	

Total (max. 24)

Notes: Strongly Agree (4), Mostly Agree (3), Partially Agree (2), Totally Disagree (1), Do Not Understand (0)

PERMISSIONS

Appendix B

Emerging Asset Management Standards in Industry*

Don M. Barry
Original contributions from Don M. Barry
and Jeffrey Kurkowski

B.1 INTRODUCTION

Over the past two decades, the focus and attention on asset management has grown substantially. Many frameworks have been developed, and the desire for a global standard seems to be a constant and revisited theme.

This appendix acknowledges that many asset management frameworks have emerged and are quite helpful as direct or supporting transformation tools. We will provide a high-level description and comparison of some of these against the Asset Management Excellence pyramid described throughout the third edition. We endorse all frameworks that help drive the behavior that supports improving asset management and operational excellence.

In reality, it does not matter which framework is embraced as much as it matters that the behavior of the affected stakeholders recognize the need to drive asset management excellence, given it will promote critical asset availability and operational excellence and reduce business, safety, and environmental risks. Environmental goals and achieving those goals are becoming a significant gate to becoming a competitive asset-intensive enterprise. Embracing IT technology and how it can accelerate the desired progress of many of the selected framework focus elements will be crucial. With the emergence of Internet of Things (IoT) devices, Machine Learning (ML), and Artificial Intelligence (AI), the opportunity to become more efficient and effective in asset management has never been more available. Asset Management fundamentals must also be understood and calibrated across the organization to align all the affected stakeholders with relevant definitions and commitments.

From the previous edition of this book, a Publicly Available Specification (PAS) 55-1: 2008 had emerged, and some efforts were made to position that effort against the framework used to organize the thoughts of that volume (2nd Edition).

Since 2008, numerous other efforts have been made to help asset-intensive organizations sort out their focus on this significant topic. Some frameworks of note include:

- Publicly Available Specification (PAS) 55-1
- ISO 55000
- (GFMAM) Global Forum on Maintenance and Asset Management
- Uptime Elements

Each of these frameworks assists an asset-intensive organization in organizing its thoughts on asset management areas that likely will need attention. Working through an offered framework should help the organization prioritize its actions. Each should help an organization develop a short and longer-term plan to implement its actions.

These frameworks should not be particularly exclusive. In other words, the Global Forum on Maintenance and Asset Management (GFMAM) may quote the ISO55000 standard as part of its framework. ISO55000 could refer to PAS55. Many consider the ISO55000 standard for an asset management plan to be a "go-to" reference.

Many organizations reference other global standards along with the organized frameworks listed above. A representative and non-exhaustive list is as follows:

TABLE B.1
Some Asset Management-Related Standards to Consider

- BS EN 15341:2007 Maintenance. Maintenance Key Performance Indicators
- CEN/TS 15331, Criteria for design, management, and control of maintenance
- CSN EN 15341
- EFQM Excellence Model 2010
- EN 1325:2014: Value Management. Vocabulary. Terms and definitions
- EN 13269, Maintenance – Guideline on the preparation of maintenance
- EN 13306, Maintenance Terminology
- EN 15341: Maintenance Key Performance Indicators
- EN 15628 Maintenance – Qualification of maintenance personnel guidelines on data interpretation and diagnostics techniques
- EN 16646 – Maintenance within physical asset management
- IEC 50(191), International Electro-technical Vocabulary – Dependability
- IEC 60300-3-11, Application guide – Reliability-centered Maintenance (RCM)
- IEC 60300-3-14-2004 Maintenance and maintenance support
- IEC 60300-3-16, Application guide – Guidelines for the specification of maintenance
- IEC 60300-3-16:2008 – Dependability management – Part 3-16: Application guide – Guidelines for the specification of maintenance support services
- IEC 60300-3-3: Dependability management – Application guide – Life cycle costing
- IEC 61703, Mathematical expressions for reliability, maintainability, and maintenance support items
- IEC Dependability Suite of standards
- International Accounting Standards (IAS)
- International Financial Reporting Standards (IFRS)
- ISO 13379, and ISO 13381-1 Condition monitoring and diagnostics of machines

TABLE B.1 (CONTINUED)
Some Asset Management-Related Standards to Consider

- ISO 14000 Environmental Management
- ISO 14224 Petroleum, petrochemical, and natural gas industries
- ISO 17021-5 Requirements for bodies providing audit and certification
- ISO 31000 Risk Management
- ISO 5500X Asset Management Suite
- ISO 9001:2008 Quality management systems
- ISO/IEC 15288 Systems Engineering
- Norsok z-008 Criticality analysis for maintenance purposes
- SAE JA1011, Evaluation Criteria for RCM
- SAE JA1012, a guide to RCM standard
- United States Department of Defense (24 November 1980) Military Standard 1629A – Procedures for performing a failure mode effect and criticality analysis

PAS55 originated in the United Kingdom and was globally recognized to some degree. The Publicly Available Specification (PAS) 55-1: 2008: Asset Management Standard was first published in 2004. The Asset Management and the British Standards Institute (BSI) worked together to develop strategies to help reduce risks to business-critical assets. This project resulted in PAS 55. This new standard culminates their latest thinking regarding leading practices in asset management systems. PAS 55 became internationally accepted as an industry standard for quality asset management. This standard was first applied in the electrical power generation, transmission, and distribution sector in the United Kingdom and has since evolved to be applied in other business sectors and geographies. The standard can be a valuable guideline for asset lifecycle management, quality control, and compliance.

Significant amounts of money and time are spent on managing business-critical assets each year, yet there is still confusion over terminology, and various management approaches are in use. In many cases, these approaches serve asset management needs well, but often they do not, sometimes resulting in high-profile failures. To the Institute of Asset Management (United Kingdom), it was apparent that there was a crucial need to provide a framework for asset management systems.

For many reasons, not every organization felt comfortable embracing the PAS55 standard. Other standards that had global co-development and support emerged, including ISO55000 and GFMAM. Private sector training and communications organizations also offered frameworks such as Uptime Elements.

Some frameworks are focused on physical assets (versus financial assets), while others include people as valued assets.

The physical assets are declared as the following five asset classes:

- Real Estate and facilities (offices, schools, hospitals)
- Plant and Production (oil, gas, chemicals, pharmaceuticals, food, electronics, power generation)
- Mobile Assets (military, airlines, trucking, shipping, rail)

Real Estate and Facilities   Plant and Production   Mobile Assets   Infrastructure   Information Technology

FIGURE B.1   Asset classes.[1]

- Infrastructure (railways, highways, telecommunications, water, wastewater, electric, and gas distribution)
- Information Technology (computers, routers, networks, software, auto-discovery, service desk)

In this appendix, one will find references to the terms "leading" and "best" practices. It is a contention that one will be presented with what some refer to as "best practices" when looking to improve a specific area or set of processes. It is universally recognized that a "best practice" or set of "best practices" that may be "best" for one enterprise or organization may not be the "best" for another. Therefore, the concept promoted here is that an enterprise or organization will review the most common, or "leading practices," when looking to improve an area or set of processes. They will adopt the ones they believe are a good fit for them and apply them as necessary. When they have completed the exercise, they will have defined their "best practices."

Similarly, picking a framework considered "leading" is to determine what is "best" for yur organization now.

B.1.1   PAS55

PAS 55 is designed to help organizations display asset management competency by meeting a particular set of requirements. Requirements address "leading practices" rather than "best practices" in each area.

All applied processes must be practical and require evidence of what is being done and why. The standard is nonprescriptive – as in standards like the International Organization for Standardization (ISO) 9001, Occupational Health and Safety Assessment Series (OHSAS) 18001 or ISO 14001. All elements of the standard framework need to be covered in the process.

The standard suggests integrating holistic, systematic, systemic, risk-based, optimal, and sustainable qualities.

Its scope manages assets across the asset lifecycle and works to optimize asset performance, capital investments, and sustainability planning while driving to the asset-intensive organization's goals.

B.1.2   ISO55000

Like PAS 55, ISO 55000 is framed around the famous PDCA (Plan-Do-Check-Act) cycle.

TABLE B.2
Plan-Do-Check-Act Description used in ISO55000[2]

PDCA	Description
Plan	Establish the asset management strategy, objectives, and plans to deliver results following the organization's policy and strategic plan
Do	Establish the enablers for implementing asset management (e.g., asset information management system(s)) and other requirements (e.g., legal requirements) and implement the asset management plan(s)
Check	Monitor and measure results against asset management policy, strategy objectives, legal and other requirements; record and report the results
Act	Take actions to ensure that the asset management objectives are achieved and continually improve the asset management system and performance

FIGURE B.2   Challenges facing asset management leadership.[2]

ISO 55000 is focused on the notion of business value creation through the prudent management of assets.

All organizations have existing strengths and capabilities in the management of their assets. Many are disjointed, short-term focused, and not connecting the day-to-day operational activities with longer-term business priorities. An objective review of the current capabilities is valuable to identify what is already well received and can be improved.

PAS 55 and ISO 55001 can provide an excellent checklist of required good practices and a standardized "Maturity Scale" to ensure consistency of definitions and benchmarking potential.

B.1.3   GFMAM

GFMAM (The Global Forum on Maintenance and Asset Management) is a nonprofit organization supporting maintenance and asset management professions by collaborating on standards, knowledge, and practices.

GFMAM was founded in May 2010 in Switzerland.

It exists to be a worldwide community, providing leadership for maintenance and asset management communities. It facilitates the alignment of maintenance and asset management practices and knowledge and encourages exchanging ideas.

Like many such organizations, GFMAM promotes and strengthens the maintenance and asset management community globally. It encourages and facilitates the alignment of maintenance and asset management knowledge and practices. Organizations like GFMAM raise the credibility of member organizations by raising the profile of the members' calibrated understanding facilitated through their Global Forum.

GFMAM supports establishing and developing associations or institutions whose aims are maintenance and asset management focused.

B.2   THE BENEFITS OF AN ASSET MANAGEMENT STANDARD

In today's economy, factors that drive the need for good asset management are becoming more apparent. Asset risks appear more often on the boardroom agenda, and there is more focus on regulatory compliance from governmental and industry institutions. Organizations clearly emphasize cost containment, price management, return on investment, and increased overall asset value. Worldwide interest in lean principles, asset management, and asset performance is raising the bar. Many sectors are seeing increased consumer expectations about quality, service delivery, and green initiatives. On top of this, there is an increased complexity of assets, tools, and equipment as assets become more interconnected, instrumented, and intelligent.

Embracing a standard, or set of standards, can benefit companies from the regulatory point of view and help them gain a competitive advantage by ensuring that they are effectively managing their assets. Using a standard methodology for comprehensive asset management can drive cost savings and service improvement.

Overall, using an Asset Management Standard encourages companies to do the following:

- Achieve asset management leading practices.
- Start processes to map the entire asset base and create the information strategy following the company's overall strategy.
- Organize around valid lifecycle asset management processes.
- Challenge and reduce current time-based work and replace it with a "risk-based" management approach.
- Position asset management-specific accountability from the "shop floor to the top floor" and create motivational performance management.

- Focus on building the asset management knowledge base.
- Understand and target the tools, and engage the entire organization.
- Adopt a genuinely holistic approach by continuously challenging reasonable, leading, or best practices.

B.3   FUTURE CHALLENGES AND DIRECTIONS

More companies are realizing the benefits of embracing asset management and the asset lifecycle approach. The drivers for adaptation include the increasing requirements of different regulators, the influence of financial and insurance companies, and the desire to improve the organization's overall image in the market.

Sector-specific application guideline projects are being launched, including guidelines for property asset management. Additionally, EAM vendors are preparing for alignment with the standard they believe works for them in a given cycle.

Specific challenges for asset management include the following:

- Integrating asset management into companies' long-term strategies by creating a Chief Asset Officer (CAO) role on the board
- Connecting and integrating asset management with financial and asset management strategies and processes
- Developing a competency-building framework of asset management educational tools
- Assuring environmental, regulatory, and legal compliance to meet sustainable manufacturing requirements

B.4   HOW ASSET MANAGEMENT MODEL FRAMEWORKS COMPARE

The original edition of this book – and arguably the original version of the Maintenance Excellence Pyramid that has been in place since the early 1990s – has displayed a complete approach to asset management that supports much, if not all, of the assertions and attributes of many of the emerging asset management standards.

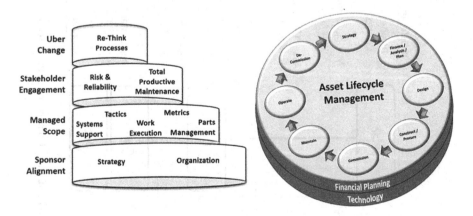

FIGURE B.3   Asset Management Excellence pyramid and the asset lifecycle models.[1]

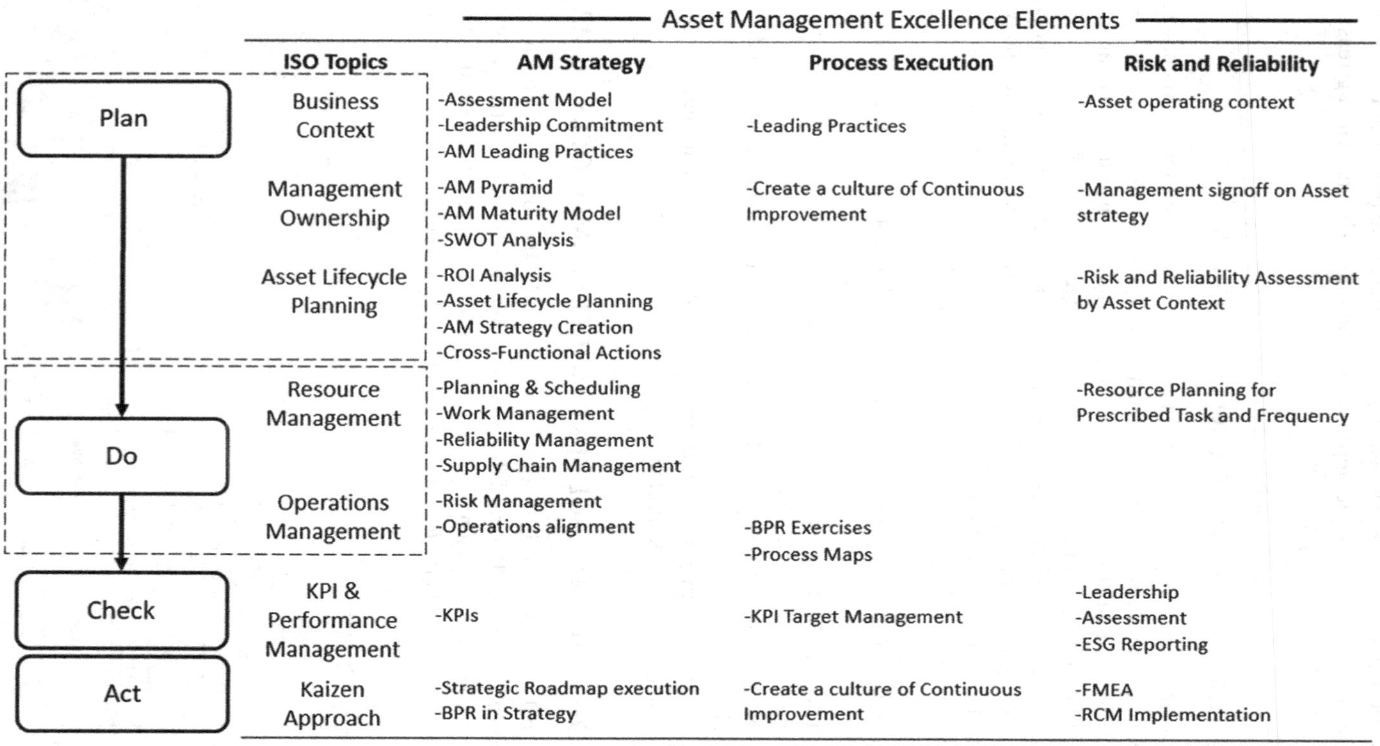

—— Asset Management Excellence Elements ——

ISO Topics	AM Strategy	Process Execution	Risk and Reliability
Business Context (Plan)	-Assessment Model -Leadership Commitment -AM Leading Practices	-Leading Practices	-Asset operating context
Management Ownership (Plan)	-AM Pyramid -AM Maturity Model -SWOT Analysis	-Create a culture of Continuous Improvement	-Management signoff on Asset strategy
Asset Lifecycle Planning (Plan)	-ROI Analysis -Asset Lifecycle Planning -AM Strategy Creation -Cross-Functional Actions		-Risk and Reliability Assessment by Asset Context
Resource Management (Do)	-Planning & Scheduling -Work Management -Reliability Management -Supply Chain Management		-Resource Planning for Prescribed Task and Frequency
Operations Management (Do)	-Risk Management -Operations alignment	-BPR Exercises -Process Maps	
KPI & Performance Management (Check)	-KPIs	-KPI Target Management	-Leadership -Assessment -ESG Reporting
Kaizen Approach (Act)	-Strategic Roadmap execution -BPR in Strategy	-Create a culture of Continuous Improvement	-FMEA -RCM Implementation

FIGURE B.4 Comparison of the Asset Management Excellence pyramid versus ISO55000 (Part 1).[2]

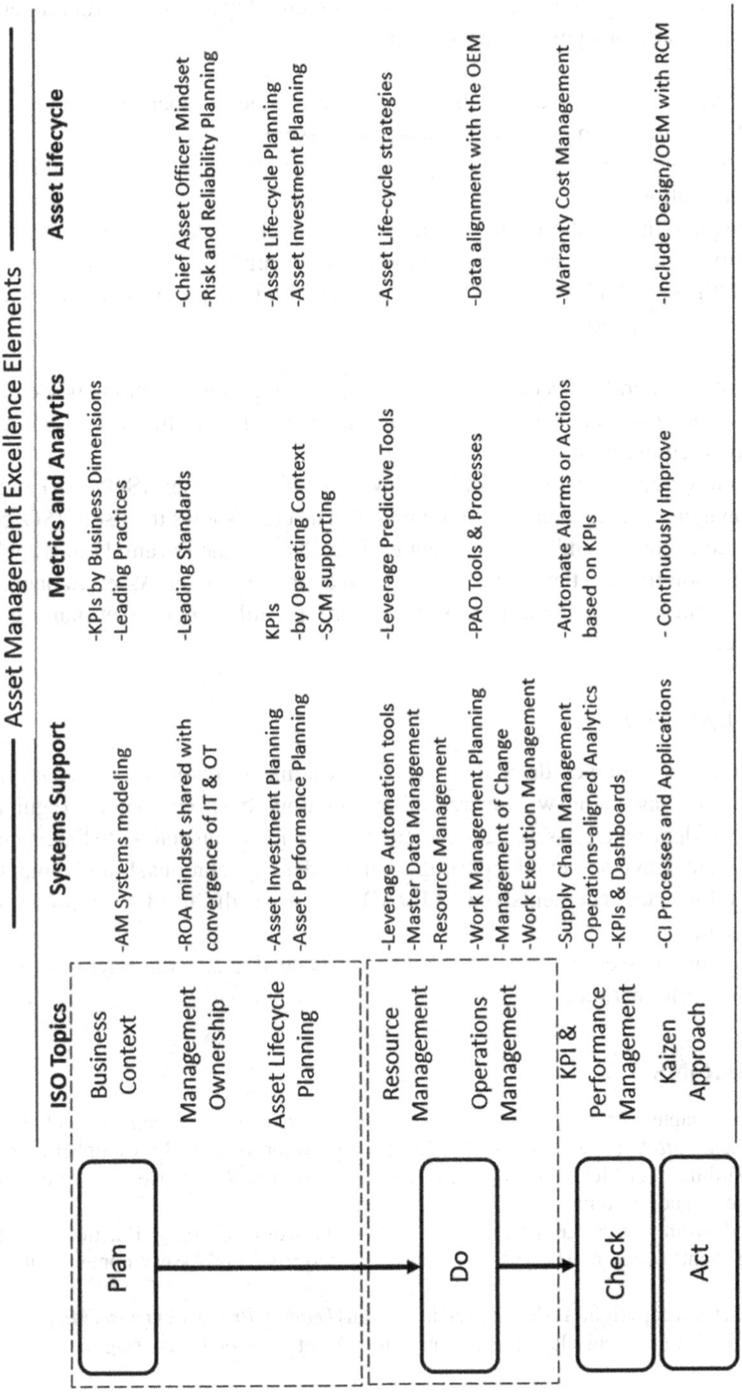

Asset Management Excellence Elements

PDCA	ISO Topics	Systems Support	Metrics and Analytics	Asset Lifecycle
Plan	Business Context	-AM Systems modeling	-KPIs by Business Dimensions -Leading Practices	
	Management Ownership	-ROA mindset shared with convergence of IT & OT	-Leading Standards	-Chief Asset Officer Mindset -Risk and Reliability Planning
	Asset Lifecycle Planning	-Asset Investment Planning -Asset Performance Planning	KPIs -by Operating Context -SCM supporting	-Asset Life-cycle Planning -Asset Investment Planning
Do	Resource Management	-Leverage Automation tools -Master Data Management -Resource Management	-Leverage Predictive Tools	-Asset Life-cycle strategies
	Operations Management	-Work Management Planning -Management of Change -Work Execution Management	-PAO Tools & Processes	-Data alignment with the OEM
Check	KPI & Performance Management	-Supply Chain Management -Operations-aligned Analytics -KPIs & Dashboards	-Automate Alarms or Actions based on KPIs	-Warranty Cost Management
Act	Kaizen Approach	-CI Processes and Applications	- Continuously Improve	-Include Design/OEM with RCM

FIGURE B.5 Comparison of the Asset Management Excellence pyramid versus ISO55000 (Part 2).[2]

A base principles model(s) for asset management is essential for successful execution. The following principles are considered essential to the successful implementation of an asset management transformation:

- An organizational structure that can implement these imperitives and internal standards with clear direction and leadership
- Staff awareness, competency, commitment, and cross-functional coordination
- Adequate information and knowledge of asset class needs, their condition, performance, risks, costs, and the interrelationships
- Definition and adoption of a Total Lifecycle Strategy for critical assets (for mature organizations)

Leveraging a generally accepted Asset Management Standards Framework or the Asset Management Excellence pyramid and the related asset lifecycle models can assist in this requirement.

The following figures provide a high-level mapping between ISO55000 and the Asset Management Excellence pyramid model. The chart shows the list of ISO55000 category requirements and the alignment to the indices in the Pyramid model, which can address solutions to the requirements. Both approaches to an Asset Management Standard framework are looking to work toward a similar goal, asset management excellence.

B.5 SUMMARY

Today's societal demands drive the need for sustainable, resilient, and more intelligent asset management, with increased expectations from companies, regulators, and shareholders when assets become much more interconnected (IoT), complex, and intelligent. Environmental reporting, shareholder expectations, brand reputation, and technology advancements (i.e., ML/AI) will drive the need to improve asset management.

Embracing an asset management framework is needed as an accelerator to compete in the world markets.

PERMISSIONS

* Chapter adapted from pages 455–465, *Asset Management Excellence: Optimizing Equipment Life-Cycle Decisions*, 2nd Edition by Editor(s), John D. Campbell, Andrew K. S. Jardine, Joel McGlynn Copyright (2011) by Imprint. Reproduced by permission of Taylor & Francis Group.

1 Adapted from copywrite (PAM Day 1, Part 1) from (Leading Practices in Asset Management) edited by (Barry). Reproduced by permission of Asset Acumen Consulting Inc.

2 Adapted from copyright (PAM Day 1, Part 2) from (*Leading Practices in Asset Management*) edited by (Barry). Reproduced by permission of Asset Acumen Consulting Inc.

BIBLIOGRAPHY

GFMAM-The Maintenance Framework

IAM – BSI – PAS55-1 2008. Asset Management. Part 1.

IBM White Paper: Enabling the benefits of PAS 55: The new standard for asset management in the industry – 2009.

John Woodhouse presentation. Process integration whole-life planning and optimization in asset management. 2009. Managing Director of TWPL (UK), Chairman, Development & Standards Committee, IAM.

Index

Pages in *italics* refer to figures and pages in **bold** refer to tables.

Printed in the United States
by Baker & Taylor Publisher Services